Construction Studies Today

Trevor Hickey

D1350326

Gill & Macmillan

Gill & Macmillan Ltd
Hume Avenue
Park West
Dublin 12
with associated companies throughout the world
www.gillmacmillan.ie

© Trevor Hickey 2006

ISBN-13: 978 07171 3955 2
ISBN-10: 0 7171 3955 7

Print origination in Ireland by Carole Lynch
Design by Vermillion, Dublin
Illustrations by Peter Bull Art Studio

*The paper used in this book is made from the wood pulp of
managed forests. For every tree felled, at least one tree is
planted, thereby renewing natural resources.*

Picture Credits
For permission to reproduce photographs the author and
publisher gratefully acknowledge the following:

4T, 19, 20R, 49, 90, 109T, 185B © Alamy; 39 © Ballymun
Regeneration Ltd; 74 © The Bridgeman Art Library/
Egyptian National Museum, Cairo, Egypt/Giraudon;
77L, 77R, 83R, 89, 108T, 109B, 110T, 112, 116B, 116T,
118, 131, 387 © constructionphotography.com;
20L © Corbis/Lowell Georgia; 3, 10, 11L, 11R, 12,
21 © Dept of the Environment, Heritage and Local
Government; 81 © Graham Hickey; 35L, 35R, 83L, 84L,
84C, 84R, 106L, 106C, 106R, 108C, 108B, 110B, 150L,
150R, 154, 153, 185TL, 185TR, 185CL, 185CR, 379T,
379B © Trevor Hickey; 13, 82T, 82B © Irish Image
Collection; 30 © The Irish Times; 4B © Lonely Planet
Images/Richard Cummins; 5, image reproduced courtesy
of the OPW; 349 courtesy of Sustainable Energy Ireland.

Contents

Acknowledgments

I would like to thank the following people for their help at various stages along the way: my good friend and mentor Donal Lynch for reading the first draft and pointing out the multitude of shortcomings, for always being willing to answer even the most mundane of questions and for his constant support and encouragement over the years; my colleagues at the University of Limerick, Donal Canty, Seamus Dolan and Declan Philips for their technical expertise; Peter Bull and his team for their hard work and attention to detail on the artwork; Chris Thomas and Carole Lynch; everyone at Gill & Macmillan, including Mairead O'Keeffe, Aoife O'Kelly and Helen Thompson. Thanks in particular to Hubert Mahony and Aoileann O'Donnell for their professionalism and patience. Finally, thanks to my wife and my three beautiful daughters for putting up with me.

Built Heritage

Our heritage is something left to us by our predecessors – that is, the people who have lived before us, like our grandparents and great-grandparents. Our built heritage is therefore the buildings and structures that were left to us by previous generations. Built heritage is also referred to as *architectural heritage.*

Categories

In the Irish context, architectural heritage can be divided into four main categories:

- Vernacular buildings: cottages, rural buildings....
- Formal buildings: country estate houses, state buildings....
- Religious buildings: cathedrals, churches, shrines....
- Fortified buildings: castles, towers, fortifications....

The traditional thatch-roofed cottage is a typical example of Irish vernacular architecture.

This courthouse in Carlow, constructed in ashlar masonry, is a wonderful example of formal architecture built in the classical style.

Fortified architecture: Bunratty Castle in Co. Clare built circa 1425.

This Church of Ireland church, which displays many Gothic features, is like many which were built in towns around Ireland.

One of the most important things to realise about the built heritage of Ireland is that it is not always consistent with international architectural trends and is often an eclectic arrangement of various styles. There are two main reasons why Irish architectural heritage varies in this way.

Firstly, many architectural styles, such as Romanesque and Gothic, originated in mainland Europe and because of the isolation of Ireland (as an island at the western extremity of Europe) these new styles were not seen here until long after they had become firmly established in mainland Europe.

Secondly, it is often the case that many buildings which survive for long periods of time (e.g. several hundred years) undergo rebuilding and renovation throughout their lifetime. When this occurs the renovators tend to include architectural details that are fashionable at the time of renovation. So, for example, a Romanesque building such as a church built during the twelfth century that is later renovated or extended in the fourteenth century, may have Gothic pointed arches in addition to the original rounded or semicircular arches. When this happens it becomes difficult to classify a building as belonging to either period. Instead we describe the style of the various elements of the building (windows, doors, arches, roof structures) rather than applying an overall label that would be less accurate. In this sense, when we explore old buildings we should begin by trying to identify elements of the building fabric that reflect certain architectural styles, rather than trying to immediately classify the building as belonging to one specific architectural style.

The Importance of Heritage Buildings

There are several reasons why these buildings are important. Most buildings are significant because they represent the centre of human activities, a connection between people and places and an expression of local customs and traditions. The reasons why heritage buildings should be protected can be summarised as follows:

- They reflect the origin and evolution of a people over many years.
- They are an important part of our social history.
- We get a sense of identity and belonging from these buildings.
- Much can be learned about the everyday lives of previous generations through the close examination of these buildings.
- The design, materials and techniques used illustrate the influence of climate, economic conditions and social structures on the dwellings of the time.

- Valuable lessons about the construction of buildings are passed on from generation to generation.
- Modern construction methods are often based on principles established during the design of these buildings.

Recognising a Heritage Building

Not all old buildings are a significant part of our built heritage. It is not realistic to say that a building should be preserved just because it is old. Many old buildings are examples of how not to build, are dangerous, and should possibly be condemned and demolished.

Our architectural heritage includes buildings of great artistic achievement: churches, courthouses, country estate houses and significant commercial buildings. However, of equal importance are the everyday buildings of craftsmen, which tell us how building designs evolved and what everyday life was like for previous generations.

Three factors to be considered when deciding if a building forms a significant part of our architectural heritage are:

1. Architectural Significance
(a) Style – an example of a particular architectural style.
(b) Association – a building associated with a particular industrial, institutional, commercial, agricultural or transportation activity.
(c) Construction – an example of a particular method of construction or use of a material.
(d) Design – an example of artistic excellence, unique design or craftsmanship.
(e) Architect/Builder – an example of a particular architect's work. In particular, an architect who has made a significant contribution to the community, county or country.

Custom House Dublin: Designed by James Gandon, an English architect, who was also responsible for the Four Courts and parts of what is now Bank of Ireland on College Green. The Custom House is often considered the most architecturally important building in Dublin.

(f) Integrity – a building that occupies its original site, has suffered little alteration, and retains most of its original materials and design features and is in good structural condition.

(g) Visual/Symbolic Importance – a building's importance as a landmark structure, or its symbolic value to the local area.

2. Cultural History

(a) Historical Association – the direct association of a building with a person, group, institution, event or activity that is of historical significance to the area; or associated with (and illustrative of) patterns of cultural, political, military, economic or industrial history.

(b) Historical Pattern – a building's direct association with broad patterns of local area history, including development and settlement patterns, early or important transportation routes, social, political or economic trends and activities.

The General Post Office on O'Connell Street, Dublin, has great cultural importance because of its historical association the 1916 Easter Rising.

3. Context

(a) Streetscapes/Environment – contributes to the existing character of the street. A building's continuity and compatibility with adjacent buildings and/or visual contribution to a group of similar buildings.

(b) An intact, historical landscape or landscape features associated with existing buildings.

(c) A notable historical relationship between the building's site and its immediate environment, including original native trees, topographical features, outbuildings or agricultural setting.

Streetscape: Enniscorthy, Co. Wexford.

(d) A notable use of landscaping design in conjunction with an existing building.

(e) Age – the age of the building in relation to the local area/other buildings.

Some Examples of Our Built Heritage

There is a wonderful variety of heritage buildings to be found all over the country. The following examples give a brief outline of the main periods of Irish architectural heritage. Many other buildings which are not described here (e.g. castles and fortified houses) are also quite common.

Pre-Christian Celtic Period

This period dates from approximately 4,500 BC to 400 AD and covers the Stone, Bronze and Iron Ages. One of the best examples of architectural heritage from this period is the passage grave at Newgrange in Co. Meath which is estimated to be around 5,100 years old.

Newgrange consists of a large mound (called a *cairn*) constructed of stone and covered in grass. A 18·7 m long, stone-lined passage leads to a chamber in which around 24 people can stand. The passage is constructed of standing stones capped with large flat stones near the entrance. The height gradually rises through the use of corbelling.

Stone-lined passage at Newgrange, Co. Meath. This photo captures the sun shining through the 'light box' at the entrance to the passage, shortly after sunrise, on the morning of the winter solstice.

This structure is a wonderful example of the corbelling technique (later used to construct the Gallarus Oratory in Dingle) and demonstrates how significant loading can be safely carried using this technique. The cairn is 11–13 m tall and 79–85 m in diameter. Newgrange was built as a pagan celebration of the return of light (the days get *longer* from the winter solstice onward).

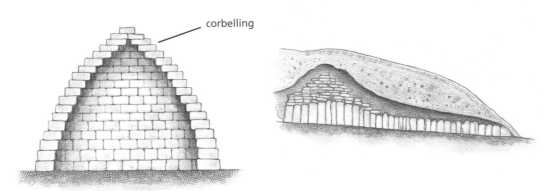

corbelling

The overlapping arrangement of the stones allows a self-supporting arch or roof to be created. This technique was used in the construction of the main chamber of the Passage Grave at Newgrange, Co. Meath.

Christian Celtic Period

This period spans the fifth to the tenth century – including the Viking Invasions (ninth and tenth centuries). Christian architectural heritage is best represented by the many monasteries that remain to be seen around the country.

These communities usually had a number of small buildings clustered within a stone wall or earth rath (ring). The daily activities of a monastic community were regulated by the ringing of a bell. The bell was often housed in a round tower.

Round Towers

The round tower could also have served as a place of refuge during attack. Although they were ineffective in this role because of the lack of defensive features (e.g. light, sanitation and means of escape). Round towers were also vulnerable to fire because of their wooden doors and floors, also because their very shape would act as a chimney in the event of a fire!

They were, however, used as a store for valuables (including religious scripts) and as a landmark for the location of the monastery. Over sixty round towers remain standing today, although many are partly in ruins.

Typical Construction Details

Common features of Irish round towers include:

Round Tower at Glendalough, Co. Wicklow.

- Free-standing.
- Cylindrical form (typical dimensions approx. Ø5 m, circumference 15 m, height 30 m).
- Entrance door positioned well above ground level.
- Four axially placed (north, south, east, west) windows at the top (bell storey).
- Conical stone cap.

Round towers vary in height but on average they are 30 m high. Their circumferences also vary but average 15 m. This leads to the suggestion that the following formula may have existed for their construction: the height of a round tower is twice its circumference. For example the tower at Glendalough is 30·48 m high and its circumference is 15·30 m.

Round towers were always built near an existing church. The round tower was usually positioned to the north-west of the entrance of the church with the tower's door facing the church.

The first step in the construction of the round tower was the excavation of a circular (doughnut-like) strip foundation. This was typically completed to a depth of one metre. A ring-like foundation of mortared stone would then be built to bring this to just above ground level. The width of the ring foundation would match the wall thickness although sometimes it would be wider.

Once the foundations had been brought to ground level, the centre was carefully marked. The centremark was the only reference point the builders would have when raising the 30 m high tower.

The walls which formed the base of the tower would be in the region of one metre to 1·15 m thick. These walls were constructed using roughly dressed stone from a local quarry. The initial course would be laid without scaffolding but as soon as they rose above one metre, a timber, ground-fast scaffold would be erected.

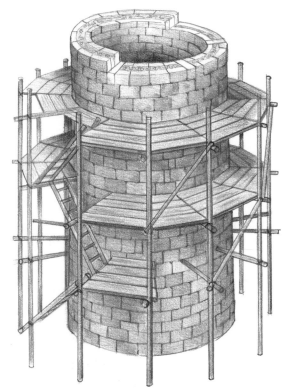

Putlog (or putlock) scaffolding was used in the construction of round towers. For round towers, the horizontal beams (putlogs), that supported the scaffold planks, were allowed to penetrate 300 to 500 mm into the walls to provide stability.

The complexity of the challenge posed by the construction of a round tower should not be underestimated. To successfully construct a 30 m high tube of stone that gets gradually thinner as it rises and yet remains perfectly circular while staying exactly vertical is extremely difficult, especially considering the materials, equipment and technology available over 800 years ago. Still, the craftsmen of the time managed to build many such towers.

Putlog or putlock scaffolding: evidence of this technique can still be seen today, as small holes in the outer surface of the walls of many round towers.

In order to keep the tower perfectly straight a plumbline would be used to keep track of the centre as marked at the base of the tower. This would act as a datum, or reference point, from which the radius of each successive course of stonework would be measured. By measuring outward from this, in gradually decreasing radii, the taper (or batter) of the tower could be achieved.

In practice a wooden beam compass would also be used to check the courses were perfectly circular and a wooden profile would be used to check the batter of the tower as it was raised.

Because the scaffolding was in the way, the outer stone surface was left undressed until the tower was complete. Then working from the top downwards, as the scaffolding was dismantled, the masons would dress each stone ensuring the batter and radius were kept precise.

Door openings were usually raised several meters above ground level. This position made the tower more secure but was also for structural reasons. Placing the door at ground level would considerably weaken the walls – creating the doorway higher up was less likely to undermine the walls. Door opes were lintelled at first, well-made arched doorways were seen in later towers.

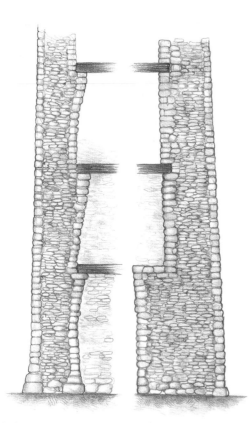

Vertical cross-section of round tower showing both construction methods used to support internal floors.

In some round towers, the floors were supported on timber beams that were built into the walls.

Alternatively, the floor beams rested on a ledge on the inner surface of the wall. To create this ledge the inner surface sloped inward more sharply than the outer surface. This created a *stacked-bucket* effect which created the ledges to support the floors.

Each floor had an opening so that it could be accessed using a ladder.

These towers typically had a stone-corbelled cap. This is a distinctive feature of Irish round towers. The towers of mainland Europe tended to have tiled, lead or slate roofs. Early Irish round towers may initially have had wooden roofs. It is thought that the stone cap evolved from the need to prevent fire caused by lightning strikes.

The stone cap was constructed using the corbelling technique. The outer surface of each stone was carved to have a curved sloped face so that the entire cap had a smooth conical form.

Medieval Period (Eleventh and Twelfth Centuries)

This period of architectural development in Ireland is represented by the Romanesque style. Romanesque architecture is usually recognised by the heavy thick walls, sturdy columns, small windows and semicircular arches. The bulky walls were necessary to support the heavy, stone, vaulted roofs.

The stone vault is perhaps the greatest achievement of the Romanesque period. Stone vaults were developed as an alternative to the highly flammable wooden roofs of pre-Romanesque structures. A typical stone vault is simply an extended semicircular arch. This type of vault is called a *barrel vault*. The outward acting forces generated by these arched stone roofs led to long, narrow thick-walled buildings.

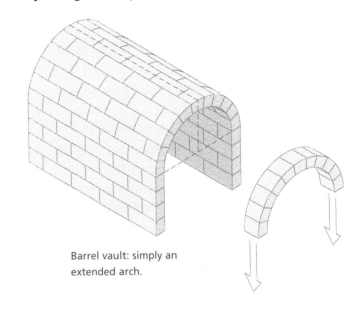

Barrel vault: simply an extended arch.

A stone vault is constructed through the use of timber scaffolding and centring. Centring is a semi-cylindrical structure used to support the voussoirs (tapered stones) during the construction of an arch or vault. Once the arch or vault is complete, it is self-supporting and the centring can be safely removed.

When two barrel vaults intersect a groin vault is generated. The use of groin vaults allowed the construction of cruciform (cross-shaped) churches. While the repetition of barrel vaults allowed wider churches to be constructed. Extra abutment was provided along the outer boundaries to resist the forces generated by this structure.

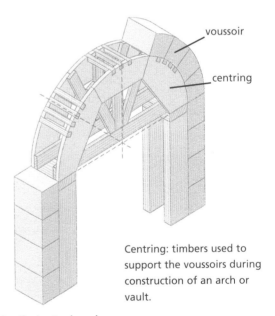

voussoir

centring

Centring: timbers used to support the voussoirs during construction of an arch or vault.

It should be remembered that the vast majority of churches built in Ireland, during and after this period, continued to have timber roof structures.

One of the most important Irish buildings constructed in this style is Cormac's Chapel in Cashel, Co. Tipperary. This church is named after its patron Cormac McCarthy, King of Desmond. It was built between 1127 and 1134. It is a rectangular building consisting of a nave (an area near the altar for clergy) and chancel (a seating area for the congregation). It is 14 m long which is large by Irish standards of the time but still quite small by international standards.

Cormac's Chapel has many of the typical features of Romanesque architecture including:

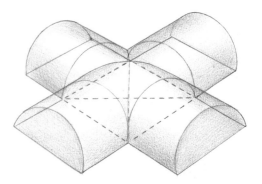

- thick heavy walls,
- vaulted roof,
- rounded arches,
- long narrow form,
- detailed sculptural decoration.

Cormac's Chapel is the earliest example of architecture found in Ireland, in the sense that it was designed completely and in detail (including internal decorative details) before construction began. Its use of stone vaulting was a completely new technology in Ireland at the time, although it was not adopted by local craftsmen afterwards.

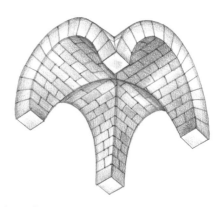

Groin vault: created when two barrel vaults intersect. A groin also provided lateral stability to the barrel by resisting the outward thrust generated by the load.

The greatest influence of Cormac's Chapel on later churches constructed in Ireland was in the area of decorative detailing. The doorways of Cormac's Chapel have richly decorated tympanums. The tympanum is the space between an arch and the horizontal head of a door or window below. The north doorway of Cormac's Chapel features a carving of a centaur and a lion.

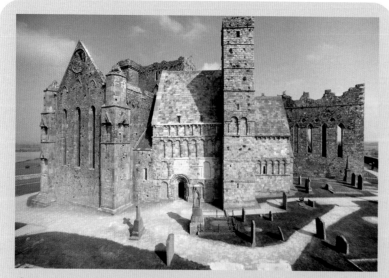

Romanesque Architecture: Cormac's Chapel, Cashel, Co. Tipperary. The chapel is the smaller building in the centre of the photo. It is surrounded by the Cathedral. The round tower is hidden from view, behind the square tower.

Tympanum above North doorway of Cormac's Chapel: note the carving of a centaur (holding a bow) and a lion.

An example of this influence can be seen in the doorway of Clonfert Cathedral, Co. Galway.

Romanesque doorway of Clonfert Cathedral, Co. Galway.

A special type of groin vault, called a *ribbed vault*, is seen in Cormac's Chapel. A ribbed vault consists of two intersecting diagonal arches with infilled panels. The arches are built first. Then the panels between the arches (called *webs*) are filled-in. This technique provides greater stability to the vault and is actually quicker and easier to construct because the most awkward voussoirs (those meeting at the groin with diagonal intersectioning edges) could be left roughly shaped. This is because the joint would be hidden when viewed from below by the presence of the ribs. The gaps generated by this technique could be filled from above with mortar.

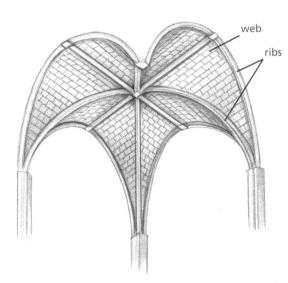

web
ribs

Ribbed groin vault: a more aesthetically pleasing and innovative version of the simple groin vault.

Web construction: a stonemason at work. (Note: some timbering has been omitted from this image for clarity.)

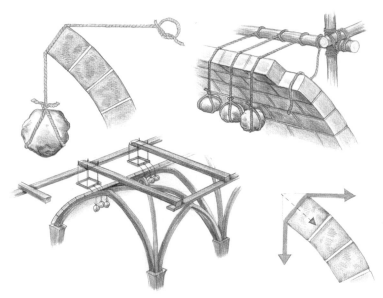

An ingenious technique (not used in Cormac's Chapel) later evolved for the construction of web sections of vaulted roofs. This technique involved the use of ropes weighted with stones to hold the voussoirs in place.

Ribbed groin vault: stone weights being used to hold voussoirs in place. This innovative technique greatly reduced the need for timber supports under the voussoirs.

Anglo-Norman Period (Twelfth Century to Sixteenth Century)

Churches and Cathedrals

The Gothic style of architecture was introduced in Ireland during this period. Although very few buildings were built in Ireland using Gothic techniques, many Irish buildings have elements, such as doors and windows, in the Gothic style. Gothic architecture is characterised by lightness. The walls are thinner, taller and have larger openings for doors and windows than Romanesque buildings. The arches of Gothic buildings are pointed rather than rounded. The flying buttress was the main technological advance that allowed this style of architecture to evolve.

Christ Church Cathedral, Dublin.

The walls of tall masonry buildings have to be designed to handle the large outward thrust exerted upon them from the roof. The first solution (Romanesque) was to design very thick walls. However, there was often a desire to build churches that were taller and grander. The problem with this, is that as the walls got higher they also had to get thicker and this limited the practical height achievable. An alternative design had to be found.

Eventually it was discovered that the load from above was thrusting outward at an inclined angle and could be dealt with by extending several relief walls (buttresses) perpendicular to the main building, rather than by thickening the entire wall. This saved space, material, construction time, and reduced the cost of construction. It was later noticed that only the outward portion of the buttress actually carried the load. The portion in the middle of the wall wasn't carrying any load except its own. The use of an arch allowed this defunct portion to be removed, which led to the much more elegant design we see today. This structural element is called a *flying buttress*.

A flying buttress can be defined as an inclined bar of masonry carried on a segmented arch that transmits outward and downward loads from a roof or vault to a solid buttress. In contrast, the solid buttress relies on its mass to transmit the load safely to the ground.

The fact that the loads are being carried by buttresses allows the use of thinner walls and larger openings. This allowed the large stained-glass windows, characteristic of Gothic church architecture, to be created. The use of flying buttresses can be seen in Christ Church Cathedral in Dublin.

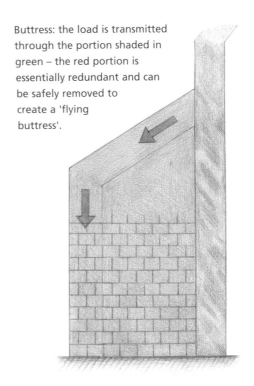

Buttress: the load is transmitted through the portion shaded in green – the red portion is essentially redundant and can be safely removed to create a 'flying buttress'.

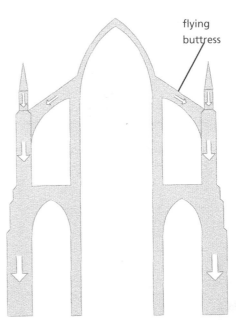

flying buttress

Sectional view of typical Gothic church showing loads being transmitted through flying buttresses.

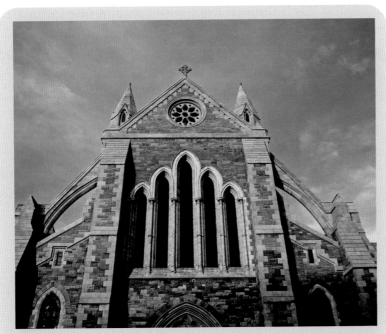

Flying buttresses at Christ Church Cathedral, Dublin.

Gleninagh Castle, Co. Clare: typical example of an Irish tower house. Note the defensive features: narrow ground-floor windows, a machicoulis above the entrance and bartisans at the corners.

The Tower House

The tower house was the typical residence of the Irish landowner in the fifteenth and sixteenth centuries. The tower house was the first dwelling to be widely adopted by both settlers and natives alike. Over 2,000 tower houses were built in Ireland and many of these can be still be visited today. A tower house is a rectangular stone tower, up to six storeys high. It usually had a tower on two or more of its corners, at least one of which contained a spiral staircase for access to the upper floors.

Access was gained through a ground floor doorway, which was protected by an iron grille (known as a *yett*) that could be pulled across using chains. Tower houses usually had a solid wooden door. On the outside of the wall, high above the doorway, a small stone chamber (called a *machicoulis*) was constructed.

This chamber, which was carried on stone corbels projecting out from the walls, had openings called *machiolations* (sometimes referred to as *murder holes*) from which arrows etc. could be fired. A small, overhanging turret (called a *bartisan*) protected the corners of the tower house.

The main design feature of tower houses is vertical planning. Each floor of the house contains one large room. The living and sleeping quarters of the owners were contained on the upper levels. These levels also had larger more decorative windows, unlike the defensive arrow loops (vertical slit openings) of the ground floor.

Tower houses also contained large stone fireplaces, usually set in the gable ends of the building, with their flues taken through the walls and carried upwards on the outside of the walls. There was no running water in these houses and the *garderobe* (toilet) was merely a wooden seat over a stone chute in the wall which exited on the

Milltown Tower House, Co. Louth. This four-storey, rectangular tower house has round projecting towers (also known as flanking turrets) at the east and west corners. These towers rise one storey above the roof of the house. The castle is vaulted above the ground floor. The original entrance in the south-east wall was protected by a murder hole on the second floor. The present entrance is through the base of the east turret. This leads to a modern flighted stairway which rises within the flanker to the second floor. This then changes to the original stone spiral stairway which rises to roof level. The west corner contains garderobes (toilets). The castle dates from the 15th/16th century and was lived in until modern times.

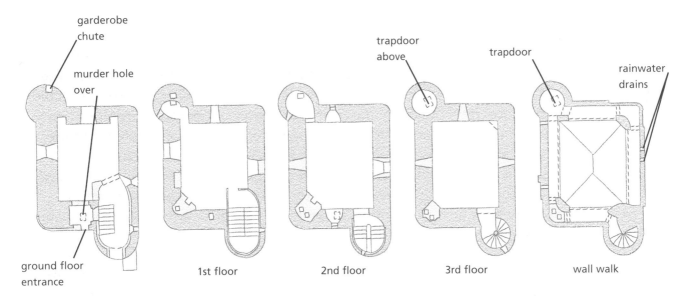

outside of the wall close to the ground. The garderobes were usually in a small chamber that also sometimes contained a slop-stone, for hand washing, which also drained through the wall.

Most tower houses would have had a defended courtyard with surrounding walls and small towers. This courtyard (or *bawn*) would have contained some free-standing wooden buildings and some buildings that leaned against the curtain walls.

Vernacular Houses (Seventeenth Century Onward)

Every culture has its own vernacular architecture. The Inuit people of Greenland have the igloo; the North American Indians have the tepee; Mexicans have the pueblo; Africans have the adobe hut. Irish vernacular architecture is exemplified by the thatched cottage. It is instantly recognised around the world as a symbol of Ireland.

The following qualities are typical of vernacular architecture:

- The builders are non-professionals.
- The buildings adapt harmoniously to the natural environment.
- The construction process is intuitive, without blueprints, and open to later modifications.
- There is a balance between functionality and aesthetic features.
- Patterns and styles are subject to a slow evolution of traditional designs suited to the region.

In vernacular architecture the contribution of the individual is always less than the overriding features of the local style – in contrast with formal architecture where the influence of the architect is clearly noticeable.

Traditional vernacular dwellings blend easily into the landscape for several reasons. The use of natural building materials is a major factor. They were usually built in sheltered positions which tended to give these homes a sense of *hugging the land*. In this way they appear as an organic part of the landscape, rather than standing out boldly against the skyline as many modern houses do. Furthermore, the simple proportion and scale of these buildings coupled with the use of bright colours (especially red) around windows and doors is pleasing to the eye. Ultimately the greatest aesthetic quality of these vernacular dwellings is the sense of respect for the landscape which they convey through their simplicity of design and humbleness of execution.

In the rural community the thatched cottage was seen as a symbol of hardship and was associated historically with poverty and a low standard of living. For many years thatched cottages were knocked down by their owners and replaced with modern bungalows. It is only in recent times that the thatched cottage has been recognised as a beautiful and sensitive building which should be treasured as an example of how architecture can work in harmony with the landscape.

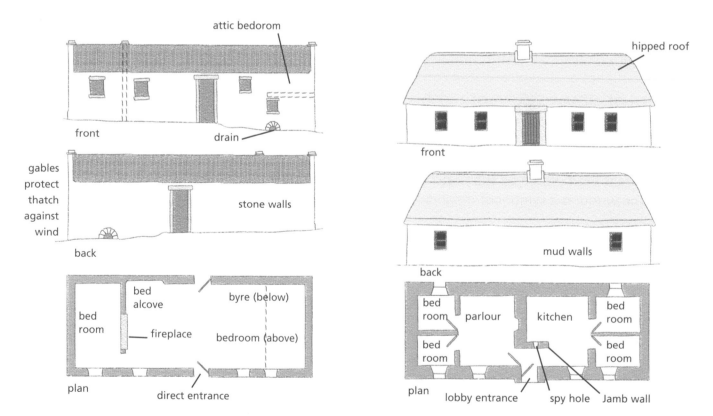

Vernacular houses: two distinctive designs evolved, influenced mainly by weather conditions and locally available building materials. Along the west coast, stone-walled houses were built. The roof was protected from strong winds by projecting gable walls and the thatch was often tied down using rope netting. In the Midlands and east, mud-walled houses had hipped roofs.

There are many examples of the functionality of vernacular dwellings. For example, early byre houses had two opposing doors, one to the front of the house and one to the rear, linked by a flagstone floor. This served several purposes: it allowed for the movement of people and animals, it made it possible to regulate the draught and smoke from the fire, it allowed the dweller to select which door would be used on a given day according to the direction in which the wind was blowing.

When you walk into one of these houses you immediately notice the lack of a doorstep at the entrance – there is no change in ground level between exterior and interior. This immediately links the house and the land in a way not experienced in today's homes. Ironically, the recent change in building regulations relating to access for wheelchair users has brought this back into being.

The half door was another functional element of these homes. It kept animals out and children in, allowed more light into the home, provided a draught for the fire and provided support for a person to lean on while chatting to passing neighbours.

The byre house (house containing people and livestock) provides a good example of functional design in that it was usually built on a slope with the byre end downhill. This ensured that the animal waste was contained at the byre end of the building and flowed out through a drain that was cut in the floor.

Furthermore, the byre house provides another example of functional design in the use of the attic space of the byre as a sleeping area. Naturally this gave the sleeping farmer a good source of heat on cold nights.

Another example of functional design is seen in the development of the jamb wall in cottages in the east of Ireland. It was common practice to position the hearth in the centre of these houses. This created a difficulty because of the proximity of the opposing doorway. To solve this problem, the jamb wall was devised. It was positioned to provide a screen between the doorway and the hearth and so reduce heat loss and provide shelter to the fire from the draughts caused by people entering and leaving the house – it also gave some privacy to the room. A small opening was usually created in the wall to allow the dweller to see visitors arriving.

Design Influences
While it is nice to think that the design of vernacular houses was driven solely by a wish to build attractive homes, the reality is that these houses evolved over the centuries as a result of particular influences. These can be classified as follows:

Climate – The influence of the climate on the design of vernacular dwellings in Ireland is most apparent in the design of roofs. Along the Atlantic seaboard the high winds led to the evolution of the *gable-end* roof, as distinct from the *hipped* roof common in inland and eastern regions. This is because the gables, which faced out to sea, protected the thatch from being lifted by the wind.

Materials – Vernacular dwellings are built using locally available materials. The use of mud or clay as a walling material in the Midlands and eastern regions was very common, this contrasts with the use of stone in parts of the West where suitable clay could not be found. Similarly, Irish cottages are always only one-room wide (3·6 m to 4·5 m) because of the scarcity of long timbers for roofing.

Technology – Advances in technologies affected the development of vernacular house designs. For example, the arrival of corrugated iron sheeting led to a decline in the use of thatch roofing, especially for outhouses and sheds where heat retention is less important.

Skills – These houses were built either solely by the home owners or with the assistance of local artisans (skilled workers who made things by hand). There were no organised construction companies or builders available to poor farm workers so the quality of the building depended very much on the level of skill and knowledge available locally.

Existing Buildings – There is some evidence in remote places, especially in West Connaught, that vernacular dwellings evolved from older structures with oval or circular plans. The beginning of this evolution may be represented by dome-shaped, stone clochain.

Economics – This is an extremely significant factor in the quality of vernacular architecture. Many of the most popular features of vernacular dwellings, such as the thatched roof and white-washed walls have evolved as a result of economic constraints. For example, thatch was used because this material was cheap and readily available, growing wild in the countryside. The poor farm labourers who lived in these homes could not afford to install a more durable slate roof. Similarly, the low cost of lime-based wash meant that it was the only option when it came to painting the exterior of the house.

Social Structure – The influence of social structure was evident in the evolution of the *byre-houses* of the west of Ireland. A byre is essentially a cowshed. A byre-house is a house that gave shelter to both people and livestock. These houses were common in the west of Ireland until the nineteenth century. The housing of animals in the family home was seen as undesirable by the poorer farm labourers who looked up to the wealthier landholders and wanted to live like them. When this influence eventually spread to the west and north, animals were housed in sheds and the byre was then converted into a bedroom or parlour (sitting room).

Culture – The influence of culture is very evident in vernacular houses. Most people know the phrase 'there's no place like home'; of course, in the Irish language we say 'Níl aon tintean mar do thinteán féin', literally 'there's no hearth like your own hearth'. This expression reflects the cultural importance of the fireplace in the Irish home. It is a reflection of the sociable nature of Irish people and this is in turn reflected in the central position of the fireplace in the thatched cottage. While there are many examples of houses whose fireplaces are positioned on gable walls, the hearth usually enjoys a central location and is always located in the kitchen, at the centre of all activities both practical and social. Many modern rural homes continue to be built with a fireplace or stove in the kitchen.

Neoclassical (Eighteenth Century Onward)

Classical architecture is the term used to describe the architecture of ancient Greece and Rome. The Parthenon in Athens and the Colosseum in Rome are both examples of classical architecture.

Proportion is a key concept in Classical architecture. Classical architects achieved aesthetically pleasing designs by devising a proportional relationship called the *golden mean* (or *golden section*). The golden section shows equal relationships between the sections and the whole. In simple terms this means that if you divide a line according to the golden mean, the ratio of the short part to the long part is equal to the ratio of the long part to the whole. The golden mean is expressed mathematically

Classical architecture: The Parthenon, Athens.

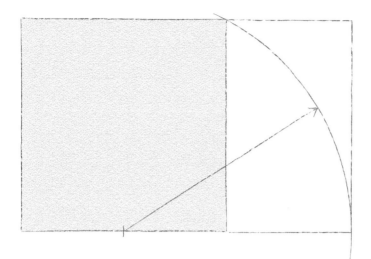

Golden section: a rectangle whose sides are in the ratio 8:5, can be devised from any square as shown.

as 1: Φ (pronounced Phi) or 1: 1/2(√5 + 1), approximately 1:1·618 or 8/5.

This relationship has been considered inherently beautiful by architects and builders for centuries. It has been used extensively in the design of many significant buildings, including the Pyramids of Egypt, the Parthenon in Athens, the United Nations headquarters in New York, and even the CN Tower in Toronto (world's tallest building, 553·34 m).

It is because of the classical awareness of the importance of proportion that these buildings were, and are, considered so beautiful and why they have had such a significant influence to this day.

A number of buildings were built in the Classical style in Dublin and in other cities in Ireland. These buildings are said to be neoclassical because they are a *modern* interpretation of the ancient classical style. The Casino at Marino, Co. Dublin is a wonderful example of neoclassical architecture. Designed by Sir William Chambers, it was constructed between 1755 and 1773. It was taken into state ownership in 1930 and restored by the Office of Public Works.

CN Tower, Toronto.

United Nations building, New York.

The Casino (meaning house by the sea) is quite a small building – only 15 m square to the outer columns. Seen from the outside, the building has the appearance of a single-roomed structure, with a large panelled door on the north elevation and a single large window on each of the other elevations. This is all an illusion, however, as it actually contains sixteen rooms on three floors. Only two of the panels in the door open to allow entrance. This change in scale between exterior and interior allows the Casino to contain many more rooms than would appear possible from the outside. The panes of glass in the windows are subtly curved, disguising the partitioning which allows what looks like a single window to serve several separate rooms.

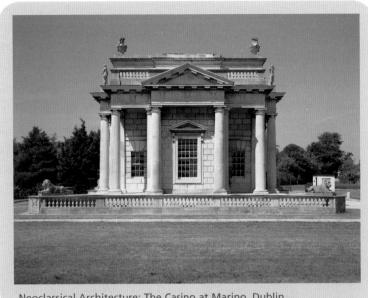

Neoclassical Architecture: The Casino at Marino, Dublin.

The exterior façades are all richly decorated with sculptures and decorative carving. The façades are dominated by the solid attic storey, statues and urns. The entrance façade has a doorway that rises almost to the height of the columns. The decoration is also functional – four of the columns which surround the building are hollow and used to drain rainwater from the roof, and the Roman funerary urns on the roof are used as chimneys – to avoid spoiling the external appearance.

Georgian Architecture (Eighteenth Century Onward)

Georgian architecture, which is based on the principles of classical architecture, flourished in Ireland during the eighteenth century. This period is specifically from 1714 to 1830 and marks the reign of four consecutive English kings (namely George I, George II, George III and George IV) – hence *Georgian*.

Georgian architecture is most important in the Irish context for its influence on house building in both the urban and rural settings. In the urban setting (e.g. Merrion Square, Dublin; The Mall, Cork; Pery Square, Limerick) a typical Georgian town house is a terraced four-storey, over-

Georgian town house: four stories over basement. Note the relationship between the parts (e.g. windows) and the whole.

Georgian door: an example of the splendid doors seen in many towns around Ireland.

basement dwelling constructed in red brick. Brick was first used in the 1700s, having been brought back from England as ship ballast. This brick was of variable quality. This is seen in the variety of texture and colour in Georgian terraces of cities such as Limerick and Dublin. In the 1800s bricks were mass-produced, rather than handmade. By the early 1900s they were available in a range of shapes and mouldings, dimensionally precise and uniform in colour. The colour of brick depends on the clay used and the temperature at which it is fired. Under-burned brick is lighter in colour and softer, while over-burned brick can be identified by its darker colour and hardness.

Georgian houses had ornate panelled doors, which were usually painted in vivid colours, with detailed fanlights overhead. Many of these doors have survived with their polished fittings intact.

The sliding sash windows were typically a well-proportioned arrangement of three panes over three in the top floor and six panes over six on lower floors. The thin glazing bars and the painted reveals of the windows give the windows a bright appearance. The delicate tracery of the fanlights and the ironwork of the railings and balconies all contribute to the overall appearance of the Georgian town house. In keeping with the classical approach, the scale and proportion of each element of the façade is in keeping with the whole. This gives the typical town house an aesthetically balanced and pleasing appearance. These houses are further enhanced by the wide streets and squares that are a typical feature of Georgian town planning.

A typical Georgian country house reflects the influence of the classically styled country estate houses of the gentry. These houses are commonly symmetrical in appearance, three bays wide, with a central doorway. They have a hallway that is flanked by two rooms which run the full depth of the house (i.e. from front to back). There is usually a fireplace in both ground floor rooms as evidenced by the centrally located chimneys. There are usually two or three bedrooms on the upper floor. Some features were adopted from formal houses such as corner *quoin* stones, large windows and a fanlight above the front door. A basement was common in the larger versions of these houses.

Exploring the Built Heritage of Your Local Community

Most villages, towns and cities have a built heritage which can be explored. Often we do not realise that the buildings we see everyday have long and interesting histories. Sometimes these buildings will have had different uses in the past, been built by famous people, or have even witnessed significant events in history such as the GPO in Dublin. The fact that these buildings have survived through generations means that they often represent the very best in local architectural endeavour. The best way to develop an appreciation of heritage buildings is to visit and explore them. It can be very rewarding to discover the history of a building and the people who lived or worked there. Exploring heritage buildings is a two-stage process involving the examination of the building's context (environment, history, people) followed by the detailed exploration of the building itself.

Georgian country house: a simpler interpretation of the Georgian style that maintains the overall symmetry and balance found in Georgian architecture.

Examining a Heritage Building's Context

Construction History
- Who built it?
- Why was it built?
- When was it started and when was it completed?
- Who was its architect and its builder?
- How has the building been used and adapted over the years?

Link with Site and Situation
- How does the building relate to the street and to its immediate surroundings?
- Has the physical geography of the site influenced the structure?
- How is the building situated on its site?
- What elements make the structure part of a group?
- Is this a landmark building? If so, what characteristics gave it this quality?
- How do passers-by relate to the building?
- What makes this building unique?
- What is the architectural style of the building?
- Do planning laws or heritage regulations affect the building or its site?

People Dimension
- Does this building suggest anything about the people who built it?
- How does the structure influence people's attitudes and behaviour?
- Are concepts of class, status, race or ethnicity useful in interpreting the building? If so, in what ways?
- Does the structure tend to isolate people or bring them together?

The Time Dimension
- How has this site been used by people in the past?
- How has the community around the building changed over time?
- Has the site or the structure reflected these changes?
- Can the history of the building or the site be divided into periods?
- To what extent was the building innovative in its design?

Examining a Heritage Building's Construction

There are several elements, which should be noted when creating a record of a building in your locality. These include:

- Building type or form. Is it a dwelling, church, castle, industrial building etc.?
- General characteristics:
  - Foundation type, if known.
  - Overall shape of plan and arrangement of interior spaces.
  - Number of storeys, vertical divisions or bays.
  - Construction materials such as brick, concrete, stone, timber and slate.
  - Wall finish such as type of bond or course.
  - Roof shape such as gabled or hipped.
  - Structural system such as solid wall, reinforced concrete, post and beam.
- Exterior features by type, number, location, material and condition including porches, windows, doors and chimneys.
- Decorative elements on exterior features such as columns, mouldings, corbelling, sculptures etc.
- Interior features such as stairways, functions of rooms, spatial relationships.
- Interior decorative elements such as wainscoting, flooring, panelling, beams, vaulting, mouldings and fireplaces.
- Number, type and location of outbuildings.
- Other man-made elements such as roads, more recent buildings and landscape features.
- Deterioration due to neglect, lack of use, vandalism, weathering and their effect on the building's historical integrity.
- Restored/reconstructed buildings: date of restoration/reconstruction, historical basis for the work, amount of original material remaining and replacement material, effect of the work on the building's historic integrity.

A written record of these details should be made during, or immediately after, the visit. It is also very useful to take photographs and make sketches of important details.

Sketching Buildings

It is reasonable to ask why we should bother to take the time and effort required to sketch a building when a photo could be taken (which would be more accurate and detailed). The answer is simple. In order to make a sketch which accurately reflects the scale and characteristics of a building the *artist* must first study the building.

When we study a building in this way we gain a much greater appreciation of the qualities of its structure, materials and finish. This level of awareness is not achieved by merely taking a photograph. When buildings are sketched regularly the artist develops a *trained* eye – i.e. the ability to note details at a glance and better understanding of architecture.

What to Use?

The materials and equipment selected for any drawing will greatly affect the results achieved. It is very important to be comfortable with whatever medium you are using and to get lots of practice. Usually a soft pencil of grade 2B is suitable for most sketching – it allows for a good range of tone to be achieved, rubs out easily and doesn't make heavy indentations in the paper.

Taking Visual Measurements

Transferring the shape and scale of a building onto paper can be made easier through the use of *pencil measuring*. To do this, the pencil is held at arm's length with the point upwards and the thumb uppermost on the pencil. The point is lined up with the top of the object and the top of the thumb is lined up with the bottom of the object. This measurement is then transferred to the sketch.

Using Verticals and Horizontals

The first step in creating an accurate drawing is to identify the main vertical and horizontal lines of

Sketching buildings: taking a visual pencil measurement.

Transferring the main vertical and horizontal lines to the sketch.

the building and to transfer these to the sketch. Put a faint horizontal line across the middle of the sheet at eye level. Now draw a vertical line down the middle of the sheet. These lines will represent the centre of the building and will act as a reference point for the transfer of measurements.

Now, taking these lines as a starting point use the pencil measuring method to draw vertical lines to represent the main vertical divisions of the building.

Repeat this method for the main horizontal lines.

Gradually build up the sketch, adding smaller divisions such as doors and windows before sketching in the details to finish it off.

Finished sketch: if you like, you can use the side of your pencil to add shadows or colouring pencils to add colour to enhance the sketch.

Sources of Information

Apart from speaking to the people who currently own or maintain the building, information can be obtained from a variety of sources, including:

- Interviewing local people (especially elderly people, historians, politicians).
- Local newspaper archives.
- Government inventories – county/city records.
- Civic trust records.
- Planning authority records.
- Historical photographs.
- Architectural plans.
- Historical maps.
- Telephone directories.
- Land registries (organisations responsible for the registration of transactions relating to land and property).

Revision Exercises

1. Write a short paragraph explaining the importance of heritage buildings.
2. Explain, in your own words, how the heritage value of an old building can be assessed.

3. Outline how the context of a heritage building is examined.
4. Generate a neat annotated sketch of a heritage building in your local community.
5. Write a short paragraph explaining why you think a heritage building in your local community should be conserved.
6. Generate a *close-up* sketch of three important exterior features of a heritage building or structure in your local community.
7. Explain, using a neat annotated sketch, the principle of corbelling.
8. Generate a simple sketch of a church built in the Romanesque style.
9. Explain how flying buttresses made it possible for thinner, taller walls with larger openings to be constructed.
10. Explain the importance of the *golden section* in Classical architecture and sketch an Irish building which employs this principle.
11. Higher Level, 2000, Question 10
    'Domestic architecture often combines one or two styles, varied or adapted locally depending on the climate, location, materials available, the skills of the builder and workers, economic status, lifestyle, social concerns or restraints and fashions.' *Hearthstones* (1993): Caneta S Hankins. Discuss.
12. Higher Level, 2001, Question 10, Option 1
    'Our countryside buildings should not be taken for granted. They deserve far more study and their appropriate use of materials, their sympathy for the landscape and their human scale deserve appreciation, for they are distinctively Irish and a significant part of our architectural and cultural heritage.' *Irish Countryside Buildings* (1985): P and M Shaffrey. Discuss.
13. Higher Level, 2003, Question 10, Option 2
    'Vernacular styles of buildings exist all over the world. These styles are characterised by their simplicity, by their use of local materials and by the ease with which they can be constructed. The knowledge required for the creation of such buildings was long regarded as common knowledge and freely available to all. The decline of the vernacular tradition with its simple forms and its accessibility to people has resulted in the loss of the knowledge and skills needed to design and construct small buildings, especially the buildings in which people live – their homes.' *Be Your Own Architect* (1992): Peter Cowman. Discuss.

14. Higher Level, 2004, Question 10
    A suburban dwelling house, built over one hundred years ago, has a large rear garden with mature trees and shrubs. A roadway provides access to the rear garden, as shown in the sketch.

Planning permission is being sought to divide the rear garden (as shown) and to erect two town houses in the divided garden.

(a) What arguments might be presented:
   (i) In support of the erection of the town houses?
   (ii) In support of the retention of the property in its original state?

(b) Make a recommendation to the planning authority on this proposal and discuss in detail three reasons in support of your recommendation.

Conservation is a broad term used to describe action taken to prevent the decay of old buildings. The goal of conservation is to sustain the life of an old building. There are two approaches that can be taken to conservation.

Restoration

One approach is to turn back the clock on a building; in other words, to restore the building to its original state using only the original materials and building techniques. This is called *restoration*. When a building is restored in this way it is sometimes used as originally intended (e.g. as a mill) or used as a heritage site that people can visit to learn about the past. There are many tradespeople and architects who are qualified to do this work – it is important that this type of work is carried out by specialists who have experience of working with old buildings.

Reconstruction

The alternative to restoration is to use new materials and techniques to add to the building in its present state. When this is done the damaged parts of the building are repaired or replaced using modern techniques and materials and the building is often adapted to a new function. This is called *reconstruction*. An example of this is when a granary building is converted into apartments or offices.

An unused granary building which has fallen into disrepair. There are many of these buildings in towns around the country.

A reconstructed granary building finds a new use as offices.

Care of the Built Heritage

The Department of the Environment, Heritage and Local Government is responsible for the conservation of both the natural and built environments. However, a number of other groups are also involved in taking care of the heritage buildings found all over the Irish countryside. Some of these groups are government organisations such as the Office of Public Works, the Local Authorities, The Heritage Council and the National Inventory of Architectural Heritage while others are non-governmental organisations such as the Civic Trusts, An Taisce and An Bord Pleanála.

Office of Public Works

The Office of Public Works (OPW) is responsible for the day-to-day maintenance and management of all national monuments and heritage buildings. Part of this responsibility includes producing information leaflets and providing guided tours of the many heritage sites around the country. The OPW has one of the largest architectural practices in the country. They design and manage large-scale conservation projects as well as providing independent advice to the government. They also manage many working buildings including Dublin Castle and Farmleigh House.

Farmleigh House, Phoenix Park, Dublin.

Local Authorities

Local authorities (e.g. city and county councils, county boroughs, borough corporations and urban district councils) are required by law to take steps to protect the architectural heritage in their areas. Local authorities have the following powers:

- To order necessary conservation work to be carried out to protect a building.
- In extreme cases of neglect, to purchase (by compulsory purchase order) a site on which a heritage building is at risk.
- To carry out the works itself and recover the costs from the owner or occupier.
- To impose a substantial fine and/or prison term (through the courts) for those found guilty of damaging a protected structure.

The owner of a protected building is also required by law to ensure that the building does not become endangered – if a person owns a site that has an old building on it they must take care of the building. When a planning authority requires work to be carried out to prevent a protected structure from becoming endangered, the owner concerned may be eligible for grant assistance to help pay for the work.

The Heritage Council

The Heritage Council was established as a statutory body under the Heritage Act 1995. Its role is to advise the government on matters relating to conservation. It does this by proposing policies and priorities for the identification, protection, preservation and enhancement of the national heritage. National Heritage is defined as including monuments, archaeological objects, heritage objects such as art and industrial works, documents and genealogical records, architectural heritage, flora, fauna, wildlife habitats, landscapes, seascapes, wrecks, geology, heritage gardens, parks and inland waterways. The Council has a particular responsibility to promote interest, education, knowledge and pride in the national heritage.

National Inventory of Architectural Heritage

The National Inventory of Architectural Heritage (NIAH) is a section within the Department of the Environment, Heritage and Local Government.
The work of the NIAH involves identifying and recording the architectural heritage of Ireland, from 1700 to the present day, in an organised manner. Its main functions are to:

- provide a source of guidance for the selection of structures for protection.
- supply data to local authorities, which helps them to make informed judgments on the significance of buildings in their area.
- foster greater knowledge and appreciation of Ireland's architectural heritage.

An Taisce

An Taisce (meaning: the *Store House* or *Treasury*) is an independent environmental advocacy group. An Taisce is probably best known for its role under planning legislation. The planning laws require that An Taisce is consulted by local authorities on a vast array of development proposals especially in relation to how they will impact on the natural and built heritage. This means that sensitive planning applications must be referred to them by planning authorities for their comments. An Taisce members, both local and national, make observations on around 3,000 planning applications each year. They also make around 300 submissions annually to An Bord Pleanála. They also maintain a record of buildings at risk and campaign to have them conserved. They also manage and maintain a number of heritage sites around the country. The nature of the work in which An Taisce is involved means they are often at the centre of controversial issues such as the construction of interpretative centres, single one-off houses in the countryside or developments that impact on important habitats. People with an interest in the environment or in property development can have very different opinions on the role of An Taisce. The organisation itself states that it has no vested interests and that it takes the public interest standpoint in all planning matters.

Civic Trusts

There are a number of Civic Trusts established in various cities and towns around the country. Each Civic Trust is an independent self-funding group, which engages in conservation projects in co-operation with local authorities, state agencies and other interested parties. Civic Trusts are non-profit-making voluntary groups and are usually registered charities. The Civic Trusts specialise in carrying out exemplary work in the restoration of heritage buildings.

An Bord Pleanála (The Planning Appeals Board)

An Bord Pleanála functions as a *court of appeal* for individuals or organisations who are not satisfied with the decision of a local authority in relation to a planning application. It is an independent body whose sole responsibility is to decide whether the decision of the local authority was correct and fair to all parties involved.

The appeal system is designed to be independent, fair, impartial and open. The common good and the principles of proper planning and sustainable development guide the decision-making process of the board. One example of the openness of the system is that the entire file relating to any appeal can be viewed by any member of the public just three working days after the appeal is determined. The file remains available for viewing for five years. Much of this information is also published on-line.

Principles of Conservation

Conservation is a very broad concept which can usually be thought of as either restoring a building to its original state or reconstructing it to a new design. Regardless of the approach taken, the guiding principles are the same. The overall goal is to highlight the historical significance of the building as much as possible. It is very important that when reconstruction, in particular, is being carried out that the future use of the building is compatible with the fabric and structure of the original building.

For example, if a developer wishes to build a multistorey car park where an old granary building currently stands, the old building might not be suitable because it may not be able to sustain the loads applied by hundreds of parked cars. However, that does not mean that the granary should be knocked down. Instead the developer could be encouraged to use the existing building to develop something more suitable such as apartments or offices.

If old buildings are to be respected and valued as a part of our heritage (see chapter 1, Built Heritage, for reasons why our built heritage is important) we must take the needs of these buildings into account when deciding what future use they will serve. It is important that all construction activity is sustainable. New uses should be found for old buildings, owners of old buildings should be encouraged to maintain their properties properly, and developers should be encouraged to purchase old abandoned buildings and adapt them to new uses.

The main principles of conservation are:

- **Retain or restore the historical significance of the building** – This means future generations should be able to see for themselves the features of the building that make it an important part of our built heritage. Features such as floorboards, lime plaster cornices and window shutters all contribute to the overall feel of a heritage building.
- **Carry out proper research before beginning construction work** – Before any work is carried out on a heritage building it is important that a clear picture of the building's history is built-up. This mainly involves finding out when the building was constructed and what materials were used to construct it. There is a wide variety of sources of this type of information including:
  – the building itself,

- the drawings and documents relating to the building,
- local craftspeople with traditional building skills,
- national archives such as the Irish Architectural Archive,
- other voluntary organisations such as the Irish Georgian Society.
- **Make the minimum physical intervention possible** – This means that wherever possible the original fabric and character of the building are kept. To do this the golden rule is: *repair rather than replace*. In other words the building's owners need to strike a balance between solving the type of problems that come with old buildings (leaking gutters, rotted window frames, draughty doors) and protecting the authentic sense of history those old buildings convey.
- **Maintain the visual surroundings and setting** – If a building was a painting then the visual setting would be the picture frame – changing the frame can have a significant visual impact. To maintain the visual setting the owner must take into account the surroundings of the building. For example, a country estate house would usually have large planted grounds surrounding it. If a developer was permitted to build on these grounds the setting would be lost and the large house would seem out of place among the new buildings.
- **Reversibility** – Ensure all changes made to a heritage building can be undone without damage to the original fabric of the structure. Any new work that is carried out on a historic building should be reversible. Important architectural details should be protected and detailed records of changes made during construction should be kept. As little damage as possible should be done to the original building during reconstruction.

Procedure

The conservation of historic buildings is properly carried out as a series of six steps:

1. Research and analysis of building history.
2. Survey of building.
3. Planning of conservation works.
4. Work is carried out.
5. Work is recorded.
6. Maintenance plan is put in place.

Following this procedure will ensure that the original value of the building is both protected and revealed for future generations. A number of areas of old buildings are often dealt with incorrectly during conservation works – especially elements of the external envelope, including windows and external walls.

Examples

Windows

The sliding sash window was introduced to Ireland in the eighteenth century and was the most common type of window used for the next two hundred years.

The use of sliding sashes is seen in Georgian town houses in many of the country's towns and cities. These sashes consisted of three, six or nine panes of glass. The limited availability and high cost of glass meant that many small panes rather than one large pane was used – this changed around 1850 when plate glass became more widely available. The proportion of these buildings and their elements (such as doors and windows) are based on the golden section. This proportion guides the design of many parts of the building including the panes of glass. If this relationship is not maintained the appearance of the building is badly affected. It is essential during conservation work that timber sashes are not replaced with uPVC. This is because a typical uPVC replacement window will lack the essential proportion of the original timber window. The frame and glazing bars of a uPVC replacement will be thicker and wider than the original timber elements. This *chunky* look ruins the aesthetic balance of the window. Also, there is a tendency to use *internal glazing bars* (i.e. bars between the panes of double-glazing) in uPVC windows to create the appearance of a Georgian window.

Traditional timber sliding sash windows. The proportion of these windows makes them an appropriate and elegant component of Georgian style buildings. The size of the panes and the thin glazing bars are particularly important.

Original timber Georgian sliding sash.

An unattractive uPVC copy of a Georgian sliding sash: note the internal glazing bars and the chunky frames.

However, when light falls on these windows the window tends to look like one large pane of glass and the Georgian effect is lost.

Mortars and Rendering

The single biggest danger when repairing mortar is that the incorrect material will be used. When dealing with old buildings it is important to remember that many of them have thick solid walls that absorb moisture during wet weather and release it afterwards. To do this the wall must be able to *breathe* – in other words the wall must be able to allow the moisture it absorbed during heavy rainfalls to evaporate. In order to do this, the joints between the bricks or stonework must be of lime mortar. Lime mortar, as opposed to cement mortar, is more flexible and porous. This allows old walls to respond to changes in the weather. The golden rule is that the joints should always be weaker than the stone or brickwork surrounding them. The joints should accept the weathering and not force the stresses (expansion/contraction) into the stone/brickwork.

If the lime mortar joints in a red brick house are repointed using a cement mortar the wall will lose its flexibility and breathability, and will become damp and possibly crack.

Mortar joints: jointing occurs as the blockwork is laid and includes flush, bucket handle, tooled and struck or weathered styles.

Pointing: is done after the blockwork is laid and can be similarly finished. Poor joint treatment can reduce the structral integrity of the wall and give it a terrible appearance.

A render is an external coating applied to a building to protect its walls from the effects of weathering. There were many types of renders used but the most common approach was to use lime-based render applied to rubble stone walls. Unfortunately it has become fashionable to remove renders to expose the stonework underneath. This is not a good idea for a number of reasons:

- It exposes the surface of the wall to the rain, making it more likely to become damp.
- It shows poor quality stonework which was never intended to be seen.
- It detracts from the appearance of a group of buildings (e.g. in a street) if one of them is stripped in this way.
- Renders are often of a bygone era, using a mixture of materials (such as lime and animal hair) that are no longer available.
- Once removed an original render is lost forever.
- It conveys a lack of respect for the craft and traditions of the time in which the building was built.
- It disrupts the aesthetic balance between the whitewashed walls and the natural, dark, slate roof.

Original renders should only be removed in the case where they have decayed so badly as to be ineffective. Any damaged render removed for this reason should be replaced with a lime-based render.

Removing the he external render from a typical verncaular dwelling can be very damaging to the walls and exposes stonework that was never meant to be seen.

Revision Exercises

1. Explain, in your own words, the various approaches that can be taken to the conservation of old buildings.
2. Describe briefly the role of the Local Authorities in relation to the care of heritage buildings.
3. Generate a neat sketch of a building you are familiar with which has been conserved.
4. Outline briefly the main principles of conservation and explain in your own words the importance of each.
5. Generate a neat annotated sketch showing the correct timber window and an incorrect uPVC replacement in a typical Georgian building.
6. Explain the importance of pointing in relation to the conservation of heritage buildings.

7. Higher Level, 2001, Question 8, (a) and (b)
   A single-storey traditional dwelling house, over 100 years old, has thick stone walls, a slate roof, wooden doors and windows, and solid floors. The house is in need of repair and it has been decided to undertake essential renovations.
   (a) Make a checklist of four renovations you expect would be needed in a house of this age.
   (b) Describe, using notes and sketches, how each of these renovations could be carried out in a manner which would respect the age and character of the original house.

8. Higher Level, 2004, Question 3, (b) and (c)
   (b) The original external render of an old house is to be removed to reveal solid stone walls of random rubble construction, as shown on the sketch. The owner has the option of either leaving the external stonework exposed or of re-plastering the walls. Outline two reasons in favour of each option listed above. Recommend a preferred option and give two reasons to support your recommendation.

   (c) If the house is to be re-plastered, a 1 lime : 3 sand mix is recommended for the external render.
        Give two reasons why such a mix is recommended for this house.

Development

Development is the term used to describe construction of all kinds, including the building of houses, shops, factories and roads. Ireland has over 600 towns and small villages. These towns are centres of social, cultural and commercial activity used not only by local people but visited by people from the surrounding areas and by tourists. All new building work must be carefully controlled, both in terms of the design of the buildings and their location, so that the qualities of the natural and built environments are maintained and enhanced.

Without planning and controls there is a risk that unsafe, unattractive or environmentally damaging structures would be built. There is also a risk that structures would be built in unsuitable places. For example, without development laws there would be nothing to prevent a factory being built in the middle of a housing estate.

Regeneration of Ballymun by Dublin City Council. This photo shows Coultry Park (including soccer pitch, basketball court and play areas) currently being developed. It also shows a variety of new housing and some of the original towers which will be demolished.

Planning is the term used to describe the procedures in place to control development. These statutory procedures are managed by the local planning authority, such as the local County Council, City Council, County Borough, Borough Corporation or Urban District Council.

The planning process is based on development plans. There is a development plan for each county as well as a more detailed local area development plan for specific parts of the county (usually where there are large numbers of people living in a particular area). The local planning authority, in consultation with the local community through its public representatives (e.g. county councillors), prepares the development plans. Development plans are prepared in accordance with national policy guidelines issued by the Department of the Environment, Heritage and Local Government. For example, the Sustainable Rural Housing Guidelines (2005) describe various issues which must be addressed by local planning authorities in relation to the development of single houses in the countryside.

The development plan provides guidance on:

- The type of developments that are acceptable in different parts of the town/suburb (i.e. zoning).
- Areas that are sensitive to development.
- The buildings to be protected.
- Control of future development.
- Traffic safety and road improvements.
- Water supply and sewerage disposal.
- Environmental issues.

Sometimes the development plan also outlines what types of developments are needed, and will be welcomed, in a particular area. This is often referred to as *zoning*. An area might be zoned for residential, commercial or educational development. This means that any development in that zone must be for the purpose outlined (e.g. only housing may be built in an area zoned as residential). When a developer submits an application to construct a new building, the development plan is consulted to check whether the proposed building is in keeping with the plan. If the new building does not fit in with the local area development plan, it will not be allowed to go ahead. Development plans are reviewed every five years.

Planning Permission

With a few exceptions, everyone wishing to carry out any sort of building work must first apply to the local planning authority for permission to do so. The planning authority has the power to have any building that is constructed without the correct planning permission demolished without compensation to the owner.

Part of the process of applying for planning permission involves letting the general public know about the proposed development. This is done by putting a notice in widely circulated newspapers and by putting up a notice on the site. By doing this the developer is informing the general public about the plan to develop the site. In this way, anyone who wishes to obtain more information can visit the local planning office and examine the application. Any member of the public may also comment on a planning application. This involves letting the planning authority know, in writing, what concerns exist about the proposed development. For example, a person might object to a neighbour's plan for an extension on the grounds that it would reduce the amount of natural light in their home or garden. The planning authority is obliged to consider all points of view when deciding on an application.

Types of Planning Permission

Planning permission can be sought in either one or two steps. The one-step method involves applying for full planning permission at the beginning of the project. The two-step method involves applying for outline permission initially and then applying for approval when the decision to proceed with the project has been taken. Benefits of the two-step method include: detailed plans and documents are not required for outline planning permission so less investment is lost if the plan is rejected, and having outline planning permission for a site increases its sale value.

Essentially, there are four types of planning permission:

- Outline – this is used to establish whether the planning authority agrees in principle with the proposed building plan, without the need to draw up detailed plans and documents.
- Approval – this involves seeking permission to construct the building suggested when outline approval was received. Detailed drawings and documents are required at this stage.
- Permission (sometimes called *full permission*) – this involves seeking permission in one step. Detailed drawings and documents are required.
- Retention – this involves seeking permission to retain or keep a building that has been constructed without planning permission. Detailed drawings and documents are required and if the application is rejected the building will have to be demolished.

Pre-planning Meetings

A pre-planning meeting is a meeting held between the developer and officials of the local planning authority to provide an opportunity for an exchange of information before the design stage begins. This usually involves the developer providing a general outline of the desired development and the planners offering feedback and advice. This helps to avoid expensive delays later in the planning process. For example, if the developer wishes to build a two-storey house in an area where all the other houses are single storied then it is likely that a two-storey house will not be approved. Finding this out beforehand will ensure that time and money are not spent designing a house that will never be built.

Submitting an Application

An application for planning permission is made on a specific form provided by the planning authority. This form requires details about the land and the person proposing the construction. Planning permission can only be applied for with the landowner's consent. The documents required vary depending on the type of development. The documents need to show clearly and in sufficient detail, what the development will look like when it is finished and how it will relate to the site and any nearby buildings. In general the following documents are required (in addition to the required fee):

Document	Number of copies
Site location map (scale 1:2500)	6
Site layout plan (scale 1:500)	6
House plans, elevations and sections (1:100)	6
Copy of site notice	1
Copy of newspaper notice	1
Letter of consent from landowner	1
Completed application form	1

A site location map shows:

- The land involved and the location of the proposed development.
- Other land in the area which is owned by the applicant.

Rural PLACE Map

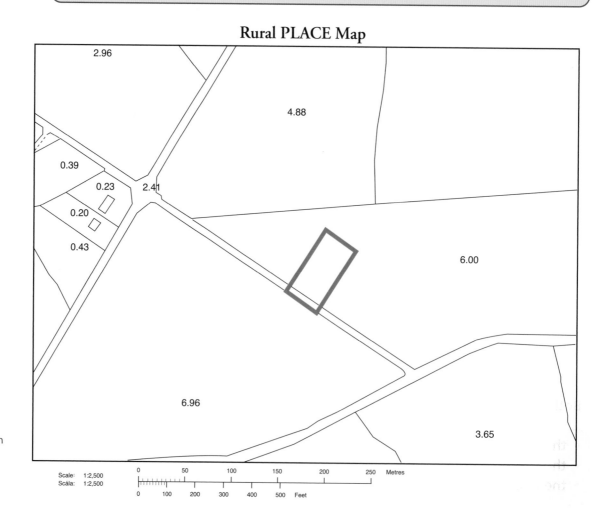

Scale: 1:2,500
Scála: 1:2,500

Typical site location map showing the proposed site outlined in red.

A site layout plan shows:

- Proposed buildings, structures and works.
- Proposed alterations or extensions to existing buildings clearly marked in a different colour outline to distinguish proposed from existing buildings, roads and site boundaries.
- Distance from existing and proposed structures to roads and boundaries.
- Approximate height of ground floor level relative to nearby road levels.
- Buildings and structures on adjoining lands.
- Existing and proposed water supply and sewage disposal systems.
- Existing and proposed access to public roads.
- Position of site notice.

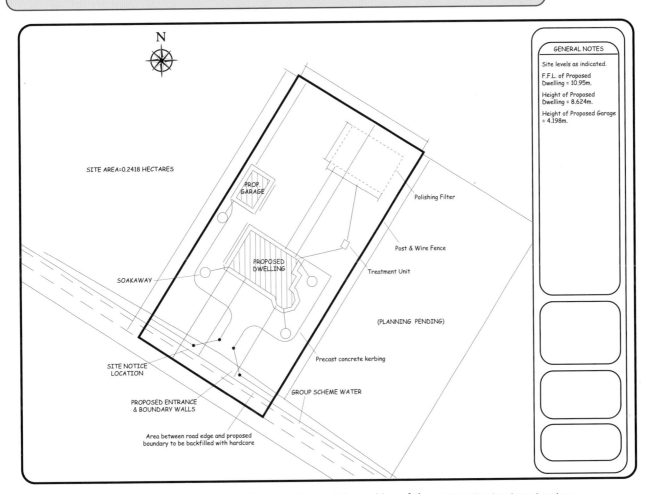

Typical site layout plan showing site access, dwelling location and the position of the wastewater treatment system.

Both the location map and the site layout plan should also show:

- the ordnance survey sheet number of the map,
- the north point and scale of the map,
- the name and address of the person who prepared the map.

For applications to build in rural areas the following information must also be included:

- Information about how the site will be accessed with emphasis on visibility and road safety requirements.
- Site suitability assessments (trial hole and percolation test) and specific design details of wastewater treatment facilities.
- Site location and setting showing how the proposed house is consistent with any landscape and design guidelines.
- Information about the measures that have been taken to ensure that the proposed site has been chosen to avoid or minimise any impacts on the landscape and heritage (protected buildings, trees etc.).

Planning Process

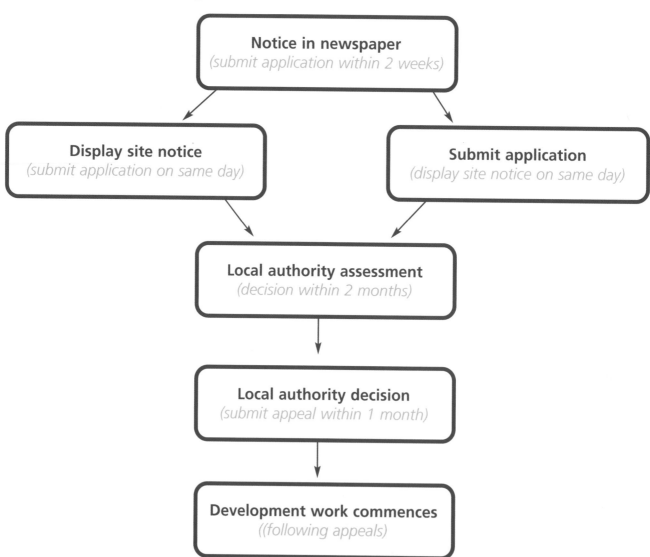

Planning process: planning applications for residential construction usually follow this sequence, provided the application is correctly completed.

The groups most involved in the planning process include:

- the developer who needs planning permission,
- the local planning authority,
- the general public, including individuals, residents' associations etc.,
- An Bord Pleanála (decide on appeals),
- The Environmental Protection Agency (when an integrated pollution control licence is required).

Objections/Appeals

An objection or observation may be made, by any member of the public, within five weeks of the day on which the application is submitted. When a planning application has been processed, the decision made (whether in favour or against the application) may be appealed to An Bord Pleanála. For example, if an application is rejected the developer might appeal if it is felt that the decision was unfair in some way. An appeal must be lodged within one month of the initial decision.

Integrated Development

Integrated development is development where a licence, approval or permit is needed from specific state government agencies (as well as requiring the usual local authority consent. These applications are referred to the relevant state body which then provides the local authority with terms and conditions for inclusion in planning permission consent). For example, a special permit is required if you are intending to build within forty metres of a stream, river or other waterway.

Environmental Impact Assessments and Statements

Environmental impact assessments are carried out by the Environmental Protection Agency. The aims of an environmental impact assessment include:

- to examine the likely effects on the environment of a proposed development,
- to ensure that adequate consideration is given to any likely effects,
- to avoid, reduce or offset any significant adverse effects.

Most large developments and all developments requiring an integrated pollution control licence require an environmental impact assessment. In these cases the developer is required to produce an environmental impact statement. An environmental impact statement contains the developer's scientific analysis of the likely effects, good and bad, of the project on the environment. It should set out any measures required to avoid or reduce adverse environmental effects. When an environmental impact statement is filed with a planning application, an environmental impact assessment is then carried out on behalf of the Local Authority. Recommendations may then be made on any extra steps to be taken by the developer before, during, or after the construction process.

Sustainability and the Environmental Impact of Building

Sustainable development is development that meets the needs of this generation without compromising the ability of future generations to meet their needs. It is essential that the construction and use of buildings is sustainable.

The concept of sustainable development recognises that development:

- has environmental, economic and social dimensions which together can contribute to a better quality of life
- must strike a balance between these three dimensions to be sustainable
- should allow future generations to enjoy a quality of life at least as high as our own
- should respect our responsibilities to the wider international community.

While it is important that everyone is aware of the effects of man-made structures on the natural environment, sustainable development is more than an environmental concept. Social structures and local economies must also be maintained for communities to survive. People living in small towns and villages have an important role to play in supporting a dynamic rural economy and social structure. For example, rural areas experiencing population decline risk losing the level of population necessary to sustain essential services (such as schools, local shops and sporting clubs) which enable a vibrant social structure to be retained. While approximately forty per cent of Irish people currently live in rural communities, it is important that development is planned to prevent depopulation. Reversing population decline by accommodating new development contributes to sustainability by helping to deliver social and economic benefits to rural areas. At present, around one in every three new homes built in Ireland are single houses built in rural areas.

Even so, individual houses should not have a negative impact on the environment. Similarly, the natural environment must not suffer lasting damage or depletion as a result of building activity. The benefits of permitting development in a rural area can be maximised by clustering new rural housing close to essential local services and community facilities. This also ensures efficient use of services.

It is important to realise that sustainable development doesn't just apply to
new construction. The adaptation of old buildings to new uses is a core
principle of sustainable development. An example of this would be to
reconstruct and possibly extend a vernacular cottage instead of demolishing it
and rebuilding on the site. Developments should be clustered to ensure
efficient use of services (i.e. electricity and water).

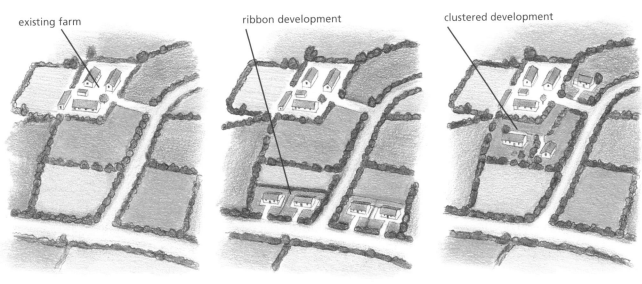

existing farm   ribbon development   clustered development

Ribbon development along roads in the countryside is inefficient and encourages dependence on the car.
Strong, sustainable rural communities can be developed by clustering houses together.

Ireland's exceptional economic performance and
the property boom since the early 1990s have
contributed to major environmental problems,
including a waste crisis and rampant urban sprawl
around many of our towns and cities. Long-
distance commuting, as a result of increasing
urban house prices, contributes to traffic
congestion, pollution, stress and increased costs.
The emphasis during this period has been on
building as many houses as possible – not on
sustainable development. This has led to a loss of
natural habitats, a reduction in green spaces and
high levels of waste.

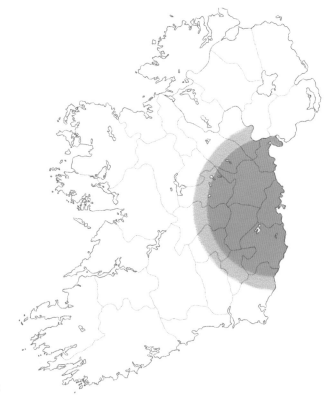

Dublin's commuter belt continues to expand as more and more
people are driving into Dublin every morning from surrounding
counties. This has a negative effect on their quality of life and the
environment. It also creates overdependence on the car and
compromises the vitality of the small towns and rural communities
that are abandoned every morning.

This increased development brings greater areas of hard or paved surfaces, which in turn have led to other environmental problems, such as flooding, in areas where there were none previously. Tax incentives aimed at coastal and amenity areas have led to the construction of many holiday homes, which sit empty for most of the year. Also, inflated land prices have led to the construction of isolated homes in rural areas which can have an adverse impact on the environment. For these reasons, sustainable development should take into account the following factors:

- appropriate use of land,
- environmental impacts of the construction process,
- environmental impacts of the materials used in construction (manufacture and disposal),
- ongoing environmental impacts of the development,
- social and economic impact of the development.

It is important that a balance is achieved between preserving the natural environment and providing adequate space for accommodation. In theory, everybody could live in small apartments in the cities and the countryside would remain untouched – but this would create a poor quality of life. On the other hand, if everybody lived in a house on a large site in the countryside there would be no countryside to speak of! Often the impression created is that a site is transformed from green field to landscaped gardens – it is important to realise that much environmental damage may have been done during the construction process. For example, the type of diesel powered machinery that is often used in construction adds to pollution of the atmosphere. Building waste such as paper, cardboard and timber is often seen burning on-site. Many builders do not segregate building waste for recycling. Construction is a very noisy process and can lead to high levels of noise pollution.

Sustainable Building Materials

Sustainable building materials are materials which have the following qualities:

- They come from renewable resources (e.g. timber from managed forests).
- They have low embodied energy – their production, transport, use and disposal does not consume large quantities of energy (i.e. gas, oil, electricity).
- Harmful waste products (e.g. carbon dioxide, sulphur dioxide) are not generated in their production.
- Packaging is eco-friendly and kept to a minimum.
- They can be reused or recycled.

Examples of Sustainable Building Materials	Examples of Less Sustainable Building Materials
• Timber • Cellulose (recycled paper) insulation • Brick products • Lime mortar • Cork	• Reinforced concrete • Cement mortar • uPVC

Where alternatives are available the most environmentally friendly option should be chosen. For example, when selecting a window frame the following options are available:

- durable hardwood and unpreserved softwood (most preferable),
- preserved softwood,
- aluminium with preserved softwood,
- uPVC (least preferable).

One way to measure the ongoing impact of a building is to consider the life-cycle cost. The life-cycle cost of a building is the total cost over its operating life. This includes the cost of construction, maintenance and day-to-day running for the lifetime of the building as well as the cost of disposing of the building if it were to be demolished. The initial costs of construction are only a small fraction of the life-cycle cost. However, the use of inferior and cheaper alternatives during construction can lead to increased maintenance and running costs. For example, the installation of low-emissivity glazing (see chapter 14, Windows) instead of standard double-glazing will greatly reduce the heating costs of the building over its lifetime, so the initial extra cost of installing low-emissivity glazing is more than covered over time. The time taken to recoup the extra cost of using a more expensive but energy efficient product is referred to as *payback years*.

One everyday example of sustainability in the home is the use of compact fluorescent lamps (cfl). A compact fluorescent lamp produces a similar amount of light as an incandescent lamp (typical *light bulb*) but consumes significantly less energy. The figures vary, depending on the manufacturer, but a typical 18–21 W cfl will produce a similar amount of light as a typical 100 W incandescent lamp – this represents an energy saving of around eighty per cent.

Compact fluorescent lamp (left) and typical incandescent lamp (right).

Cost in Use Analysis: Home Lighting

	Incandescent 100 W Bulb	Low-energy 21 W Bulb
Initial capital cost	€0·80	€4·99
Life expectancy	1,000 hrs (1 yr)	10,000 hrs (10 yrs)
Running cost at 1,000 hrs per annum and €0·14 per kWh (incl. VAT)	(1,000 hrs × €0·14 × 0·1 kw) = €14·00	(1,000 hrs × €0·14 × 0·021 kW) = €2·94
Replacement costs over 10 years	= (€0·80 × 10) = €8·00	= zero
Running costs over 10 years	= (€14.00 × 10) = €140·00	= (€2.94 × 10) = €29·40
Total life-cycle cost	€148·00	€29·40

From this analysis it is clear that low-energy bulbs are more economical in the long term. However, this example does not take interest rates, inflation or the cost associated with the loss of heat from the incandescent bulbs into account. When these are factored in, the saving is less but still significant. Also, it should be noted that the electricity saved by installing a single low-energy (cfl) bulb avoids the generation of 0·75 tonnes of carbon dioxide and 8 kg of sulphur dioxide. So, it is clear that the use of cfls saves money and is good for the environment.

In the future we will be able to assess and classify a building's environmental characteristics based on set criteria, much as we can classify electrical appliances at present (see chapter 21, Heat and Thermal Insulation). This will allow the energy performance of different buildings to be easily compared.

Roles in the Construction Industry

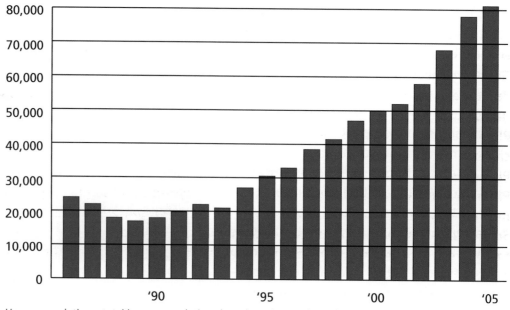

House completions: total house completions based on the number of ESB domestic connections.

In the period from 1985 to 1995, around 20,000 (average) homes were completed each year. However, in the period from 1995 to 2005 this number is closer to 60,000 (average) each year. Today, over a quarter of a million people are employed in the construction sector.

This boom in the construction of domestic dwellings has led to many changes in the industry. One of the main changes is the way in which most of the people working on large building sites are now private contractors working independently for the main builder. This means that the builder has fewer direct employees, making it easier to manage the business.

There are a wide variety of roles in the construction industry including:

- **Architect** – responsible for the design of the building and oversees the entire construction process, ensuring that everything is built according to the specifications.
- **Consultant engineer** – structural engineers are responsible for ensuring that the building can safely transmit all the loads which it will experience throughout its long lifetime. Services engineers are responsible for the design of the services systems such as: electrical, wastewater disposal and air conditioning.
- **Quantity surveyor** – estimates the cost of every element of the construction process and ensures that all necessary materials are available when needed throughout the build.
- **Building contractor** – has overall responsibility for the hands-on work on-site. The main contractor will usually employ a small number of people directly and use subcontractors for everything else.
- **Bricklayer** – builds masonry walls and other elements.
- **Carpenter and joiner** – does all woodwork construction and installation on-site such as: (first fix) stud partitions, first floors, roofing; (second fix) hanging doors, fitting skirting boards and architraves.
- **Concrete operative** – does concrete work such as foundations and footpaths.
- **Electrician** – carries out all electrical wiring of houses.
- **Floor and wall tiler** – installs tiles in kitchens, bathrooms and other areas.
- **General construction operative** – does general work around the site wherever needed.
- **Painter and decorator** – paints internal and external elements such as walls, doors, railings and gates. Also, hangs wallpaper and installs other decorative features.
- **Plant operator** – operates heavy machinery such as excavators, dumpers, telescopic lifts and cranes.
- **Plasterer** – plasters internal walls, finishes plasterboard, renders external walls.
- **Plumber** – installs water supply, central heating systems and sanitary appliances.
- **Roofer** – installs roof covering (e.g. underlay, flashing, battens and tiles or slates).
- **Scaffolder** – assembles scaffolding and certifies it is safe for use.
- **Steel fixer** – on larger sites, is responsible for the installation of steel reinforcement for concrete.

Planning to Build a Single House in the Countryside

Choosing where to build a house in a rural area is about striking a balance between a number of factors including: the impact the house will have on the landscape, the comfort and quality of life of the home owners and essential construction details such as access to roads and connecting to a water supply. A well-sited house in a rural setting will save the owner money over time if it is well sheltered from the weather because it will retain more heat and require less energy.

There are four stages to consider when planning a new house for the countryside:

1. Choosing a site.
2. Assessing a site's potential.
3. Selecting an appropriate house design.
4. Devising a suitable landscape design.

Choosing a Site

Choosing a site involves checking the county development plan for information relating to the design and construction of single houses in the countryside. Requirements vary from county to county and these need to be checked first. The next step is to consider the type of landscape in which the house will be sited. This is essential because it will influence the design of much of the house. The final step is to assess the potential of the proposed site to ensure that it is suitable for development.

Checking the Development Plan

As we learned in chapter 3, every county (and many towns and urban areas) has a development plan to provide guidance in relation to the types and locations of future construction projects. Many development plans provide detailed information on the siting, location and design of dwellings permitted in rural settings. The purpose of this guidance is to ensure that newly built houses are appropriate to their location, contribute to the appearance of the countryside and respect and reflect local traditions. The development plan can be viewed or purchased at the local planning office. The development plan will include any requirements under national policy documents (e.g. Sustainable Rural Housing Guidelines 2005).

Reading the Landscape

Ireland has a beautiful and diverse landscape including mountains, plains, valleys and coastal landscapes, each providing striking scenery in which to place a home. There is always debate about the many benefits and the considerable costs of building in the countryside, however, one thing on which everyone is agreed is the need to maintain the beauty and integrity of our countryside. To do this every home that is built in the countryside must be carefully designed and constructed to ensure that it is appropriate to its location and does not harm the natural environment in any way.

Reading the landscape means looking at and appreciating a wide variety of natural and man-made factors, such as:

- natural contours or lie of the land,
- rock types,
- trees, hedges and other plants,
- traditional house types and styles in the area,
- traditional building materials used in the area.

Assessing a Site's Potential

When a potential home owner visits a site for sale, s/he usually tries to *picture* their dream home there. When they do this they are assessing the site's potential to accommodate their home, cars, garden, clothes line and everything else involved with a home. While many of the things associated with a home are flexible, some are not, and these must be considered when assessing a site's potential.

The following factors influence the siting of a house in a rural setting:

- access to the site from the road,
- shelter and privacy,
- orientation of the house to the sun.
- existing boundaries (hedges, trees, streams),
- location of the wastewater treatment system.

Most local planning authorities provide detailed guidance regarding safe access to the home entrance from the road. These describe various requirements such as clear sight lines along the road in both directions, the width of entrance, the distance that the entrance is set back from the roadside and so on.

Once the landscape has been considered, the next step is to decide where in that landscape it is best to locate a new home. The primary factors influencing this decision are shelter and orientation to the sun. The goal is to seek natural protection from the wind and rain while avoiding exposure and prominence. Seeking shelter goes hand-in-hand with integrating a dwelling with the land. In other words, if you build a house in a natural hollow in the land, the house will automatically look more appropriate in the landscape. On the other hand, if you build on the crest of a hill the house will stand out more against the surrounding landscape. It also makes financial sense to *hug the land* when siting the house – a well-sheltered house will retain more heat than an exposed one.

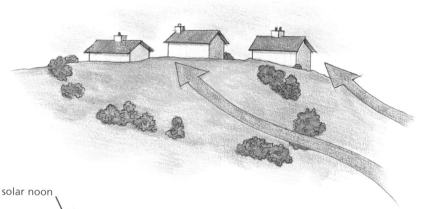

Building a house on a hill is a bad idea because it exposes the house to the wind and to the driving rain – causing the house to suffer huge heat losses. It also makes the house stand out like a sore thumb on the horizon.

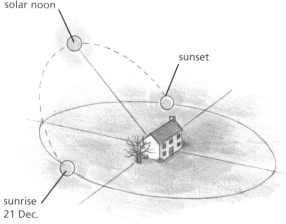

Winter sun path: the sun rises late, sets early and stays low in the sky.

This principle also applies to the orientation and design of the house. There is a long-standing tradition in Ireland of facing the gable end of the house into the wind in order to reduce the cooling effect of the wind. In order to do this it is important to use official and local sources of information to establish the prevailing wind direction.

It is also essential to make the most of the available sunlight. Orienting a house to within 15° of south can lead to energy savings of up to thirty per cent. This is because the home can absorb free heat energy from the sun during the day, instead of having to burn fuel to provide heat. The position of the sun is generally due south but it does vary with the seasons. Also, the side of the house with the largest amount of glazing (often the kitchen/dining area) should face south to capture solar heat. (See chapter 20, Light, for more detail.)

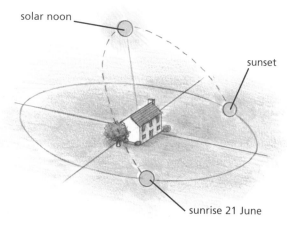

Summer sun path: the sun rises earlier and reaches higher into the sky during summer, providing more heat and light energy.

Existing natural boundaries and site features, such as trees and hedgerows, should be retained to help the newly constructed house blend in with its surroundings and provide privacy and shelter.

Finally, the wastewater treatment system must be located downhill and usually to the rear of the house. On a sloped site this tends to limit the options for where the house can be situated.

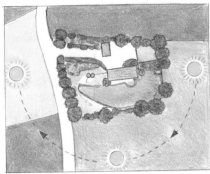

Top left detail west-facing slope (site above road): septic tank to the front (near the road) pushes house further uphill, house located near northern boundary for shelter, privacy screening (hedges/trees) needed to south and west sides of house will reduce solar gain if not planned properly.

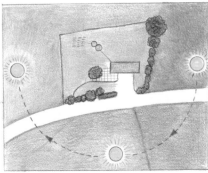

Top right detail north-facing slope (site below road): septic tank behind house and furthest from road. To maximise light (and solar gain) large windows will have to face the roadside, also patio/deck area will have to be to the front; both measures reduce privacy.

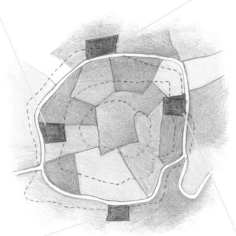

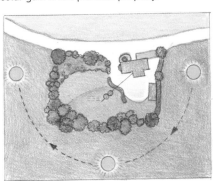

Bottom left detail south-facing slope (site below road): septic tank behind house to the rear of the site, screening used to create privacy from road, layout allows for patio/deck area and car parking to be screened from road and provides excellent privacy.

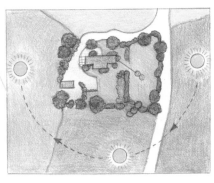

Bottom right detail east-facing slope (site above road): septic tank in front of house, closest to road. Garage and utility area located to rear of house and screened from road by landscaping. Evening patio/deck area private but smaller due to garage and driveway.

Site selection: four possible sites located around a hill in a rural setting. Factors to consider include: the septic tank is always located downhill from house, orientation to the sun to maximise solar heat gain, the relationship of the house to the road (privacy) and the protection of existing natural boundaries (hedges, trees).

Selecting an Appropriate House Design

There is a long tradition of building homes in the countryside. Unfortunately, over the last thirty years many houses have been built that are alien to the Irish landscape. If a new house is going to sit well in the Irish countryside it will need to echo past designs. This means the design should be simple and follow the example of older conventional houses in the area. Traditionally, rural houses have relied on their good proportion, careful use of colour and quality of materials to make a positive impact. Frivolous extra details (especially uPVC windows and fascias) and loud colours seen in many houses today do not fit in well in a rural setting. Very large houses should be broken down into smaller components that are more appropriate to the rural setting.

If the appearance of a house is not in keeping with the styles traditionally built in an area or uses materials which are out of place, it will make a negative impression on most people who see it. When this happens we say that the visual amenity value of the area has suffered. This means that the enjoyment people get from just looking at the hillside (for example) on which the house is built is reduced because the house is an eyesore. Of course, the appearance of a house depends on individual taste and opinions vary greatly. However, there are a few simple guidelines illustrated below that can be followed to help ensure that most people feel the house blends in reasonably well with its surroundings.

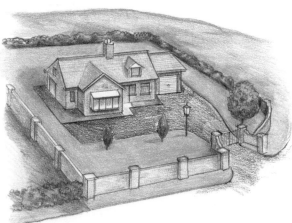

Failing to link with the land: inappropriate features include: urban house design and detailing, house cut into landscape, urban features (wall, lanterns), non-native/ornamental trees, too much lawn and tarmacadam, cars parked to front, no privacy for home owner.

Linking with the land successfully: rural house designs using local materials, house designed to suit site, natural stone walling, native planting, landscaping used to blend house into countryside, gravel driveway, cars parked to rear, privacy for home owner.

Devising an Appropriate Landscape Design

One of the ways in which the visual impact of a newly built house can be reduced is through the use of landscape design. This involves selecting suitable planting to link the newly built house to the natural landscape. A large area of perfectly mown lawn will look out of place on the side of a typical hill in the countryside. Native hedges and trees should be planted to screen the house so as to provide shelter and privacy but also to reduce the visual impact of the house especially when seen from the road. This also involves planting around the base of the house, forming a connection between the house and the site rather than laying a concrete path which has the effect of lifting the house up on a pedestal.

Planning to Build in an Urban Area

Building in an urban area differs from building in a rural area. In this case, blending with natural landscape features is not as relevant. This is because when a residential development is being planned it usually involves clearing a large site and starting from scratch. There are three main stages to consider when planning a large-scale housing development in an urban setting:

- contextual design features,
- visual design features,
- physical design features.

Contextual Design Features

When designing a residential development in an urban area it is important to think about the *big picture* – for example, will the people who are going to live there enjoy a good quality of life? When a large number of people are living in a relatively small space there is a need for a variety of convenient local services and facilities, including:

- Retail – newsagents and grocery shopping.
- Recreational – sports (gym, swimming pool), cinema, pubs, café, parks and other open public spaces.
- Educational – primary and secondary schools and a library.
- Spiritual – churches and other places of worship.
- Transport – adequate public transport links to urban centres (bus, light railway).

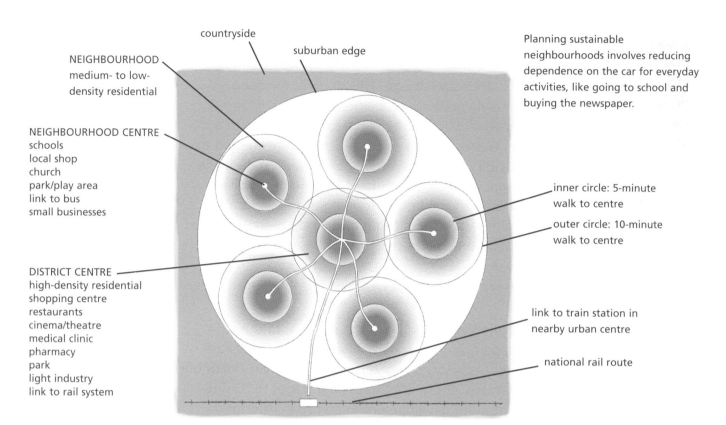

countryside

suburban edge

NEIGHBOURHOOD
medium- to low-
density residential

Planning sustainable
neighbourhoods involves reducing
dependence on the car for everyday
activities, like going to school and
buying the newspaper.

NEIGHBOURHOOD CENTRE
schools
local shop
church
park/play area
link to bus
small businesses

inner circle: 5-minute
walk to centre

outer circle: 10-minute
walk to centre

DISTRICT CENTRE
high-density residential
shopping centre
restaurants
cinema/theatre
medical clinic
pharmacy
park
light industry
link to rail system

link to train station in
nearby urban centre

national rail route

For example, the home owner should be able to walk to the local shop to buy a litre of milk and not have to travel by car. Children should not have to trek long distances to get to school in the morning. There should also be adequate social and recreational facilities for both children and adults, including somewhere to play football, walk the dog or practice skateboarding skills. These spaces should be well located and accessible to everyone living in the area.

This type of approach to the design of housing developments is sustainable because it eliminates the use of the car as much as possible and it creates communities where people get to know each other, develop friendships and enjoy a better quality of life.

Another important consideration is the type of dwellings included in the development. If the scheme includes a variety of dwelling types (e.g. detached, semi-detached and apartments) there will be a greater social mix in the area. Taking this approach means that everyone has the opportunity to live in well-designed areas with good facilities and services. If only one type of dwelling is constructed then only people at a particular income level will live there. This can lead to the creation of slums and a socially divided and unfair society.

It is also important that the dwellings take advantage of the physical features of the site. This involves:

- protecting local ecology (mature trees, waterways, wildlife areas),
- orienting dwellings to capture natural light and solar heat gain,
- including terraced dwellings that minimise the loss of heat,
- designing dwellings that can be extended in the future as families grow,
- including appropriate storage areas for car parking, bicycles and bins (especially important for apartments).

Visual Design Features

The purpose of considering the visual impact of a development is to decide whether the overall appearance is appropriate and balanced. This involves looking at each element of the development (e.g. houses, apartment blocks, roads, footpaths, planting etc.) and also the relationship between these elements.

One of the first things to look at is the landscape structure. This involves consideration of:

- the amount of open space in the development,
- the system of roads,
- the height-to-width ratio of the streets,
- the variety of spaces created (long winding streets, parking squares, high-density blocks),
- the use of planting to provide scale and to soften the streetscape,
- the position of houses and the topography of the site.

The way in which these elements are arranged will determine the character of the development and can make it a dull impersonal place or a vibrant interesting neighbourhood.

Limiting the visual length of the street makes it more interesting visually and encourages walking. Street length can be limited by building a taller building at the end of the street or by gently curving the street.

Another important consideration is the building forms and materials used. The buildings should ideally reflect the traditions of the area (if they exist). Similarly, a local selection of materials should be used in the construction of the homes in order to connect them to the area. It is often said that human beings are creatures of habit – in this way we often like a new home to reflect tradition in a up-to-date contemporary way.

Physical Design Features

The most important element of any residential development is the houses. After all, this is where we spend most of our time. The structure of the house (envelope) and its immediate surroundings (curtilage) impact greatly on the quality of life of its occupants. Other essential elements include the provision of services and the control of vehicular movement.

Envelope

The internal space of the house should be of adequate size and well laid out. Natural light should brighten the living spaces and make them positive places to inhabit. The design of the windows will greatly influence not only the amount of light entering the building but also the external appearance of the home (especially in terms of the proportion and layout of the windows and the materials used). The home should lend itself to extension in the future (e.g. attic conversion) and there should be adequate acoustic insulation to allow privacy and enjoyment.

Curtilage

Most home owners have a high expectation of privacy to the rear of their home. The back garden is usually the place where urban dwellers sit in the sun, barbeque and spend time with friends and family. For this reason, it is important that there is a reasonable amount of separation between dwellings and that rooms are not overlooked. It is

In new developments a minimum distance of 22 m should be maintained between houses that are 'back to back'. Screening that is above eye level should be provided at the boundary.

also important that gardens are of a good size or that private open space is provided as an alternative (e.g. for apartments). The storage of essential items such as wheelie bins should be also planned for.

Services

The supply of electricity and water are essential to every development. Other important services include the supply of natural gas, the removal of wastewater (sewage and rainwater) and access for waste disposal vehicles. The provision of services is normally routed underground although certain elements such as substations and gas regulators will be located above ground. It is desirable that these are accommodated in a way that blends in with the surrounding homes. Similarly, refuse collection points should be accessible by the usual 15 m trucks that collect domestic waste. Where access is limited a communal refuse collection point should be used.

Vehicles and Pedestrians

The parking of cars and vehicular movement is an essential element of housing development design. Many neighbours have disagreements over parking and speeding in residential areas. Well-designed roads, footpaths and parking spaces can solve these problems. The use of communal parking spaces that are overlooked by a number of dwellings will prevent the street being dominated by the appearance of parked cars while also providing security. A maximum speed of 30 kph should be applied to residential areas. Chicanes and speed ramps will help to regulate the flow of traffic.

underpass access to parking avoids gaps in streetscape

parking behind and between houses

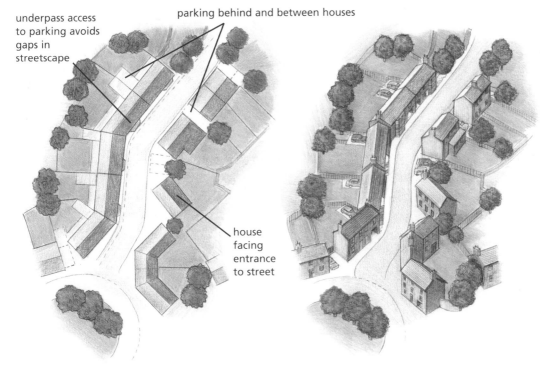

house facing entrance to street

Urban design: informal urban street. A curved street is used to control traffic and to prevent wind funnelling. It also adds visual amenity.

parking for houses facing square

focal point

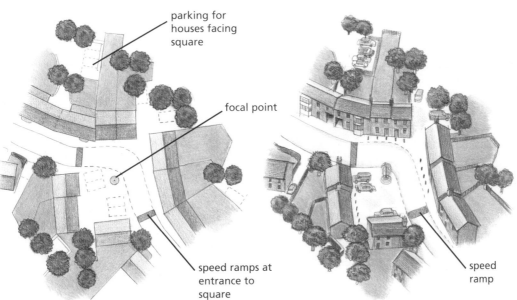

speed ramps at entrance to square

Urban design: parking square. Small parking squares are used to slow down traffic and to provide a focal point (e.g. water feature) in residential developments.

speed ramp

Urban design: higher density block. A block like this would usually be located in the centre of a residential development; it includes apartments and surrounds a communal parking area, which is overlooked for security.

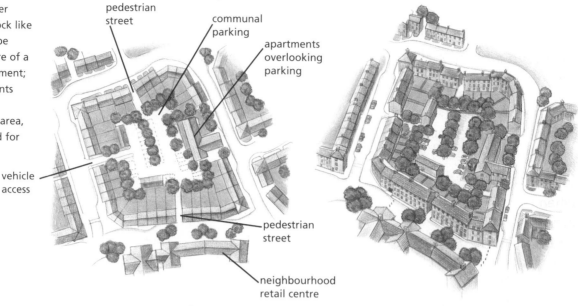

pedestrian street

communal parking

apartments overlooking parking

vehicle access

pedestrian street

neighbourhood retail centre

Urban design: lower density area. Lower density areas can be included in larger developments to provide an opportunity to incorporate a mixture of detached and semi-detached homes with larger gardens that would suit bigger families.

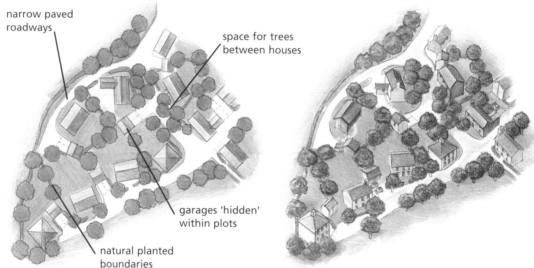

narrow paved roadways

space for trees between houses

garages 'hidden' within plots

natural planted boundaries

Urban design: formal square. A formal square provides interesting contrast to the other less formal curvy layouts. It is modelled on the Georgian square, with a common green space in the centre of the square surrounded by shared parking spaces.

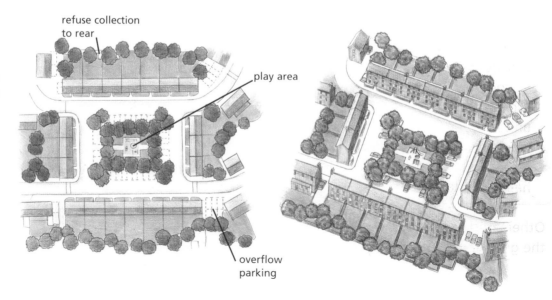

refuse collection to rear

play area

overflow parking

It is essential that the design of the development encourages walking for short trips. Wide footpaths that are separated from cycle paths are necessary to encourage people to walk, rather than drive, short journeys (e.g. a trip to the local shop).

Planning to Build for Lifetime Use

Every building should be designed for access and use by all members of society. A good designer will picture herself or himself in various situations when considering how best to create a building. The designer will imagine himself or herself as a child or as an elderly person, on crutches or in a wheelchair in order to create a building that works for everyone throughout their lifetime. This approach to housing design is known as *lifetime use* house design. Lifetime use housing design requires us to think about the needs of everybody at all stages of their lives when designing a home. Much of this approach to building design isn't just best practice, it is a requirement of the building regulations.

Changes to the building regulations were made to achieve this goal in housing and the new provisions apply to all new homes built since January 2001. These changes specifically apply to the approach, main entrance, circulation and toilet facilities of a house. These requirements are designed in particular to make it possible for a wheelchair user to visit a typical home comfortably.

Public Buildings

The building regulations also apply to public buildings. Some of the provisions for public buildings include:

- at least one entrance accessible to wheelchair users,
- an internal layout which allows wheelchair users to circulate freely,
- a passenger lift in high-rise buildings,
- a proportion of guest bedrooms in hotels which are suitable for wheelchair users,
- unisex toilets which are suitable for wheelchair users,
- wheelchair spaces in theatres, cinemas, concert halls, and sports arenas,
- facilities for people with hearing impairments in theatres, cinemas, concert halls and places of worship.

Other considerations include car parking spaces and the slope (gradient) of the ground outside the building.

Lifetime Use Housing Design Features

Choosing a Location for a Home

In terms of choosing a location for a house, the approach must allow for:

- a site boundary access (gateway) of minimum width 800 mm,
- an entrance route with a firm non-slip surface, with a minimum clear width of 900 mm,
- level or gently sloping approach,
- a 1,200 mm deep canopy and landscaping/planting at the entrance to protect people from the rain.

All newly built houses should be readily accessible by everyone. This example shows adequate space for a wheelchair user to get out of the car or to wheel past it, a smooth driveway surface, low-level solar-powered lighting for night time access and protection from the weather at the front door.

Entrances and Corridors

The front door of every house should be accessible to wheelchair users. In exceptional circumstances, where this is not possible, a side or rear entrance should be accessible. An accessible front door should have a maximum threshold height of 15 mm and a minimum clear opening of 800 mm (see chapter 15, Doors). Ideally, the ground approaching the entrance should be level. Where this is not possible a ramp (slope 1:20) may be used. The doorbells and letterboxes should be at height of 900–1,200 mm above ground level. The door number and house name (if any) should be clearly visible from the street.

Areas of the home normally used by visitors should be accessible by people with disabilities, including wheelchair users, and the elderly. Design features necessary to achieve circulation include:

- turning spaces of minimum 1,500 mm diameter,
- corridors and hallways should have a minimum unobstructed width of 1,200 mm,
- internal doors to have a minimum clear opening of 800 mm,
- door openings should not incorporate saddles,
- clear space of 500 mm at the leading edge of doors to allow wheelchair users to reach the handle,
- door handles to be easy to operate.

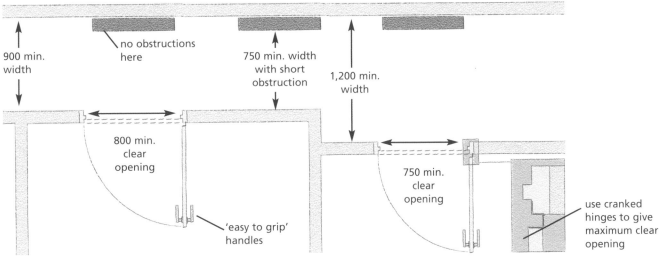

900 min. width

no obstructions here

750 min. width with short obstruction

1,200 min. width

800 min. clear opening

750 min. clear opening

'easy to grip' handles

use cranked hinges to give maximum clear opening

Internal corridors should allow wheelchair access to the main living area and downstairs toilet of every newly built home. Where the corridor is narrow the door opes need to be wider to allow 'angled' entry.

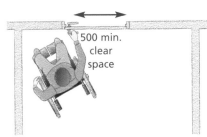

500 min. clear space

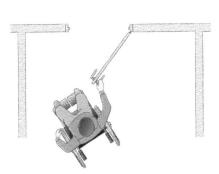

Wide doors are used because they are easier to negotiate and reduce the risk of the wheelchair user grazing his/her knuckles. Light switches in hallways should be located at a height between 900 mm and 1,200 mm. Small permanent obstructions in corridors, such as radiators, should not reduce the unobstructed width to less than 750 mm and should not be positioned opposite doorways.

It is essential that space is provided, beside internal doors, to allow the door to swing past the wheelchair. This space would also be required for someone with a broken leg who is on crutches.

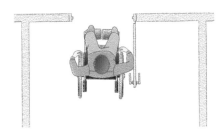

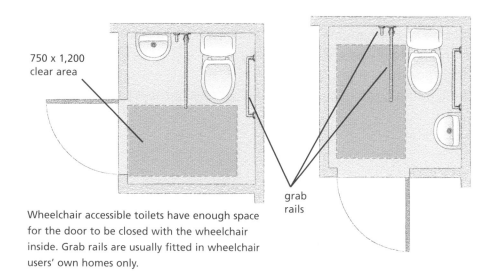

750 x 1,200
clear area

grab
rails

Wheelchair accessible toilets have enough space
for the door to be closed with the wheelchair
inside. Grab rails are usually fitted in wheelchair
users' own homes only.

Ground Floor Toilets

An accessible toilet facility should be provided at ground floor level in every new house. A clear space 750 mm by 1,200 mm, accessible by wheelchair users, should be available beside the toilet to allow safe transfer from the chair to the toilet seat.

In general, the size and layout of a bathroom or toilet facility, and the positioning of the door, should allow for the door to be easily closed with the wheelchair inside.

Bathrooms

Specific bathroom design features to facilitate safe and easy use by wheelchair users include:

Toilet
• fold-up grab rail fitted 400 mm from toilet centre line, on transfer side of toilet,
• vertical grab rail 600 mm long, 500 mm from toilet centre line, on transfer side of toilet,
• horizontal grab rail 1,200 mm long, on wall beside toilet,
• lever-type flush handle fitted to transfer side of cistern,
• toilet seat to finish 450–460 mm above floor.

Sink
• full-size wall-mounted sink 750–900 mm above floor,
• allow 700 mm clear knee space underneath,
• locate within easy reach of toilet,
• lever taps fitted,
• wall space for long mirror (900–1,800 mm above floor level) for use from seated position,
• vertical grab rail 600 mm long, 200 mm from tap centre line.

Bath

- bath 1,600 mm long with 450 mm rim height,
- bath to have built-in slip resistance,
- long horizontal grab rail fitted to inside wall,
- vertical grab rail fitted to wall,
- taps positioned at midpoint of inner side (not at end) for easy access,
- a 400 mm storage ledge at one end of bath.

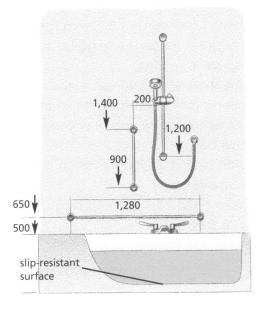

Accessible bathtub: includes lever-type taps at the side, grab rails and longer shower head rail, so the shower head can be lowered to suit (this design is very suitable for children too).

Window

- direct access to window to facilitate ease of opening and closing,
- low cill height to allow opening and closing from seated position,
- lever handles for ease of use.

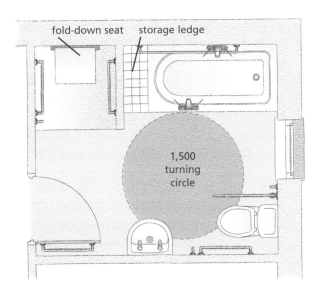

Wheelchair accessible bathroom: this design incorporates both a bath and a shower. Grab rails are provided to all appliances and the window is easy to reach. There is also a 1,500 mm turning circle provided.

Floor

- slip-resistant floor covering,
- drainage to internal gulley with floor gently sloped to gulley,
- gulley situated so as not to cause an obstruction,
- stainless steel grid to gulley.

Shower

- shower 1,000 mm × 1,000 mm for ease of access,
- lay entire floor to fall to gulley or use non-slip, flush finish shower tray,
- fabric shower curtain easily opened and closed and allowing maximum space for chair,
- lever temperature and flow controls 900–1,200 mm above tray,
- flip-up seat, 450–500 mm wide, fitted to finish 450 mm above tray,
- shower head adjustable in height 1,200–2,200 mm above tray,
- vertical and horizontal grab rails in cubicle.

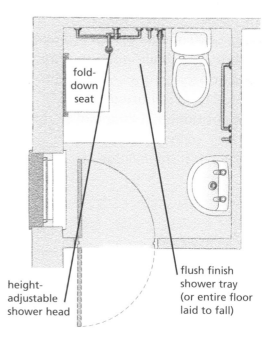

fold-down seat

flush finish shower tray (or entire floor laid to fall)

height-adjustable shower head

Wheelchair accessible shower room: includes a 'wheel in' shower space with a fold-down seat and grab rails. Once the person has transferred to the seat, they can push their wheelchair aside and draw around the shower curtain.

Many of these features can be incorporated into standard bathrooms to make them more universally accessible. Any person who has ever had a sore back or a broken leg would agree that grab rails and non-slip floors can be useful to everyone. Where certain features (especially grab rails) are not incorporated initially, fixing should be provided in the required areas to permit retrofitting.

Kitchens

Lifetime use kitchen design focuses on safe and easy access to storage spaces and appliances. Many of the activities associated with the preparation of food can be hazardous (e.g. boiling water) so the design features need to be well thought through to limit these hazards as much as possible. Typical design features would include:

- adequate unobstructed space (wheelchair turning circle Ø1,500 mm),
- frequently-used appliances (e.g. microwave) to be located in the accessible zone (450–1,300 mm above floor level),
- counter surfaces should be at a height of 800 mm generally,
- clear knee space underneath key items (e.g. sink),
- dishwasher, washing machine, dryer to be built-in under work surfaces,
- split oven and ceramic hob are most easily used,
- sockets accessible from wheelchair (900–1,200 mm) above floor level,
- lowered light switches (900–1,200 mm) above floor level,
- open shelving and carousel corner units for ease of storage,

- circular table for ease of circulation in small kitchens,
- reduced cill height to allow view from the chair and access to open and close windows,
- unobstructed spaces at doorways to allow ease of entry and exit,
- non-slip floor surfaces.

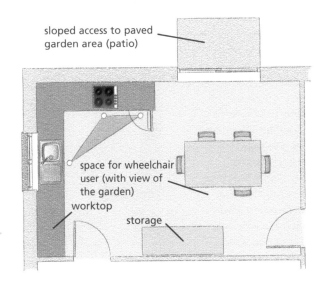

Kitchen designed for lifetime use: the relative positions of the sink, cooker and fridge is very important because they are repeatedly used when preparing a meal – they should form a reasonably small triangle in plan.

Before: typical floor plans often reduce the accessibility of the home for all users, particularly wheelchair users. In this example, a wheelchair user will have difficulty opening doors, accessing the toilets, opening the kitchen window and turning around both upstairs and downstairs.

After: the same house with a small number of minor changes which make it suitable for lifetime use. The doors have been moved slightly to allow easier opening by wheelchair users. The hallway and landing have been made slightly bigger to create turning circles and the toilets have been made accessible. These simple changes give this house a more sustainable design and make it more user-friendly for everyone.

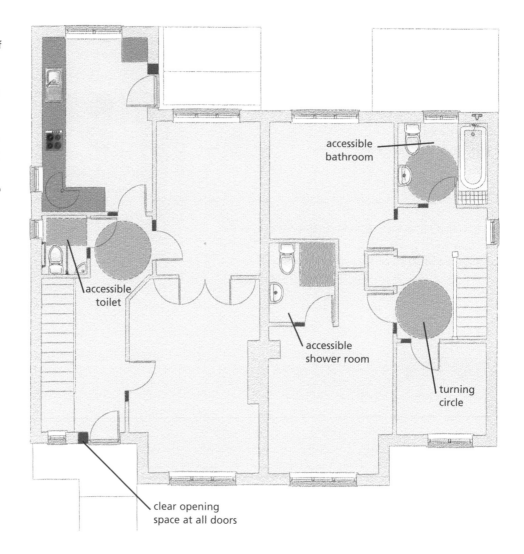

accessible bathroom

accessible toilet

accessible shower room

turning circle

clear opening space at all doors

Revision Exercises

1. Explain in your own words the consequences of poorly controlled development.
2. Explain why it is necessary for every county to have a development plan.
3. Explain briefly, the following types of planning permission:
    (a) Full permission.
    (b) Outline permission.
    (c) Approval.
    (d) Retention.
4. Explain the procedure involved in making a planning application. Include in your answer a list of the documents required.
5. Explain briefly the stages involved in planning to build a single house in the countryside.
6. Write a short paragraph on the effect of the construction industry on the environment.

7. Outline four design features that should be considered when planning a housing development in an urban setting.

8. Explain, using neat freehand sketches, the design considerations necessary for:
   (a) A bathroom designed for lifetime use.
   (b) A kitchen designed for lifetime use.

9. Ordinary Level, 2005, Question 8
   Planning permission is required before a dwelling house can be erected.
   (a) State two reasons why it is necessary to apply for planning permission to erect a dwelling house.
   (b) Explain what is meant by full planning permission.
   (c) Describe the purpose of any three of the following as they apply to an application for planning permission:
      • site notice, • newspaper notice, • site location map,
      • site layout map, • Percolation Test.

10. Higher Level, 2000, Question 5
    (a) Outline five main considerations in choosing a site for a dwelling house.
    (b) Discuss the importance of each consideration you have listed in (a) above.
    (c) Discuss in detail two ways in which a new house can be made to harmonise with the surrounding landscape.

11. Higher Level, 2003, Question 2
    A proposed combined kitchen and dining space in a new house is 6 m long by 4·5 m wide and has two adjoining external walls. This space is to be user-friendly for a person in a wheelchair.
    (a) Using a well-proportioned line diagram or freehand sketch, propose a design layout for the space indicating the positions you would choose for the following:
       • doors, • windows, • sink, • work surfaces, • storage, • fridge,
       • electric cooker, • dining table.
    (b) Using notes and detailed freehand sketches, outline two specific design considerations that would make the proposed layout suitable for a person in a wheelchair.
    (c) Discuss in detail three other design considerations that influenced the proposed layout.

12. Higher Level, 2005, Question 2
    Current building regulations require that new dwelling houses be suitable for all, including wheelchair users.
    (a) Using notes and freehand sketches, outline three areas in a dwelling house that need specific consideration to ensure that the house is suitable for a person in a wheelchair.
    (b) Select one of the areas outlined at (a) above and using notes and detailed freehand sketches, show three specific design considerations that ensure that the space selected is suitable for a wheelchair user.

Construction Materials

There is a wide variety of materials used in the construction of new houses. Some of these materials have been used for generations such as timber, stone, slate and glass while others such as uPVC and polythene are relatively new. Each material is used because it has properties that particularly suit the application. For example, concrete is very good at withstanding heavy loads so it is used for foundations, whereas polythene (plastic) is resistant to water so it is placed in floors and walls to prevent dampness. We will look at these materials under two broad headings, namely *natural* and *manufactured* (or composite) materials.

Natural Construction Materials	Manufactured Construction Materials
• Timber	• Cement
• Stone	• Concrete
• Aggregates	• Brick
• Lime	• Steel
	• Polymers (plastics)
	• Wood-based panel products

Timber

Timber can be classified as either softwood or hardwood. Softwood is the timber of evergreen (coniferous) trees such as Sitka Spruce (*whitewood*) and Scots Pine (*redwood*). Hardwood is the timber of broad-leaved (*deciduous*) trees like Oak and Beech.

Coniferous trees (e.g. Douglas fir) are typically evergreen, have needles instead of leaves and are 'cone-bearing'. Most construction grade timber comes from conifers.

Deciduous trees (e.g. Beech) have leaves which turn brown in autumn and fall off in winter. These trees produce the hardwoods commonly used inside houses for components such as stair balustrades and doors.

Softwood is very widely used in the construction of houses. This is because it is readily available, relatively cheap and strong and easy to use on-site. Whitewood is commonly used for structural work (joists, rafters, stud partitions) in the construction of houses. Timber is classified in accordance with various EU standards. When a company sells timber for use in the construction industry it must stamp certain information onto the timber. The stamp includes all of the required information in a number of codes, including the company's registration number, species, source code and the strength class.

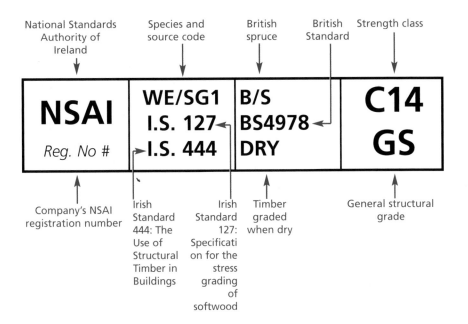

National Standards Authority of Ireland →

Species and source code →

British spruce →

British Standard

Strength class →

| NSAI *Reg. No #* | WE/SG1 I.S. 127 ← I.S. 444 → | B/S BS4978 ← DRY | C14 GS |

↑ Company's NSAI registration number

↑ Irish Standard 444: The Use of Structural Timber in Buildings

↑ Irish Standard 127: Specification for the stress grading of softwood

↑ Timber graded when dry

↑ General structural grade

Strength grading stamp.

Structural timbers are classified according to their strength. The European standard by which timbers are classified provides for a number of strength classes, each designated by a letter and number. The letter tells us whether the timber is a softwood (C) or a hardwood (D). The number indicates the value of bending strength in newtons per square millimetre (N/mm^2). There are twelve categories of strength for softwoods and six categories for hardwoods. An example of a classification would be C16 – this is softwood with a bending strength of 16 N/mm^2. A strength class of D35 refers to hardwood with a bending strength of 35 N/mm^2.

There are specific regulations governing the use of structural timber in the construction of houses in Ireland. These regulations provide guidance on the species, strength, sizes and marking of timbers for various applications. Once the strength of the timber is known, this information is used to meet the standard IS444:1998 The Use of Structural Timber in Buildings. This standard ensures houses are built using timbers that are capable of safely carrying the applied loads. The standard outlines the required size and strength of timbers to be used for various applications. For example, if a floor joist is to span a distance of 4 m, a length of 75 × 150 mm, C16 timber could be used at 350 mm spacing or a length of 44 × 175mm C27 could be used at 400 mm spacing.

As well as the bending strength, a number of other mechanical tests are carried out on the timber, including tension, compression and shear tests. When these tests are complete a visual inspection of each timber is carried out. This is called *grading*. It involves visually inspecting the timber for defects such as knots and fissures (shakes and checks). The standards provide guidance on the level of defects allowable for various applications.

It is essential that all structural timbers have a moisture content of less than twenty per cent. Timber processing companies season (dry out) timber to ensure that it has an acceptable moisture content. This improves its strength and reduces the potential damage by decay from fungi. A number of options are available, from air drying which takes several months, to high temperature kiln drying that can be done in less than a day. Dry timber is lighter and easier to handle, fastens and paints better, and is dimensionally more stable – it does not warp or split. If the moisture content of the timber is above twenty per cent it must be stamped *WET* and is not suitable for structural use.

The strength of timber depends on a number of factors including:

- species of tree,
- growing conditions (climate),
- rate of growth (slower growth = higher strength),
- seasoning process,
- handling (accidental damage),
- storage conditions.

Typical species grown in Ireland include: Sitka Spruce, Norway Spruce, Lodgepole Pine, Douglas Fir (species and source code: WE/SG1), Larch and Scots Pine (species and source code: WE/SG2).

Timber Decay

When talking about the degradation of wood, it is important to remember that in the right environment wood can last for thousands of years – it does not decay with time. This is clear from the perfectly intact wooden artefacts that have been found in the tombs of Egyptian Pharaohs.

Wood will only degrade due to external factors, including:

- fungi,
- bacteria,
- insects,
- marine borers,
- mechanical,
- chemical,
- heat.

Wooden chair from Egypt, approx. 1400 BC.

Of these external factors, fungal attack is the most relevant to the selection and use of structural timber. There are two major categories of fungal attack. These are called brown rot and white rot. They are given these names because the rotted timber takes on a mainly brown or white appearance after attack. The terms dry rot and wet rot are also commonly used to describe fungal decay of timber but these terms are misleading (and should be avoided) because all fungal decay requires moisture.

Fungal attack will occur in timber which is exposed to damp, stagnant conditions. In fact, fungal attack will only occur in timber with a moisture content above 20%. This is why structural timber must have a moisture content below 20% when purchased. This low moisture content must be maintained throughout the lifetime of the building. This is why, for example, attic spaces are ventilated. The fresh air blowing through the attic space helps to keep the roof timbers dry and cool. Fungal attack is most likely in temperatures between 20°C and 30°C.

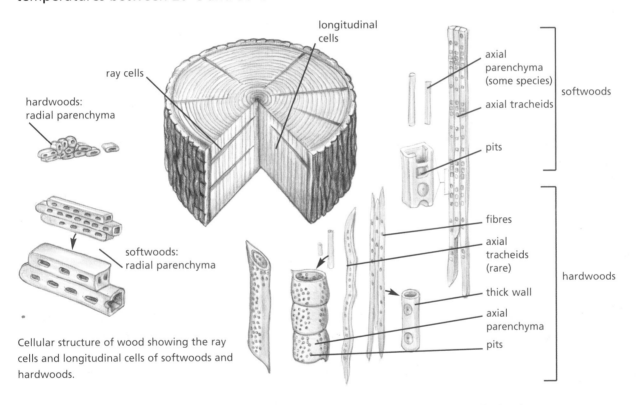

Cellular structure of wood showing the ray cells and longitudinal cells of softwoods and hardwoods.

Wood basically consists of boxes and tubes made of sugars which are linked together to form cellulose, the basic building material of plants. Chains of cellulose are arranged in different orientations and bonded by another material, hemicellulose. A further material, lignin, adds rigidity and strength to the timber. It is the arrangement of cellulose with the other two materials which give wood its characteristic properties and its *cellular* structure.

Timber constituents are shown in the table.

Softwood % oven-dry weight of wood		Hardwood % oven-dry weight of wood	
• cellulose	40–50	• cellulose	40–50
• hemicellulose	20	• hemicellulose	15–35
• lignin	25–35	• lignin	17–25
• other substances*	1–10	• other substances*	1–10

(Softwood: cellulose, hemicellulose, lignin = structural; Hardwood: cellulose, hemicellulose, lignin = structural)

*i.e. non-structural substances such as resins and tannins

When brown rot occurs in timber, the cellulose is consumed leaving the lignin behind. Lignin gives wood its rigidity and also its colour. When the cellulose is consumed, the lignin left behind gives the wood a distinctively brown appearance. This inability to substantially degrade lignin distinguishes brown rot fungi from white rot fungi. When white rot occurs lignin is consumed in addition to cellulose and hemicellulose. It is because the lignin is consumed that the rotted wood looks distinctively white.

Rot is prevented by ensuring that untreated structural elements (e.g. suspended timber ground floors, attic spaces) are well ventilated and by correct use of damp-proof membranes and damp-proof courses. Where rot has occurred, the infected timber (e.g. floor joist) must be cut out and replaced. All of the timbers should then be treated with a preservative. It is also very important to deal with the cause of the rot (i.e. dampness) by removing the source of moisture (e.g. rising damp, leaking pipes, gutters etc.) and by providing adequate ventilation.

Blue stain is another type of fungal attack which is also recognised by the colour it causes. Blue stain fungi cause bluish or greyish discolouration of the wood but do not cause decay. Blue stain has no effect on the strength of the timber. Blue stain usually appears as specks, spots, streaks or patches of colour, especially in whitewood (e.g. pine). Blue stain is only important in the selection of structural timbers where the timbers will be seen in use (e.g. exposed roof beams).

Certain blue stain fungi are responsible for the spalted appearance of some hardwoods. Spalted beech, birch and maple are much sought after for decorative artefacts (especially wood turning and furniture) for their pleasing aesthetic qualities. The spalting lines are actually *zone lines* caused when competing fungi meet.

Brown Rot

Fungi:
- Serpula lacrymans
- Coniophora puteana

Attacks:
- cellulose

Appearance of infected timber:
- brown in colour
- cracks across and along grain giving cuboid appearance
- crumbles when touched

White Rot

Fungi:
- Phanerochaete chrysosporium
- Phellinus contigus
- Donkioporia expansa
- Asterostroma spp.

Attacks:
- lignin, cellulose and hemicellulose

Appearance of infected timber:
- white/greyish in colour
- cracks along grain
- stringy fibrous texture

Brown rot.

White rot.

Penetrating Treatment of Timber

Penetrating treatment is commonly referred to as pressure treatment, because the liquid preservative is driven into the timber in a pressure chamber. Penetrating treatment provides a higher level of resistance to decay than superficial (surface) treatment, because the preservative penetrates deep into the timber.

The level of treatment can be varied, depending on a number of factors including:

- **Timber Species** – pressure treatment is usually applied to softwood timbers. This is because softwoods are more suitable for pressure treatment. The microstructure of hardwoods coupled with the high levels of natural oils and chemicals present in many hardwoods, reduces the effectiveness of (and need for) pressure treatment.

- Hazard Class – term used to describe the level of exposure of the timber when in use:
  - Hazard Class 1 – timber is under cover, fully protected from the weather and not exposed to wetting (e.g. roofing timber).
  - Hazard Class 2 – timber is under cover and fully protected from the weather but where high environmental humidity can lead to occasional wetting (e.g. sauna).
  - Hazard Class 3 – timber is either continually exposed to the weather or is protected from the weather but subject to frequent wetting (e.g. windows and doors).
  - Hazard Class 4 – timber is in contact with the ground or freshwater and permanently exposed to wetting (e.g. fence post).
  - Hazard Class 5 – timber is permanently exposed to saltwater (e.g. floating structures).
- Service Life – this is the amount of time the timber component is expected to function properly. For example, structural components (e.g. wall plates, joists, rafters) are expected to have a 60-year service life, while external joinery (e.g. windows, doors, fascias and soffits) are expected to have a 30-year service life.
- Preservative Type – a number of preservatives are commonly used, including:
  - Copper-based Preservatives – these preservatives give the timber a distinctive green colour (e.g. copper-azole and copper-quat), for this reason copper-based preservatives are not used on timber that will be used for aesthetically sensitive functions (e.g. windows and doors). Copper-based preservatives are recommended for hazard class 4 situations. Fixtures used in conjunction with copper-treated timber should be made of galvanised or stainless steel to prevent corrosion.
  - Organic Solvents or Micro-emulsions – these preservatives contain one or more fungicides or insecticides (resist attack by insects including beetles, termites and marine borers) and a dye.
  - Boron – these preservatives leave no colour on treated timber (e.g. boric acid, disodium tetraborate (borax) and disodium octaborate).
  - Note: The use of creosote and arsenic-based preservatives in residential construction is banned by EU Directive (since June 2003 and June 2004 respectively).
- Treatment Process – the penetrating treatment of timber can be carried out at low pressure (hazard class 1–3) or high pressure (hazard class 4 and 5). Organic solvents and micro-emulsions are normally applied at low pressure, while copper-based preservatives are applied at high pressure.

The double-vacuum pressure process is a commonly used method. As mentioned above, this process uses pressure to drive the preservative into the timber. This is necessary to overcome the natural resistance of the timber to penetration and to ensure a reasonable depth of protection is achieved. It is important to remember that when a plank of pressure-treated timber is cut, the newly exposed end grain is susceptible to decay and should be treated with a brush-on preservative.

Treatment process:

timber is wheeled into
chamber on a bogie

vacuum is applied to pull
air out of timber cells

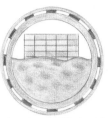

chamber is filled with
preservatives

Penetrating treatment of timber
(pressure-treating).

chamber is pressurised forcing
preservative into the timber

preservative is drained
and another vacuum is
applied to remove
excess preservative

treated timber is
removed for drying

Stone

Stone is an essential material in the construction industry. Every house, school, hospital, shopping centre, sports arena, road or bridge needs stone of one type or another for its construction.

There is a long tradition of using stone in the construction of houses in Ireland. Old stone buildings usually reflect the geology of the area in which they stand. This is because stone is a heavy material and is expensive to transport over long distances. Houses, byres and other everyday buildings were often constructed using rough stone (rubble) which was later covered with a lime-based render. For more formal buildings, such as cathedrals, the stones were shaped (dressed) into squared pieces (ashlar) that sat together neatly giving a higher standard of finish.

The recycling of stone and aggregates is becoming more important, successful and cost-effective. Many companies now specialise in the crushing, grading and resale of salvaged stone.

Stone everywhere can be classified as follows:

- igneous,
- sedimentary,
- metamorphic.

Each of these categories contains stones that have differing properties such as appearance, strength, durability and ease of use.

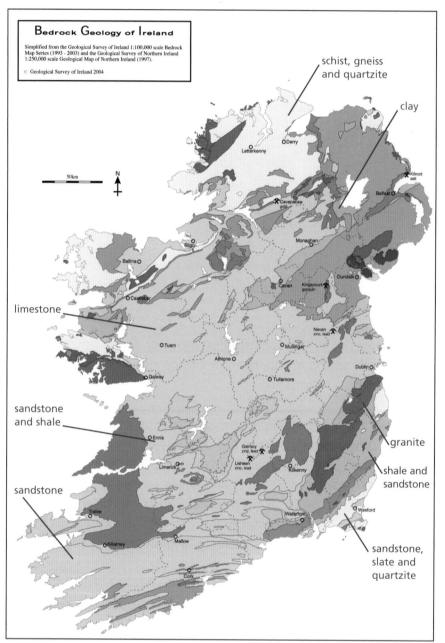

BEDROCK GEOLOGY OF IRELAND: this map shows the type of rock that would be exposed at the surface if all soil were removed. Visit www.gsi.ie for more details.

Igneous

Igneous rocks form when molten rock cools and becomes solid. Molten rock is called *magma* when it is below the Earth's surface and *lava* when it is above.

Granite is an example of a magma-based stone, whereas basalt is a lava-based stone. It is because magma takes a very long time to cool below the earth's surface that granite is a high-quality, dimensionally stable and extremely hard type of stone.

Granite

Granite is the main igneous stone used for building. It contains the following minerals:

- Feldspar – main mineral in granite, gives granite its colour (white, grey, pink, green etc.), as hard as steel.
- Quartz – pure silica, harder than steel, extremely durable, colourless.
- Mica – black biotite and white muscovite are shiny crystals that are often found in granite. They make the surface of the granite sparkle in sunlight.

Granite is usually found in mountainous regions. In Ireland, granite was traditionally quarried from a large mass which stretches from South County Dublin through Wicklow, Carlow and Wexford. It was also quarried in County Down in the Mourne Mountains. (See preceding map illustrating the Geology of Ireland – shown in red.)

Granite is used in building as dimension stone (high-quality and accurately cut) or as rubble (rough irregular shape). It is often used for paving or kerbstones because it is hard-wearing and has a naturally slip-resistant surface. Granite has always been a popular stone for construction because of its hardness

Granite building: St Joseph's Redemptorist Church and Monastery, Dundalk.

and durability. However, these properties mean that it tends to be very hard on cutting tools and machinery. Granite dust is harmful to the lungs so dust masks are required when cutting.

Sedimentary

Many sedimentary rocks are made from the broken bits of other rocks. These broken bits of rocks are called *sediment*. Sediment is the sand you find at the beach, the mud in a lake bottom or the pebbles in a river. Sediment is transported by the action of rivers and wind and deposited in beds in sea water, lakes and deserts. The sediment may, in time, form rock if it becomes compacted by weight and cemented together.

Limestone and sandstone are commonly used examples of sedimentary stone. Both are softer than granite and therefore easier to work with.

Limestone building: St Finbar's Cathedral, Cork.

Limestone

Limestone is formed from sediment laid down in either saltwater or freshwater. The most important mineral in limestone is calcite. Calcite comes from the shells and skeletal bones of marine life. This is dissolved by the action of acidic water and re-deposited among the sediment where it acts as a cement, binding the sediment together. Calcite is a soft mineral and dissolves with acid rain. Sulphur in the atmosphere near cities combines with calcite to form calcium sulphate which causes decay in limestone buildings.

Irish limestone dates from the carboniferous period (354 to 290 million years ago) and is well compacted and relatively hard. Generally the older the stone the more compacted it is.

Sandstone

Sandstone can be formed on land in wind-blown desert dunes, in freshwater or saltwater. Sandstone is classified by its binding agent. The durability of the sandstone is also determined by its binding agent. For example, siliceous sandstone has silica as a binder which forms durable sandstones, such as gritstones. Calcareous sandstone has calcium carbonate as a binder and is susceptible to acid decay from polluted atmospheres. Ferruginous sandstone has iron oxide as a binder. This produces a variety of colours (including red and brown) which are generally good weathering.

Sandstone building: St Patrick's Cathedral, Armagh.

Metamorphic

Metamorphic stones come from igneous or sedimentary stones that have changed due to high temperatures and pressures within the Earth. Metamorphism causes the recrystallisation of existing minerals and compaction (increased density and reduced porosity). Igneous, sedimentary and metamorphic rocks can all be metamorphosed. Examples of this include:

- limestone to marble,
- shale to schist,
- mudstone to slate,
- sandstone to quartzite,
- granite to gneiss.

Marble

Marble is used for decorative purposes such as fireplaces and floors. Carrara marble from Northern Italy is the most famous marble in the world. It was used by Michelangelo in the sixteenth century. It is snow white and translucent, metamorphosed from pure limestone and is still used all over the world. A well-known Irish marble, Connemara marble, is green in colour and very hard. Marbles do not weather well and are best used indoors.

Slate

Slate is traditionally used for roofing. It was split using hand tools and graded by thickness so that the heaviest slates were laid first at the eaves (gutter) level, then the medium and finally the lightest slates at ridge level. Slate was traditionally quarried in Donegal, Kerry and Tipperary.

Natural slate recovered from an old roof.

Natural slate roof.

Aggregates

Aggregate is a general term used to describe granular stone material used in construction. Sand and gravel are examples of aggregate.

Sand used in the production of mortar and concrete.

Gravel used in the production of concrete.

Hardcore used as a fill material under floors.

Types of Aggregate

There are three main types of aggregate:

- Natural aggregate – aggregate quarried from mineral sources which has been subjected to nothing more than mechanical processing.
- Manufactured aggregate – aggregate resulting from an industrial process such as thermal treatment.
- Recycled aggregate – aggregate resulting from inorganic materials previously used in construction.

When natural sand is taken from a pit for use in concrete it is washed to remove any clay or silt that may be mixed through it. This ensures that the strength of the concrete is consistent and reliable. Sand is classified by particle size: fine (less than 0·25 mm), medium (0·25 to 0·5 mm), coarse (0·5 to 1·0 mm) and very coarse (1·0 to 4·0 mm).

In addition to being clean it is important that the moisture content of aggregates is known. This is because when sand is moist it occupies a greater volume than when it is dry. This phenomenon is

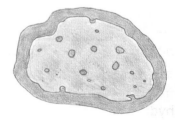

Bulking of sand: because the grain of sand is so small, the film of moisture adds significantly to its volume.

known as *bulking*. A sample of moist sand typically occupies twenty-five per cent more volume than it does when dry. This has a significant impact on the measuring of constituents when preparing concrete.

Classification of Aggregate

Aggregates are generally classified by their size. This is done by passing the material through a series of sieves. Each sieve collects granules of a specific size.

As larger particles are caught in the sieves they are removed and stored by size. Aggregates are available in both single and graded sizes. Generally, aggregates from 80 to 4 mm in size are referred to as coarse aggregate (gravel), while aggregates from 4 mm to 0·063 mm in size are referred to as fine aggregate (sand). Any material that passes the 0·063 mm sieve is referred to as 'fines'.

Aggregates are supplied in single size and graded forms. Single size aggregate is material of a single size only, such as 10 mm gravel. Aggregates that are made up of a mixture of sizes are referred to as *graded*. These include a mixed material from one size to another. An example would be 4/20. This aggregate contains granules that range in size from 4 mm to 20 mm in size. *All-in* aggregate contains granules in the 0/40 range.

Hardcore is a specific type of unbound (i.e. no cement) granular fill used as a sub-base below a concrete floor or footpath. It is supplied as 0/32 to ensure that when it is compacted, the smaller particles fill the spaces between the larger particles giving a very dense and stable fill.

		particle size in mm
coarse aggregrate		80·000
		63·000
		40·000
		31·500
		20·000
		16·000
		14·000
		10·000
		8·000
		6·300
fine aggregate	very coarse sand	4·000
		2·800
		2·000
	coarse sand	1·000
		0·500
	medium sand	0·250
	fine sand	0·125
fines		0·063
		0·000

Lime

Lime is a versatile material which has a long tradition of use in the construction of houses in Ireland. It is particularly important in the conservation of heritage buildings because of its flexible, porous characteristics. Lime comes in a variety of forms including:

- non-hydraulic lime (air lime).
  - quicklime (dry powder, lumps),
  - hydrated lime (dry powder, slurry, putty).
- hydraulic lime.
  - natural hydraulic lime (dry powder),
  - artificial hydraulic lime (dry powder).

Non-hydraulic lime is produced from limestone which is very pure in form. Hydraulic limes are produced from limestone which contains mud-bearing minerals like alumina and silica. Non-hydraulic lime, when used in construction, hardens by reaction with carbon dioxide in the air whereas hydraulic lime has the ability to set in damp conditions in a similar fashion to cement.

The Lime Cycle

Lime is made from limestone. By heating the limestone (calcium carbonate, $CaCO_3$) in a kiln at about 900°C, carbon dioxide (CO_2) is driven off. This is called *calcining*. The product is a dirty white lumpy material called quicklime.

$$(CaCO_3 \rightarrow CaO + CO_2)$$

This is then added to large quantities of water (hydration/ slaking). This causes a chemical reaction to take place which gives off heat and the lime expands to about 3 times its original volume. This change is gradual and takes several days to complete and when finished the lime is said to have been hydrated. It is called hydrated lime or slaked lime, i.e. calcium hydroxide with the chemical formula $Ca(OH)_2$.

$$(CaO + H_2O \rightarrow Ca(OH)_2)$$

The excess water is then drawn off and the hydrated lime is allowed to mature for at least three months in a covered container. During this time it thickens to the consistency of toothpaste (pure white in colour) and is called *lime putty*. Lime putty is usually delivered to the building site covered by a thin film of water in airtight plastic tubs. Lime-putty-based products harden, when mixed with water, by absorbing carbon dioxide from the air to revert back to calcium carbonate (limestone). This process is called *carbonation*.

$$Ca(OH)_2 + CO_2 \rightarrow CaCO_3 + H_2O$$

Carbonation is most effective if lime is used in thicknesses up to a maximum of 12 mm in renders or plasters and in weak thin coats when limewashing.

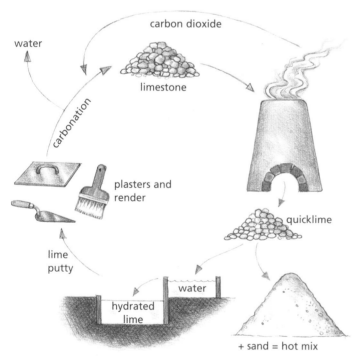

The lime cycle.

Lime mortar, suitable for rendering and pointing, is made by mixing lime putty with coarse sand (hot mix). The addition of water is not normally required because the lime putty tends to have enough water content to produce a workable plastic consistency when mixed well. Plaster can be made by mixing lime putty with fine sand. Haired lime plaster is made by adding animal hair (usually goat hair) to the mix. This improves the tensile (stretching) performance of the plaster and is very important when plastering lathed internal walls in heritage buildings. Lime wash is made by diluting lime putty with water to create a wash. Adding suitable pigments such as red or yellow ochre produces pleasing pink or yellow tones of colour.

Lime has the advantage of being cheap, reusable and readily available. However it is messy, laborious to mix and it gains strength slower than cement. Lime mortar is not suitable for use in persistently damp conditions because it readily absorbs moisture – over time the effect is to reduce the adhesion of the lime to the sand and the mortar falls out of the joints between the stonework or brickwork. Protected from persistent damp, a lime mortar will serve effectively for the lifetime of most buildings. Lime renders are more flexible than cement-based renders. This allows for moisture and temperature movements – leading to less drying shrinkage and cracking. This makes them particularly suitable for use in the conservation of old buildings. The porosity of lime render causes greater suction (absorption of water) and this helps the application of further coats.

Lime-based products are naturally white in colour. This gives many of the vernacular cottages scattered around the Irish countryside their traditional white appearance. Lime also has antiseptic properties: this helped to keep traditional cottages healthy by killing fungi, insects and other unwelcome creatures.

In modern construction, lime is used where flexibility is desired in a mix such as in the packing around the flue in a chimney. It is also added to external renders to give a white, rather than grey, appearance.

Hydraulic Limes

Hydraulic lime is made by burning limestone that contains clay. Hydraulic lime is stronger than ordinary lime and will harden in wet conditions. Nowadays, hydraulic lime has been replaced by Normal Portland Cement for most applications.

Cement

Cement is a binding agent used in mortars, renders and concretes. When cement is mixed with water a chemical reaction (called hydration) takes place.

The paste-like material formed binds itself very firmly to most materials. When mixed with aggregates it forms a very solid, stone-like, material called concrete.

Cement is manufactured in a number of locations around Ireland. The process involves heating a mixture of pulverised limestone and shale (in the ratio 4:1) to a very high temperature (about 1,500°C) in a rotating kiln. This produces a white-hot mass which cools to form *clinker* – a solid, black, glistening substance similar in size to gravel). This is then cooled and ground into powder in a ball mill (a steel cylinder containing man tonnes of steel balls). A small amount (five per cent by weight) of gypsum is added which increases the setting time of the cement. The finished product is then either shipped in special plastic-lined paper bags of 25 kg or shipped in bulk by train or truck.

Cement hardens, not by drying out, but by a chemical reaction between cement and water. This reaction is called *hydration*. When this reaction takes place crystals are formed which interlock giving the cement paste its strength.

raw materials (limestone: shale, 4:1)

mixed and ground to powder

heated to 1,500°C in an inclined rotating kiln

Manufacture of cement from limestone and shale.

clinker

clinker ground in ball mill (5% gypsum added)

cement

Setting is the term used to describe the initial stiffening of the cement. This is not to be confused with *hardening*. Setting takes places after about 90 minutes, whereas hardening takes places over a much longer period of time. In fact, cement continues to harden throughout its life. It gets progressively stronger when measured after 2 days, 28 days and 72 days. The strength after 28 days is usually taken as the working strength for structural design and engineering purposes. The strength described here is the compressive strength – the strength of the material when crushed. The compressive strength of Normal Portland cement measured on mortar prisms is at least 20–30 N/mm^2 at 2 days and 50–60 N/mm^2 at 28 days.

There are numerous cement types available for different applications. The most common type is Normal Portland Cement (NPC) also known as CEM I. The traditional name comes from the fact that the colour of the cement is similar to that of Portland stone – a limestone found in England. Normal Portland cement is suitable for most applications including the construction of foundations, walls, footpaths and so on.

However, when a high early strength is required, for example when carrying out urgent repair work, Rapid Hardening Portland Cement is used. RHPC is simply cement which has been very finely ground to allow an increased rate of hydration. Smaller particles have a greater surface area and so react with water more quickly. This means that RHPC mortar can achieve a 2–day compressive strength of at least 28–34 N/mm². However, the 28-day strength remains 55–62 N/mm². The main application of rapid-hardening cements is in the manufacture of precast concrete components where the high early strength of the concrete allows efficient reuse of moulds and formwork.

Another type of commonly used cement is Sulphate-Resisting Portland cement. SRPC is made by fusing together a blend of very finely ground limestone, shale and iron oxide at high temperatures to form cement clinker. This cement is used when there is a high level of sulphates present in the soil or groundwater. This can be naturally occurring or due to fertilizers or industrial effluents. Sulphates are corrosive to concrete and cause it to decay over time, which could prove disastrous for a foundation. Sulphate resisting cement is more expensive to produce because the chemical composition of the cement has to be adjusted during manufacture.

The strength and performance of cement products is governed by Irish and EU regulations. Cement manufacturers continually test samples in their laboratories to ensure that the cement meets the required standards. Cement mortars are usually made by mixing one part of cement to three parts of sand with one-half part of water, measured by weight. Each batch used for these test specimens consists of (450 ± 2) g of cement, (1,350 ± 5) g of sand and (225 ± 1) g of water. According to Irish Standards, the minimum compressive strengths must be 10 N/mm² at 10 days and 42.5 N/mm² at 28 days for class 42.5 N cement.

Cement should be stored in a dry, weatherproof, frost-free, enclosed shed or building with a dry floor. If the floor is concrete, it should be stored on a timber platform. Any contact with moisture will cause it to deteriorate through premature hydration. Cement can absorb moisture from the air as well as direct moisture, so once opened, a bag should be used completely. Bags should be stacked closely together away from walls, not more than eight bags high and covered with a tarpaulin or polythene sheet. Cement should also be stored so that it can be used in order of delivery (first in, first out). Cement should be checked for deterioration

Silo containing dry mortar mixture. When delivered, the silo contains 16 tonnes of dry material, complete with mixing equipment. Once power and water are connected, a constant supply of mortar is available at the push of a button.

when taken out of storage and should not be used if it is lumpy. Cement is supplied in moisture-resistant paper bags that give it a shelf-life of approximately three months.

Cement-based mortars are generally very strong and durable and perform well when exposed to weathering. Cement mortars can be coloured (by the addition of a powder pigment to the cement and sand mixture before the water is added) to suit the brickwork or roofing tiles being used. Cement is also used in the rendering of blockwork (see chapter 24, Rendering and Plastering). For large developments a premixed mortar in dry powder form can be delivered to the site in a large silo. This saves time, reduces waste, improves the consistency of the mix and cleanliness on-site.

Sydney Opera House, designed by Jorn Utzon and built between 1953 and 1971.

Concrete

Concrete is an artificial stone-like building material made by mixing cement, aggregates and water. Concrete will take the shape of the form or mould into which it is placed, allowing for very creative shapes to be designed.

Properties

Concrete has a number of properties that make it particularly suitable for use in the construction industry. It is strong in compression. It can readily withstand significant loading. It can be formed into any shape. Its surface can be finished with a smooth or rough texture. The colour of concrete is usually the grey of Portland cement but colour pigments can be added. It is relatively cheap and easy to work with. Its main shortcoming is that it is weak in tension. This means that when it experiences a pulling action it tends to crack and break. Tension is common in horizontal beams.

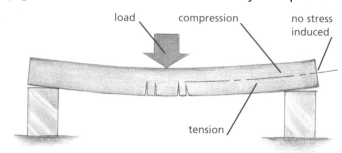

Failure in a concrete beam due to tension in the lower half of the beam.

Types and Applications

Concrete can be made to suit a variety of applications. The use of fine and coarse aggregates combined in the right proportions produces a strong concrete with very few voids or spaces. The ratio of constituents used is traditionally expressed as a ratio in the form – cement: fine aggregate (sand): coarse aggregate (gravel). For example, a mix of 1:2:4 would be traditionally used for floors. However, changes in regulatory requirements and in the speed at which houses are being constructed means that large-scale concrete mixing is best done in carefully controlled conditions off-site by specialist companies. The fresh concrete is then delivered to the site by truck in volumes of up to 8 m^3 when needed.

The manufacture of concrete is controlled by strict standards and involves the use of lettering and numbering to identify particular types of concrete – in a similar way to the stamping of wood that was discussed earlier. There are four types of concrete suitable for use on small sites where the concrete is site-batched or obtained from a local ready-mixed concrete producer. These mixes are described by their compressive strength, using the letter 'C' followed by two numbers. The first number is the cylinder strength (not used in Ireland), the second number is the cube strength (see the Cube Test following). Concrete mixes vary in compressive strength and are used for a variety of applications as shown in the following table.

Strength Class	28-day Cube Strength (N/mm^2)	Minimum Cement Content (kg/m^3)	Application
• C8/10	10	Not applicable	Pipe and kerb bedding/backing
• C12/15	15	200	Strip foundations
• C16/20	20	200	Floors
• C20/25	25	240	Footpaths/garage floors (wearing surfaces)

Water-Cement Ratio

The proportion of water to cement in the mix has a significant impact on the performance of both fresh and hardened concrete. The ease with which freshly mixed concrete can be handled, placed in formwork, compacted and finished is technically termed a consistence and is commonly called workability. This depends largely on the amount of water in the mix. Workability is also influenced by the size and shape of the aggregate used. Small, smooth, rounded aggregate will provide a more workable mix than large, angular aggregate.

The strength of the concrete is also affected – the more water used, the weaker the hardened concrete will be. This is because only a certain amount of the water in the mix is consumed by hydration with the cement, the remainder helps to make the mixture workable – this water eventually evaporates leaving tiny pores in the concrete. The greater the amount of water used the more porous the hardened concrete will be, thereby reducing its strength and durability. The range of acceptable water-cement ratio for most construction applications is between 0·45 and 0·60 by weight. The water used should be of drinking quality.

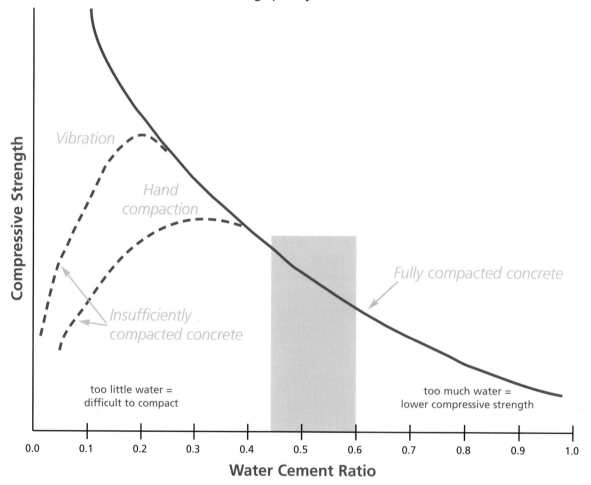

If the mixture is too wet there is also a risk of segregation, where the heavier coarse aggregate sinks to the bottom of the mixture. Generally speaking the drier the mix the stronger the hardened concrete will be, however, if the mix is too dry it will be difficult to compact it into place. This could lead to voids or spaces in the concrete. If voids occupy just 5% of the volume of the hardened concrete, there will be a reduction in strength of approximately 30%. Concrete must be sufficiently workable to allow full compaction after placing. The workability of a sample of concrete can be established using the Slump Test, which is explained in the following text.

Batching and Mixing

Batching is the term used to describe the careful and accurate selection of the quantities of cement and aggregate to be used in a mix of mortar or concrete. There are two methods of batching: batching by volume and batching by weight.

The batching of concrete by volume involves measuring the quantity of fine and coarse aggregate relative to a standard 25 kg bag of cement. Batching by volume is only recommended up to C10/15 strength. For these mixes the recommended volumes of fine and coarse aggregate are:

Strength Class	Cement	Fine Aggregate	Coarse Aggregate
C8/10	25 kg	50 litres	75 litres
C10/15	25 kg	45 litres	60 litres

Note: these mixes will provide a Slump Test result of 0–40 mm.

Therefore to volume batch C8/10 concrete a 25-litre gauge box (a cube measuring approx. 292 mm internally) would be required. This would then be filled twice with fine aggregate and three times with coarse aggregate, to which would be added one 25 kg bag of cement to achieve the correct mix. This produces approximately 0·1 m³ of concrete despite the fact that the combined volume of the constituents is much greater. This is because the fine aggregate and cement paste occupy the spaces between the coarse aggregate, thereby reducing the overall volume generated.

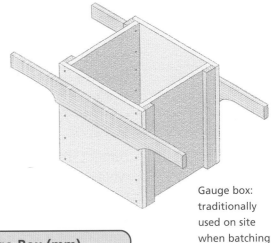

Gauge box: traditionally used on site when batching by volume.

Volume of Gauge Box (litres)	Internal Size of Gauge Box (mm)
15	246
25	292
30	310

Note: A cube whose sides are 100 mm, occupies a volume of one litre. The *shovelful* is not a recommended method of volume batching. For water, a volume of one litre weighs one kilogram.

Batching by volume is commonly used for preparing mortar and concrete on small projects such as DIY jobs around the home – ideally, it should not be used on larger projects such as house construction. Batching by volume is usually appropriate when mixing by hand. When mixing by hand, the mix ingredients should be placed on a clean board and mixed to a uniform colour throughout before adding water to obtain the necessary workability. Each batch should be inspected before using to check that it has been thoroughly mixed (uniform appearance).

If a project requires more than 0·5 m³ then the batching-by-weight method should be used. This is because batching by volume is inherently inaccurate. For example, a given sample of sand will occupy twenty-five per cent greater volume when damp, compared to when it is dry or saturated.

Batching by weight is used by concrete suppliers and professional builders on large-scale projects. This involves the use of specialist equipment which can accurately weigh and deliver a constant supply of mix in dry form.

On medium-scale projects, rotating-drum mixers with integral weighing mechanisms are used. As each constituent is added to the drum the weight is noted until the correct quantity of each material is added.

Each batch should be completely emptied before refilling the mixer. A mixing time of one to two minutes is usually necessary with rotating mixers. This method is commonly used for mortar preparation. However, with the large number of ready-mix suppliers in operation around the country today most builders choose to use ready-mixed concrete supplied directly to the site by truck.

Ready-mixed concrete is batched by weight at the production facility. The cement is stored in a large silo, while the aggregates are stored in large weighing hoppers or bins. The required quantity of each material by weight is released from the hoppers onto a conveyer belt, at the same time the silo releases the required amount of cement onto another belt. The measuring equipment is accurate to within ±3%. These computer-controlled systems automatically adjust for additional weight due to moisture. The water content by weight of aggregates is typically: sand 6–7%, coarse aggregate 1–2%, all-in aggregate 4–5%. The conveyer belts carry the ingredients to the chute at the back of the mixer truck for transport to the site. The mixer truck gradually adds water to the dry mix during transport so that the concrete always arrives on-site freshly mixed.

Mix proportions for standardised concretes, batched by weight, are as follows:

	Cement (kg)	Aggregate (30–45% fine aggregate) (kg)
C8/10	230	1,960
C10/15	265	1,930
C16/20	310	1,900
C20/25	350	1,870

Placing

Fresh concrete is placed into formwork on-site. The formwork is simply a mould that holds the concrete in place until it hardens. Formwork is traditionally made from timber or steel although specialist reusable materials are available. A release agent should be applied to the formwork before the concrete is poured to make it easier to remove it after the concrete has hardened – just like greasing a cake tin. Formwork should be clean and well secured because concrete is a very heavy material and will cause flimsy formwork to buckle. When timber formwork is used during dry weather it is best practice to hose the formwork down before placing the concrete. This prevents the over-absorption of water from the mix by the timber.

Compacting

When the concrete has just been poured it can contain up to twenty per cent trapped air. To remove this air the concrete must be consolidated or compacted. This is done by vibration – the concrete settles into place allowing it to flow into corners, around steel reinforcement and flush against the form face. Concrete can be vibrated either internally or externally using specialist equipment – smaller jobs of shallow depth can be compacted by hand using a timber board, called a screed.

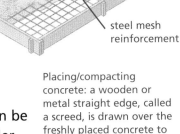

Placing/compacting concrete: a wooden or metal straight edge, called a screed, is drawn over the freshly placed concrete to bring it to the correct level.

Curing

Curing is the term used to describe the maintenance of concrete at the desired temperature and humidity (usually for the first seven days) following placement to ensure adequate hydration and hardening. It is important to remember that concrete does not harden by drying out, in fact if the concrete dries too quickly there is a risk that the cement will not have fully hydrated. This would greatly weaken the hardened concrete. The mild, damp climate in Ireland means that curing is not usually a problem. However, if the concrete is poured on a hot summer day it is important to cover it with polythene to prevent the evaporation of water. Similarly, if the weather is particularly cold it is important not to allow the concrete to freeze as this will also prevent hydration.

Admixtures

Sometimes it is necessary to adjust the properties of fresh or hardened concrete. An admixture is a substance added to concrete (or mortar) to adjust or alter the properties of the mix or of the hardened product. For example, if an extension to the rear of a terraced house is being constructed and the only access is through the house, then it might be necessary to slow down the setting of the concrete to allow enough time for placement. Typical admixtures include:

- Accelerators – used to hasten hardening and strength development.
- Retarders – used to slow down setting and hardening.
- Water-reducing agent. Used to reduce the water-cement ratio while maintaining workability.
- Superplasticisers – a special type of water-reducing agent that achieves very high levels of workability. Used when pouring concrete into formwork containing a lot of steel reinforcement. It allows concrete to be placed without compaction. Alternatively, superplasticisers are used to produce concrete of very high strength due to the reduced water-cement ratio.
- Water-resisting – admixture which reduces (by at least fifty per cent) the capillary absorption of hardened concrete. Used in structures below the water table or in water-retaining structures.

Reinforced Concrete

The use of steel to strengthen concrete is called *reinforcement*. It is usually necessary to use steel bars to strengthen concrete in situations where tension will be applied. A typical example of this is the lintel (concrete beam) above a doorway. The lower portion of the lintel experiences tension (stretching caused by bending downward) because of the weight acting on it from above. If steel was not used the lintel would crack and eventually collapse.

There are two approaches to the reinforcement of concrete. The simplest involves placing steel bars or mesh into the formwork and pouring concrete around it. This method is suitable for applications where the tension generated in the concrete will be relatively low and where the concrete will be well supported, e.g. the ground floor of a domestic dwelling.

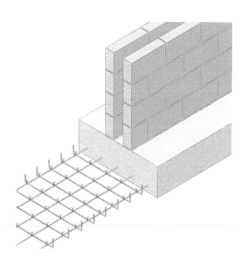

Reinforced concrete strip foundation: in this example the steel bars prevent the strip from bending due to the force exerted by the load-bearing wall.

Reinforcement is usually placed so that there is at least 50 mm of concrete covering it. This prevents air and moisture from coming into contact with the steel, preventing possible corrosion of the steel.

Prestressed Reinforced Concrete

Where the concrete will experience higher tensile stress, such as in a bridge or a beam supporting the roof of a sports stadium, it is necessary to apply tension to the steel reinforcement so as to counteract the high tensile stress applied to the concrete.

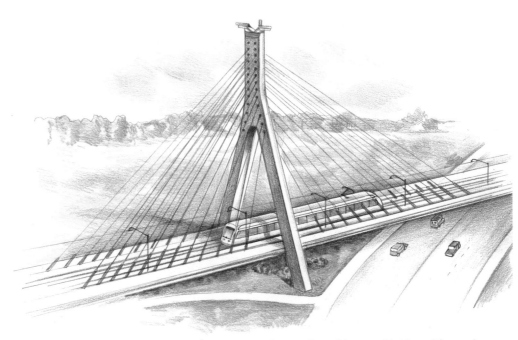

Bridge for the Dublin Light Rail System (Luas) at Dundrum. It is a cable-stayed bridge with a main span of 108.5 m. The pylon height is around 50 metres.

This can be done in two ways:

- pre-tensioning,
- post-tensioning.

Rebar: a steel reinforcing bar used in concrete. It is usually formed from mild steel, and has a patterned surface for better adhesion by the concrete.

These methods involve stretching the steel reinforcement bars (tendons) either before or after the concrete hardens.

In pre-tensioning the steel is anchored at one end and stretched by hydraulic jacks from the other end until the required tension is achieved. The sides of the formwork mould are then positioned and the concrete is placed and allowed to harden around the tensioned steel. The casting is then usually steam-cured for twenty-four hours to rapidly obtain a typical 28 N/mm^2 compressive strength. The steel has a patterned surface, which allows the concrete to achieve a firm bond to it.

When the concrete has sufficiently hardened the steel projecting out at each end is cut flush with the end of the beam. Pre-tensioning is commonly used for precast components such as lintels and suspended floor slabs.

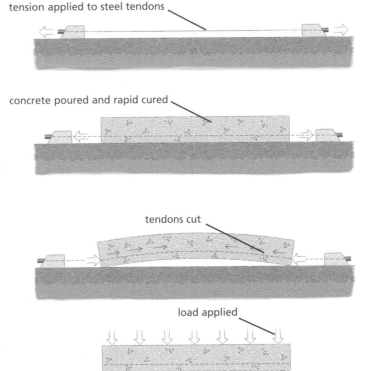

tension applied to steel tendons

concrete poured and rapid cured

tendons cut

load applied

Pre-tensioning concrete: commonly used for residential construction.

tendon wrapped in ducting

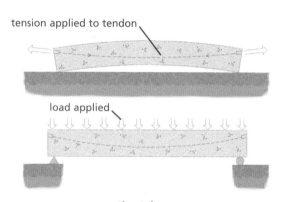

concrete poured and rapid cured

tension applied to tendon

load applied

Post-tensioning concrete: commonly used for larger-scale projects.

Post-tensioning is an alternative method that involves allowing the concrete to harden before applying tension to the steel tendon. For this to be possible a void must be created in the concrete through which the steel can be threaded afterwards. This can be done in a number of ways including placing a flexible steel duct in the formwork before pouring the concrete. The concrete is then placed around it and allowed to cure. When the concrete has cured the tendon is threaded through the duct and secured at one end.

A hydraulic jack is used at the other end until the required tension is achieved. The anchors used vary but a typical approach involves using a large nut which can be secured on the threaded ends of the tendon. As the nut is tightened by the jack the tendon is stretched until the required tension is achieved.

grouting tube

hexagon nut

threaded rebar

corrugated duct

bearing plate

hydraulic jack

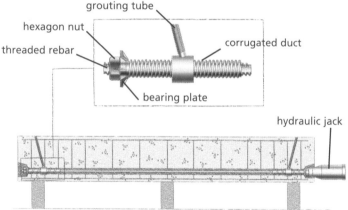

The remaining space within the void is injected with cement grout to exclude moisture. Post-tensioning is often employed in large-scale construction when a number of precast components must be linked together on-site. This is usually done using curved tendons.

Anchoring of straight tendon in post-tensioned concrete.

curved cap tendon

pre-cast component (e.g. floor)

straight tendon

intermediate support

Curved tendon used to join floor panels in large-scale construction.

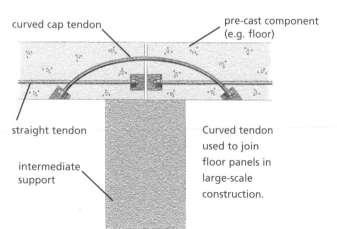

Advantages of prestressing:

- Prestressed concrete can carry a greater load and span a wider opening than a conventionally reinforced member of the same size, proportion and weight.
- Smaller members can be used (because of their increased strength) this reduces the weight of the overall design.
- Individual members can be joined together to act as a single member.

Disadvantages of prestressing:

- high level of workmanship and control required,
- special alloy steels are required,
- more expensive than traditional reinforced concrete,
- special equipment needed.

For spans up to 6 m traditional reinforced concrete is usually the most economic method. Above this range prestressed concrete is recommended.

Problems with Reinforcement

The greatest danger to reinforced concrete is the penetration of air and water causing corrosion (rust). When this happens the corroded steel occupies a larger volume than the uncorroded steel and pressure is exerted on the concrete. This can lead to cracking, spalling and delamination. In order to prevent this, concrete must be properly compacted when placed. This will eliminate voids where corrosion could occur.

Concrete Testing

Various tests are carried out to ensure that the quality and uniformity of concrete products remain consistent. Any inconsistencies in the materials or processes used will affect the quality, strength, appearance and reliability of the hardened concrete. These tests involve checking a number of elements including:

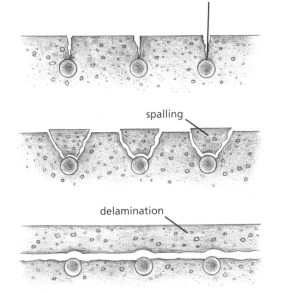

Cracking, spalling and delamination, caused by corrosion of the steel, in reinforced concrete.

- cleanliness of materials,
- bulking of sand,
- moisture content of aggregates,
- workability and consistency of concrete,
- compressive strength of concrete.

Silt Test

The Silt Test is used to determine the quantity of silt present in a sample of sand. Silt is a loose sedimentary material consisting of fine particles between 0·002 mm and 0·05 mm in diameter. The presence of silt in sand used for the manufacture of concrete will reduce the strength of the hardened concrete.

Equipment

- sample of sand,
- saline solution (one per cent common saltwater),
- graduated cylinder,
- calculator.

Method

1. Place 50 ml of saline solution into a graduated cylinder.
2. Add sand gradually until the 100 ml mark is reached.
3. Add more saline solution until the 150 ml mark is reached.
4. Shake the cylinder vigorously.
5. Allow to settle for 3 hours.
6. Measure the height of sand and silt.
7. Calculate the amount of silt as a percentage of the amount of sand.

Calculating the percentage of silt present using the following formula:

$$\frac{\text{height of silt layer}}{\text{height of sand layer}} \times 100 = \text{\% of silt present}$$

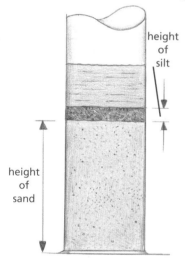

height of silt

height of sand

Silt test: measuring the result of the test.

Bulking of Sand Test (Saturation Method)

When aggregate is being measured by volume the quantity of sand must be adjusted if the sand is damp. This is because sand increases in volume when damp or moist. This occurs because every moist grain of sand is coated in water thereby increasing the size of each grain.

The Bulking of Sand Test is carried out to demonstrate this phenomenon. The test is based on the principle that sand that is completely saturated with water occupies the same volume as dry sand.

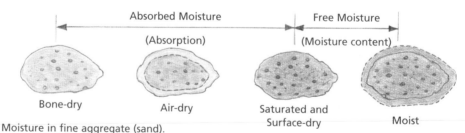

Absorbed Moisture (Absorption)   Free Moisture (Moisture content)

Bone-dry   Air-dry   Saturated and Surface-dry   Moist

Moisture in fine aggregate (sand).

Equipment

- sample of moist sand,
- graduated cylinder,
- clean water,
- stirrer,
- calculator.

Method

1. Loosely pack a graduated cylinder with moist sand to a height, H.
2. Remove the sand.
3. Half fill the graduated cylinder with water.
4. Slowly add the sample of sand, stirring continuously to expel all air bubbles.
5. Allow to settle for five minutes.
6. Measure the height of the saturated sample of sand, h.
7. Calculate the bulking factor.

Calculate the bulking factor using the following formula:

$$\frac{H - h}{H} \times 100 = \text{\% bulking caused by moisture}$$

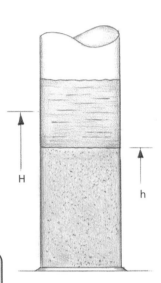

Bulking of sand test: measuring the result of the test.

Therefore, when mixing a batch of concrete using the sand tested, an extra volume of sand equal to the percentage of bulking must be added.

Water Content of Aggregate

The moisture content of aggregates must be taken into account when adding mixing water during concrete manufacture. If the aggregates are wet, less water will be required for the mix. This is very important in terms of the water-cement ratio. Remember that a water-cement ratio of between 0·45 and 0·60 is appropriate for most tasks and damp aggregates risk increasing the water-cement ratio to above this level – thereby reducing the strength of the hardened concrete.

Equipment

- 2 kg sample of moist sand,
- metal container,
- heat source (gas burner),
- weighing scale,
- stirrer.

Method

1. Weigh a 2 kg sample of sand to be tested.
2. Place this in a metal container and heat.
3. Stir sand continuously to ensure it is thoroughly dry.
4. Weigh the sample again.

Calculate the percentage of moisture that was present using the following formula:

$$\frac{\text{wet weight} - \text{dry weight}}{\text{dry weight}} \times 100 = \%\ \text{moisture content}$$

Slump Test

The Slump Test is carried out to determine the consistence workability of a fresh batch of concrete and the consistency of various batches.

Equipment

- sampling scoop,
- cloth,
- slump rod,
- buckets,
- small shovel,
- mixing tray,
- steel rule,
- rigid metal plate,
- slump cone.

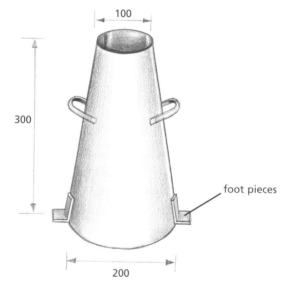

Method

1. Select three buckets of concrete from different parts of a freshly prepared batch of concrete.
2. Empty the buckets into the mixing tray.
3. Mix together thoroughly.
4. Ensure that the slump cone is clean and damp.
5. Place the metal plate on a level surface.
6. Place the cone on the metal plate and stand on the foot pieces.
7. Fill the cone in three equal-depth layers.
   - Rod each layer 25 times after adding it.
   - Spread the blows evenly over the area.
   - Make sure the rod just penetrates to the layer below.
8. Level off the top of the cone using the rod in a rolling and sawing motion.
9. Carefully clean off spillage from sides and base plate.

10. Carefully lift the cone straight up and clear, to a count of between 5 and 10 seconds.
11. Place the cone beside the sample of concrete.
12. Lay the rod across the top of the cone so that it reaches across the top of the sample.
13. Carefully and gradually measure the distance from the underside of the rod to the top of the sample to the nearest 5 mm.
    - This is the slump.

Results

Slump Class	Slump Range (mm)	Workability
• S1	10–40	Low
• S2	50–90	Medium
• S3	100–150	High

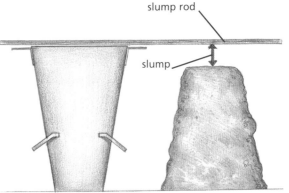

Slump test: measuring the distance from the underside of the rod to the top of the sample to the nearest 5 mm.

Conclusions

Plastic Mix

- Concrete mix that flows sluggishly without segregating and is readily moulded.

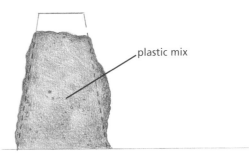

Dry (Stiff) Mix

- Concrete mix containing too little water or too much aggregate in relation to the other components and having little or no slump.

Wet Mix

- Concrete mix having a relatively high water content and runny consistency, yielding a product that is low in strength, durability and watertight properties.

Cube Test

The Cube Test is carried out on hardened concrete to determine its 28-day compressive strength.

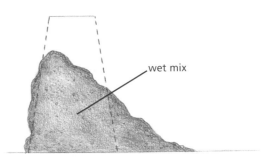

The workability (also known as, consistence) of the sample can be assessed visually as: plastic, dry or wet.

Equipment

- cube mould,
- mould oil,
- oil brush,
- scoop,
- steel float,
- mixing tray,
- small shovel,
- tamping bar.

Method

1. Using the brush, lightly oil the inner surfaces of the mould.
2. Using the scoop, select at least 6 small samples of concrete from different parts of the batch.
3. Place the samples into the mixing tray and thoroughly remix into one heap.
4. Fill the mould in three layers, tamping each layer at least 35 times.
5. Using the steel float, smooth off the top layer flush with the mould.
6. Store the filled mould for 24 hrs at 20°C, humidity 90%.
7. Then remove and store the samples under water until tested.

Cube test: the mould is filled in three layers, tamping each layer at least 35 times to ensure all air pockets are removed.

The cubes are tested in a laboratory, by compressing them in a hydraulic press. The compressive strength is determined when the sample fails.

Concrete Blocks

A concrete block is a precast unit of normal Portland cement, aggregates and water.

There are two main categories of concrete block – solid and

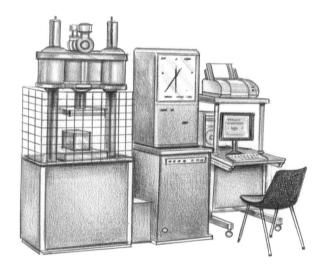

Cube testing equipment: the sample is compressed until it fails (breaks) and the results are analysed on the computer.

hollow. Solid blocks are used in a wide variety of applications including cavity walls, internal walls and boundary walls. Hollow blocks are often used in single leaf residential constructions in the Dublin area, where exposure to wind-driven rain is less severe, as well as being used for garages and other uninsulated buildings.

Both of these types of blocks are usually supplied in two lengths, 390 mm and 440 mm. With a 10 mm mortar joint they have a co-ordinating length of 400 mm and 450 mm respectively. The surface of standard concrete blocks is coarse which allows for good bonding with render (see chapter 24, Rendering and Plastering).

Solid block: used in cavity wall construction.

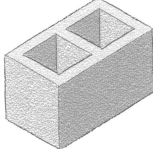

Hollow block: traditionally popular around Dublin for house construction.

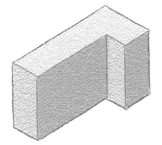

Cavity closer: an 'L' shaped block used to close the cavity around window and door opes.This prevents the spread of fire and improves thermal insulation.

A wide range of specialised blocks are also available, including:

- Fine texture blocks – have a finer finish than standard blocks. Generally for internal applications such as offices, factory units and schools.
- Painting quality blocks – specially developed for projects where painted blockwork is required. These blocks have a finer finish than standard and fine texture blocks.
- High-strength blocks – coarse in texture, they have a higher cement content. These blocks are colour-coded for identification purposes (10 mpa Red, 20 mpa Black, 30 mpa Yellow).
- Aerated blocks – the raw materials used in aerated blocks include sand, cement, lime, aluminium and water. The aluminium in the mix reacts with the lime and forms hydrogen gas. The released gas causes the mixture to expand to twice its original volume, forming extremely small air bubbles. This gives the blocks their characteristic low density. Aerated blocks are light, have high thermal properties and are easily shaped using normal woodworking tools. They must be bonded using a (1:1:6) cement, lime and sand mortar.

Brick

Brick is manufactured in plants which use raw materials extracted from nearby pits. This leads to a wide range in the texture and colour of bricks available. Raw materials commonly used include fireclay and shale. These are milled (crushed) to a fine consistency and then screened and mixed with water. The soft, putty-like mixture formed is then extruded (shaped), perforated and wirecut to length. The individual bricks are then dried in drying chambers before firing (1,000°–1,200°C) in a kiln.

Brick samples.

Steel

Steel is a type of ferrous (iron-based) metal which has a carbon content of less than two per cent. There are various types of steel available:

Low Carbon (Mild Steel) 0·05–0·32% Carbon
• used for structural frames.

Medium Carbon 0·35–0·55% Carbon
• used for machinery (e.g. excavators).

High Carbon 0·60–1·5% Carbon
• used for machine tools (e.g. chisels).

Steel is used in a variety of applications in house construction, including:
• door and window fittings (e.g. hinges),
• galvanised steel lintels in red brick houses,
• wall ties in cavity wall construction,
• boxes for light switches and sockets,
• sanitary fittings (e.g. taps, flush handles).

Other elements can also be added to produce special purpose steels. For example, chromium is added to produce stainless steel. Stainless steel, however, tends to be too expensive for general use in construction and is only used in kitchens (e.g. sinks) where its rust resistant properties are desirable.

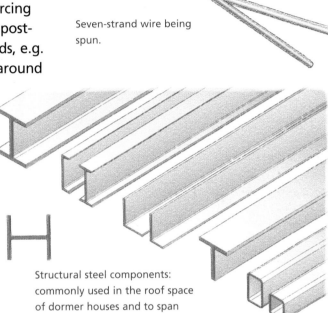

Seven-strand wire being spun.

Very high-strength steels are produced for the reinforcing wires in prestressed concrete. The steel wires used in post-tensioning are usually made up of a number of strands, e.g. a seven-wire strand consists of a single straight core around which six wires are spun in a single layer helical pattern. A standard seven-wire strand would be approximately 15 mm in diameter and have a tensile strength of 1,670 N/mm². This means it could withstand a load in the region of 230 kN.

Most steel structures use standard structural sections. A wide range of standard cross-sectional shapes are available for use in construction. These are produced by passing red-hot billets of steel through a series of rollers until the shape is gradually transformed to the desired profile. Hence the term 'RSJ', rolled steel joist. These are then cut to

Structural steel components: commonly used in the roof space of dormer houses and to span large window opes.

transportable lengths for distribution. Steel tends to oxidise (rust) in the presence of air and water and so it must be covered to prevent this. Steel may be galvanised (coated in zinc), encased in concrete or coated in a suitable paint to prevent corrosion.

Polymers (Plastics)

Plastics are correctly referred to as polymers. Polymers are often called plastics because they exhibit plastic behaviour before failure (i.e. they stretch before breaking). However, not all plastics behave like plastics, so really stretchy plastics are called *elastomers* (e.g. synthetic rubber). The word *polymer* derives from Greek and means 'many parts'. Polymers are chains of large molecules called *monomers*. Monomers are generally composed of combinations of hydrogen, oxygen, carbon and nitrogen. Many plastics are made from petroleum (paraffin) and natural gas or coal derivatives. For example, acetylene is produced from coal – acetylene is used to make polyvinyl chloride (PVC).

Polymers may be categorised as either *thermoplastic* or *thermosetting*. Thermoplastic polymers can be heated and moulded into shape and then reheated and remoulded. Thermosetting polymers do not have this property – once heated and moulded they cannot be remoulded. Examples of thermoplastic polymers include:

- Acrylic – sheet material substitute for glass commonly used in garden sheds or moulded to form bath tubs.

Acrylic bath.

- Polystyrene
  - Foam – thermal insulator used to retain heat in concrete ground floors and concrete cavity walls.

Expanded polystyrene insulation (EPS).

- Polythene
  - Low density – used in sheet form under the concrete in the ground floor of buildings (damp-proof membrane).
  - High density – e.g. buckets and cutlery trays.

Polythene sheet.

- Polyvinyl chloride – windows and doors.

  - uPVC (unplasticised, meaning rigid, PVC) is widely used in construction – its rigidity and low flammability are useful in pipe, electrical conduit, siding, windows and doors. A high-impact strength, coloured (e.g. white and brown uPVC) is used for windows and doors. The

uPVC windows.

    heated raw material is forced through dies (shaped moulds). It extrudes as thin-walled hollow box sections, complete with rebates, grooves and nibs for beads, weather and glazing seals and for fixing hardware.

  - PVC in combination with plasticiser (sometimes in concentrations as high as fifty per cent) is used as garden hose, imitation leather upholstery and shower curtains.

- Nylon – used in cog and pulley wheels because of its resistance to wear during use.

Nylon pulley wheel.

Examples of thermosetting polymers, include:

- Polyester resins
  - Resin and hardener – mixed to form a hard brittle plastic. Common examples would be sockets and light switches found in most houses.
  - Sometimes added to glass fibres to

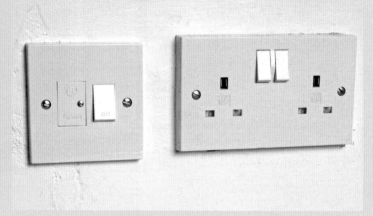

Polyester light switch and socket.

repair surfboards, kayaks, car bodies and to reinforce acrylic bath tubs.

- Phenolic resins (bakelite)
  - Strong, hard, highly heat resistant – commonly used in electrical contacts, inside switches and in pot and pan handles.

- Polyurethane
  - Used in rigid forms as an insulator for buildings, hot water storage tanks, water heaters and refrigerators.
  - Used in soft form for furniture upholstery and mattresses.
  - Commonly used in liquid form as a varnish for timber.

Polyurethane insulation (PUR).

Wood-based Panel Products

The importance of timber to the Irish construction industry cannot be overstated. The traditional use of timber in house construction, coupled with the growing popularity of timber-framed housing has seen the use of timber and timber products continue to grow. The volume of timber used in house construction ranges from 1 m³ for an apartment to over 20 m³ for a large dormer bungalow – the average falling between 6 and 10 m³. In Scandinavia and Canada 90% of houses are timber-frame, in the US that figure is 80% and in Scotland it is 60%. In Ireland in the early 1990s, less than one per cent of all construction was timber-frame. Since then, timber-frame building has risen sharply to 25% in 2005. An average timber-frame house uses 8 m³ of timber.

This growth in the use of timber has been augmented by the return to timber products in areas where timber had gone out of favour. For example, many people who are building their own homes are now returning to the use of timber for doors and windows as opposed to using non-sustainable uPVC products. A number of companies produce wood-based panel products in Ireland including Finsa (Co. Clare), Weyerhauser, (Co. Tipperary), SmartPly (Co. Waterford), Masonite (Co. Leitrim) and in Northern Ireland, Spanboard (Co. Derry).

Wood-based panel products are sheet materials in which wood is used in the form of strips, veneers, chips, strands or fibres. These panels are usually supplied in sheets 1,220 mm × 2,440 mm and in various thicknesses, although smaller sizes are available. There are four main categories of panel materials:

- plywood, including blockboard and laminboard,
- particleboard, including chipboard, flaxboard and cement-bonded particleboard,
- oriented strand board (OSB),
- fibreboards, including softboard, hardboard and medium-density fibreboard (MDF).

Wood-based panels are widely used in the construction industry for a number of reasons, including:

- high strength to weight ratio,
- ease of working, finishing, fixing,
- range of sizes and thicknesses available,
- range of types and special products available,
- environmental credentials – made from a renewable raw material, recyclable, low life-cycle costs.

Plywood

Plywood is a the technical term used to describe two categories of plied panels:

- veneer plywood,
- core plywood.

There are two types of core plywood – blockboard and laminboard. In everyday life, veneer plywood is called plywood and core plywood is called either blockboard or laminboard.

Veneer Plywood

Veneer plywood is defined as plywood in which all the plies are made of veneers orientated with their plane parallel to the surface of the panel. Veneer plywood is generally made from veneers that are peeled from a log. Standard plywood veneer is produced using a lathe, which peels a log in a similar manner to a blade pencil sharpener. Most decorative veneer is sliced from flitches (a halved or quartered log) after the log is cut into quarters. Before the veneers are cut the logs are normally soaked or steamed in order to increase the moisture content. This helps to produce a smoother veneer. The veneers are then

Various methods of veneer cutting/slicing are used to produce different grain patterns.

Plywood (veneer plywood).

Rotary cutting: the rotation of a log in a lathe to produce a continuous veneer.

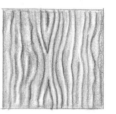 

Flat (plain) slicing: longitudinal slicing of a $1/2$ log.

 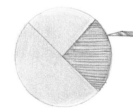

Quarter slicing: slicing of a $1/4$ log perpendicular to the annual rings.

Half round slicing: slicing across the annual rings to produce the look of both rotary cutting and flat slicing.

Rift cutting: slicing of oak and similar species perpendicular to the noticeable radiating rays so as to minimise their appearance.

dried to a moisture content of about
4–8%. In some cases, small strips of
veneer may be jointed into full-size
sheets by edge glueing, stitching or using
perforated paper adhesive tape.

Open defects, such as knot holes, are
repaired using plugs or filler to improve
the panel. The veneers are then sorted
into grades, by visual inspection.

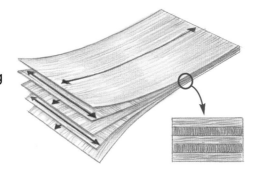

Plywood: bonding the
plies perpendicularly to
each other ensures that
each panel is
dimensionally stable and
has good strength both
along and across the
panel.

The veneers are bonded together with a synthetic resin adhesive. The type of
adhesive used during manufacture depends on the intended application.
Some adhesives are suitable only for interior use while others are designed for
panels that will be used outdoors. Typical adhesives include:

- urea-formaldehyde (UF) – panels
  intended for interior use only,
- phenol-formaldehyde (PF) – panels
  intended for exterior use only,
- melamine-urea-formaldehyde
  (MUF) – panels intended for
  exterior use only.

Each veneer is laid so that the grain of
adjacent veneers are at right angles to
each other. This ensures that each
panel is dimensionally stable and has
good strength both along and across
the panel. The assembled veneers are
known as a *lay-up*. The lay-ups are then
compressed and heated in a batch
press. The cured panel is allowed to
cool after which it is trimmed to size
and, if necessary, sanded. An odd
number of veneers is always used. This
ensures the direction of the grain of
both surface veneers match thereby
ensuring aesthetic and structural
consistency.

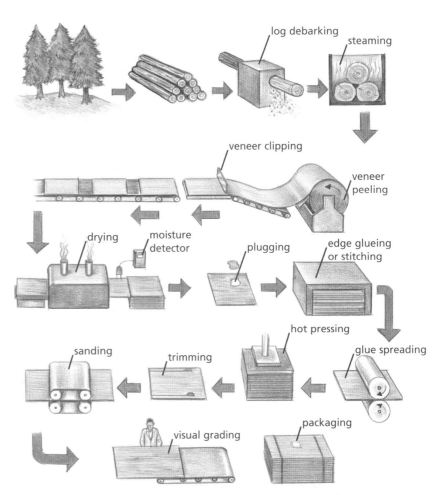

Plywood manufacture.

Plywood is produced all over the world using a wide range of timber species
including softwoods as well as temperate and tropical hardwoods. Almost

any species can be used, provided that it can be reliably peeled. Softwoods commonly used include spruce, pine and fir. Hardwoods used include birch and beech. The quality of the finished plywood depends on the quality, species and lay-up of the veneers, as well as on the glue type and bonding quality. Coatings or preservative treatments can also be used to enhance the durability of plywood panels.

Veneer Plywood Types and Applications
Some of the typical types of plywood include:

- **Structural Plywoods** – are used for floor decking, wall sheathing, flat roofs, concrete formwork and external cladding. These panels are often manufactured with tongue and grooved edges for improved jointing.
- **Marine Plywood** – was originally developed for ship and boat building and has a very high performance under severe exposure conditions. It is also commonly used in construction applications where high performance is required. Marine plywood is manufactured according to European standards using timbers having a moderate or better durability rating. Examples of such species include: hardwoods such as Mahogany, European Oak and Teak; and softwoods such as Pitch Pine, Douglas Fir, Western Red Cedar and Yew. High-quality veneers are used and must be bonded using a weather-resistant resin.
- **Utility Plywoods** – are similar to structural plywoods except that they have a surface appearance suitable for joinery and furniture manufacture.
- **Decorative and Overlaid Plywoods** – are special end-use plywoods that are overlaid with veneers, phenolic films or other finishes to give a decorative or hard-wearing finish.
- **Speciality Plywoods** - are products aimed at specific applications ranging from flexible plywoods suitable for bending into complex curves, to highly compressed bullet-proof plywoods. Lightweight panels and panels with a non-slip finish are also available.

Core Plywood

As mentioned previously, there are two types of core plywood – blockboard and laminboard. Blockboard is a five-ply panel consisting of a central core of sawn timber battens, not more than 30 mm wide, sandwiched on each side by a pair of veneers. Laminboard consists of a central core of slats laid on-edge (at right angles) to the two veneers each side of the panel. In

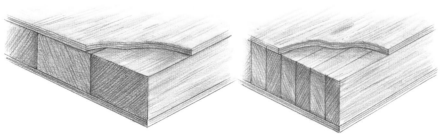
Blockboard (shown on the left) and laminboard.

both cases the direction of the grain of the core strips runs at right angles to that of the adjacent veneers and parallel to the external surface veneers.

The technique of manufacturing blockboard and laminboard developed alongside the plywood industry from the turn of the century to use up *waste* material left over following plywood manufacture. The method of production is similar to that for plywood and the wet stages of veneer manufacture are exactly the same. Core plywood panels are usually used in situations requiring a product similar to plywood in appearance but at a lower cost. They are normally restricted to interior applications such as joinery, door blanks, furniture and shopfitting.

Particleboard

Particleboard is commonly referred to as chipboard in Ireland. It is a panel product in which particles of wood are bonded together to form the panel. Particleboard is most commonly made from softwood chips such as spruce, pine and fir although hardwoods such as birch are sometimes used. The wood chips are prepared in a mechanical chipper, dried and then mixed with an adhesive such as urea-formaldehyde (UF) or melamine-urea-formaldehyde (MUF). The chips are formed into a mat and are then pressed between heated platens to compress and cure the panel. The finished panels are then cut to size and sanded. In order to achieve a smooth surface, the panel density is increased at the faces by the use of smaller wood particles with a larger percentage of resin binder compared to the core of the panel.

Particleboard is not normally suitable for exterior applications. The term *moisture resistant* is sometimes used in relation to panels classified for use in humid conditions. These panels may be resistant to periods of short-term wetting or high humidity, but are not waterproof and direct wetting will lead to failure.

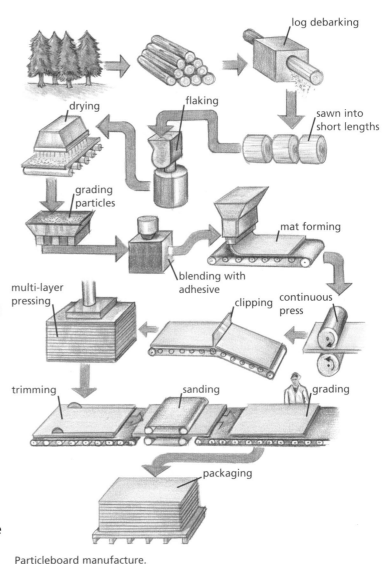

Particleboard manufacture.

Particleboard.

Generally, particleboard has a pale straw colour, but for identification purposes the whole panel, or a part of the panel, may be dyed to indicate a particular application – e.g. green for panels with enhanced moisture resistance or red for panels integrally treated with flame-retardant chemicals. Common applications of particleboard include kitchen units and worktops.

Oriented Strand Board

Compared to other types of panel products, oriented strand board (OSB) is a relatively new product. It was first developed around 1980, but has since grown in popularity and is now widely used in the construction industry.

Oriented strand board (OSB).

Approximately 300,000 m³ is produced annually in Ireland, most of which is exported. OSB is a panel material in which long strands of wood are bonded together with a synthetic resin adhesive, commonly phenol formaldehyde (PF). The wood strands are longitudinally cut from debarked logs using rotating knives. The ribbon of flakes produced is usually about 75 mm wide and this breaks up on handling to produce individual flakes about 75 mm along the grain and from 5–50 mm across the grain. After drying, these flakes are sprayed with resin to form a mat. The rest of the manufacturing process is similar to that for other wood panel products, involving a combination of applied pressure and heat. The timbers used in OSB manufacture include both softwoods and hardwoods.

OSB is usually manufactured with three distinct layers. In the surface layers the strands are generally oriented in the long direction of the panel – in the core layer the strands are oriented across the panel. This gives the panel strength in both directions in a similar manner to plywood. Not all strands are oriented exactly in the specified direction and the surface appearance can appear to be quite random. However, the overall effect is still present.

OSB is readily identified by its large, long wood strands. Panels tend to have a number of gaps or hollows on the surface due to the overlap of strands – OSB will never possess the smooth surface found in fibreboards and particleboards. Its advantage lies in its high level of mechanical performance (strength, durability), which is directly related to the use of longer and larger strands of wood. OSB varies in colour from a light straw colour to a medium brown depending on species, resin and pressing conditions used during manufacture.

OSB is primarily a panel for use in construction and is widely used for flooring, flat roof decking and wall sheathing. As with particleboard, OSB panels should generally be kept away from direct contact with water. Panels for use in humid conditions have a degree of resistance to short-term wetting and high humidity but are not intended for prolonged exposure to moisture. There are four specific types of OSB manufactured, i.e. OSB1 to OSB4. OSB2 (load-bearing for dry conditions) and OSB3 (load-bearing for humid conditions) are the most commonly used. A colour-coding system is used to differentiate between the various types. Two colours are used for each panel: the first colour indicates the purpose (white for general purpose, yellow for load-bearing), while the second colour indicates whether the panel is suitable for dry (blue) or humid (green) conditions. Common applications of OSB include flooring, wall sheathing, roof sarking, packaging, and furniture. OSB is also commonly used as site hoarding although it generally has a limited lifespan in this application.

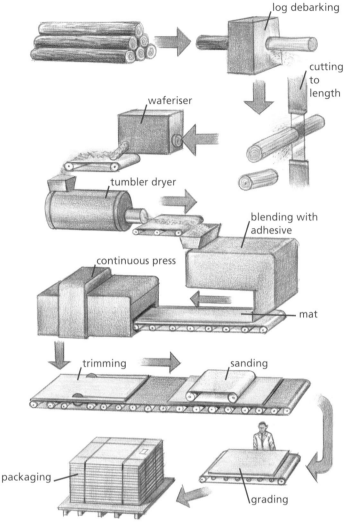

Oriented strand board manufacture.

Fibreboard

Fibreboard is produced by one of two basic processes:

- wet process (softboard, mediumboard, hardboard),
- dry process (medium-density fibreboard – MDF).

The technical difference between a wet process and a dry process fibreboard is that the wet process fibreboards have a fibre moisture content of more than 20% at the forming stage, whereas dry process fibreboards have a fibre moisture content of less than 20% at the forming stage and they are produced with the addition of a synthetic resin binder.

Wet process fibreboards can be made using either softwood or temperate hardwood species, although some low-density mediumboards are made from recycled paper fibre. Wet process boards are made by reducing steamed wood into fibres and adding water to form a slurry. This is then formed into a mat on a moving wire mesh. The mat is either rolled (softboards), or rolled and then pressed (mediumboards, hardboards), at a high temperature to the desired thickness. Depending upon the degree of pressure applied, and hence the final density of the panel, the product is termed *softboard*, *mediumboard* or *hardboard*. The manufacture of wet process fibreboards does not involve the use of an adhesive. The fibres are held together from the felting together of the fibres and the adhesive properties of the natural lignin present in the wood. Other additives such as wax, natural oil or fire-retardant chemicals are sometimes added to improve the properties of the final product.

Wet process fibreboards are used in the construction and furniture industries. Hardboards are used in furniture as drawer bottoms and unit backs, door facings and caravan interiors. Standard hardboard is not recommended for exterior use or for use in areas subject to direct wetting or high-humidity conditions. Mediumboards and softboards are used as pinboards, underlay materials, as components of partitioning systems and as an acoustic absorbent. It is important not to confuse mediumboard with medium-density fibreboard (MDF) as they are very different products.

Medium-density fibreboard (MDF).

Dry process fibreboard is usually made using softwood species. The dry process of fibreboard manufacture creates a product called medium-density fibreboard (MDF). The typical manufacturing process is similar to that for wet process fibreboards, except the fibres are mixed with a synthetic resin binder and then dried, before being formed into a mat. The mat is pressed between heated press plates to the desired thickness. Several mats may be stacked together to produce a thick board.

Confusingly, the density of MDF can be varied during manufacture to create high-density MDF (sold as HDF), light density MDF (sold as LDF) and ultra-light MDF (sold as ULDF). The constituents of a typical MDF panel manufactured in Ireland are 82% wood fibre (mainly softwood), 10% synthetic resin binder, 7% water and less than 1% paraffin wax solids. Urea-formaldehyde is the most common adhesive used, although depending on the grade and end-use of the product other binders may be used.

There is a wide range of panel types available and although some fibreboards are available in load-bearing grades, most MDF is not intended for structural applications. MDF is known for being particularly good for machining and has a smooth surface which can provide a high-quality lacquered or painted finish. It is available in a wide range of sheet thicknesses and sizes. MDF is used for interior design and furniture applications. When used for kitchen and wardrobe doors it is usually veneered with a hardwood or vacuum-covered in a vinyl or foil wrap. MDF is increasingly used in non-structural building applications such as skirting boards and architraves, windowboards, staircases and decorative façades.

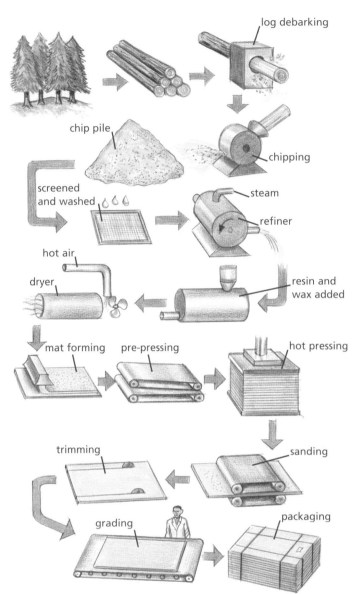

Medium-density fibreboard.

Revision Exercises

1. Explain the classification system used for identifying timber used in the construction of houses.
2. Generate a neat annotated sketch of a stone building in your local community. Indicate the type(s) of stone used in its construction.
3. In your own words, with the aid of a neat sketch, briefly describe the lime cycle.
4. Describe, with the aid of neat freehand sketches, how cement is manufactured. Explain each step involved.
5. Explain the following terms:
   (a) aggregates, (b) placing, (c) batching, (d) compacting, (e) water-cement ratio, (f) curing, (g) workability, (h) hardcore.
6. Describe, with the aid of neat freehand sketches, a test suitable for establishing the consistence workability of a sample of concrete.
7. Explain the principle involved in the prestressing of concrete. Describe, with the aid of neat freehand sketches, one method of prestressing concrete.

8. Describe how concreting materials should be stored on-site.
9. Sketch an example of the use of steel in the construction of a house.
10. Discuss briefly the advantages and disadvantages of uPVC windows.
11. Outline the manufacturing process of oriented strand board.
12. Ordinary Level, 2002, Question 5
    (a) In relation to concrete work explain in detail any three of the following terms:
        (i)   Coarse Aggregate,
        (ii)   Fine Aggregate,
        (iii) Water-Cement Ratio,
        (iv)   Slump Test,
        (v)   Reinforced Concrete.
    (b) Give two examples of where reinforced concrete may be used in the construction of a domestic dwelling. Write a short note on each.
13. Ordinary Level, 2004, Question 9
    (a) Explain in detail three of the following terms as they apply to concrete:
        • aggregates, • batching, • formwork, • Slump Test, • curing.
    (b) Describe three locations where precast concrete components are used in the construction of a dwelling house.
    (c) Explain three advantages of using precast concrete components.
14. Higher level, 2002, Question 3.
    (a) Discuss the importance of the use of steel in the manufacture of reinforced concrete, with reference to the:
        (i)   strength properties of both materials;
        (ii) design considerations to avoid deterioration over time.
    (b) Describe in detail, using sketches and notes, three methods of combining concrete and steel in the manufacture of concrete lintels.
    (c) List one advantage of each method described in (b).
15. Higher Level, 2003, Question 4
    Poor design detailing may result in the occurrence of both brown rot and white rot in a domestic dwelling.
    (a) Outline the conditions necessary for the development of each type of rot.
    (b) Select one location in a domestic dwelling where rot may occur and, using notes and sketches, show how the rot may be eliminated.
    (c) Using notes and sketches, show the design detailing that would prevent the occurrence of rot at the selected location.

Any building is essentially a combination of three types of systems:

- Structural system – designed and constructed to support and transmit loads safely to the ground.
- Enclosure system – the building envelope consisting of the floors, exterior walls, roof, windows and doors.
- Mechanical systems – provide essential services to a building including water, sewage disposal, heating, ventilation and electricity.

In houses, the structural and enclosure systems are sometimes integrated. For example, the external walls of a typical Irish house serve as both the structural and enclosure system. This chapter is mainly concerned with the structural systems used in the construction of houses.

The structural system of every house is required to support and transmit various loads. These loads can be classed as static or dynamic.

- Static Loads – loads that are applied slowly to a structure and do not change quickly. Examples of static loads include:
  - Live Loads – any moving or movable loads resulting from people, collected water and/or snow or movable equipment.
  - Dead Loads – loads associated with the building weight and any elements permanently attached to it.
  - Soil and Hydraulic Loads – ground pressure loads exerted on a rising wall and hydraulic loads from groundwater.
- Dynamic Loads – loads that are applied suddenly to a structure, often with rapid changes in magnitude and point of application.
  - Wind Loads – forces exerted by the energy of moving air. Wind exerts positive pressure on the windward side of a building and negative pressure or suction on the leeward side.
  - Earthquake Loads – cause lateral movement at the base of a building that can cause failure or collapse in extreme cases. Flexible buildings such as timber-frame structures perform well under earthquake loads.

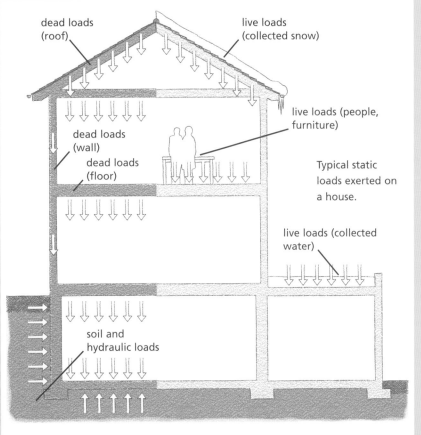

dead loads (roof)

live loads (collected snow)

dead loads (wall)

dead loads (floor)

live loads (people, furniture)

Typical static loads exerted on a house.

live loads (collected water)

soil and hydraulic loads

Structural Forces: Concepts

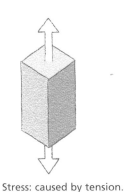

Stress: caused by tension.

Stress
Stress is defined as the internal resistance of a structural member to external forces applied to it, equal to the ratio of force to area, expressed in units of force per unit of cross-sectional area. (N/m^2)

Strain
Strain is the deformation of a structural member under the action of an applied force. Strain is expressed as a ratio of the change in size or shape of the member relative to its original size or shape.

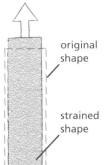

original shape

strained shape

Strain: change in shape caused by stress.

Elasticity
Stress and strain are related by the concept of elasticity. An elastic material becomes strained when stress is applied to it. While we usually associate elasticity with stretching, it is also a feature of squeezing. The elasticity of a material is important in construction because it provides an indication of how a material will perform when under tension or compression.

The relationship between stress and strain can be calculated using Young's modulus. The formula for Young's modulus is:

$$Y = \frac{Stress}{Strain} = \frac{F/A}{\Delta L/L}$$

where:
F = total applied force.
A = area over which the force is applied.
F/A = stress (force per unit area).
L = original length.
ΔL = displacement.
$\Delta L/L$ = strain (displacement per unit length).

The value of Young's modulus for aluminium is about $7 \cdot 0 \times 10^{10}$ N/m^2[1]. The value for steel is about three times greater, which means that it takes three times as much force to stretch a steel bar by the same amount as a similar aluminium bar.

Tension

Tension is defined as stress in a structural member that tends to pull or stretch. It has a lengthening effect. A structural member designed to resist tension is called a *tie*.

Compression

Compression is defined as stress in a structural member that tends to squash or crush. A structural member designed to resist compression is called a *strut*.

Shear

Shear is defined as stress in a structural member that tends to cause one part to slip over another. Shear is experienced in steel structures where plates are joined using bolts.

Torsion

Torsion is defined as the twisting of an elastic body about its longitudinal axis caused by two equal but opposite rotating forces (torques).

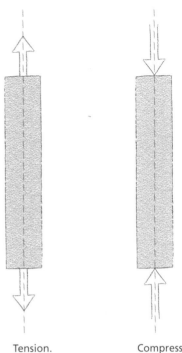

Tension.   Compression.

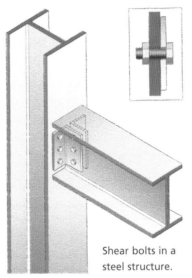

Shear bolts in a steel structure.

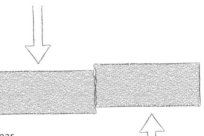

Shear.

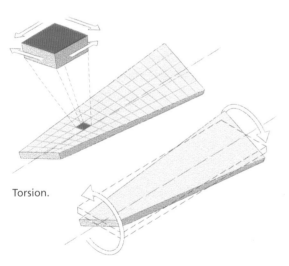

Torsion.

Eccentric Force

An eccentric force is defined as a force applied parallel to the longitudinal axis of a structural member but which is not coincident with the axis, i.e an off-centre force.

Eccentric force.

Bending

Bending is defined as the bowing of an elastic body as an external force is applied transversely to its length. Bending is the structural mechanism that enables a load to be carried perpendicular to the direction in which it is applied. Bending is experienced in lintels above window and door openings in houses.

Bending.

Slenderness Ratio

Long slender columns tend to collapse due to buckling rather than crushing. The ability of a column to resist buckling is determined by its slenderness ratio. The higher the slenderness ratio the better the column will perform. The slenderness ratio of a column is defined as the ratio of its effective length to its smallest radius of gyration. The effective length of a column is the portion

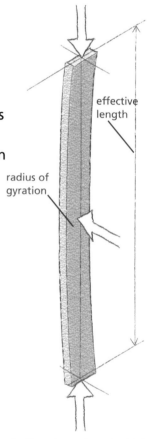

effective length

radius of gyration

Radius of gyration: the higher the radius of gyration of a column, the more resistant the column is to buckling. Square or circular columns perform best – columns of rectangular section tend to fail in the direction of the least dimension (buckling tends to occur about the weaker axis).

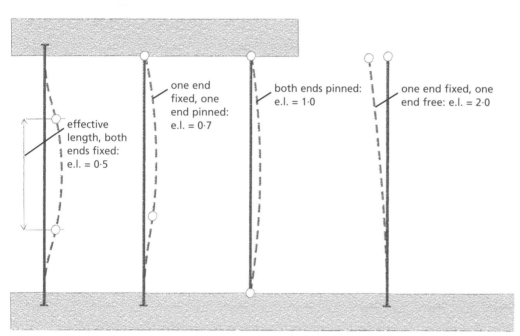

effective length, both ends fixed: e.l. = 0·5

one end fixed, one end pinned: e.l. = 0·7

both ends pinned: e.l. = 1·0

one end fixed, one end free: e.l. = 2·0

Effective length: the effective length of a column determines its load-bearing capacity – when this portion buckles, the entire column fails.

of the column's length that is likely to buckle under load. If both ends of a column are fixed, the effective length is reduced by half (because only the middle portion will buckle) and its load-bearing capacity is increased by a factor of four. The radius of gyration is the distance from the longitudinal axis at which the mass of the column can be assumed to be concentrated.

Structural Forms and Elements

A structure is defined as an assembly of elements designed and constructed to function as a whole in supporting and transmitting applied loads safely to the ground.

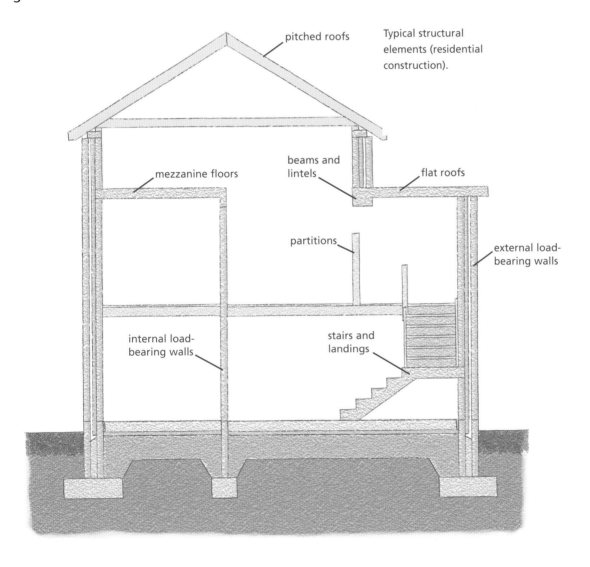

pitched roofs

Typical structural elements (residential construction).

beams and lintels

mezzanine floors

flat roofs

partitions

external load-bearing walls

internal load-bearing walls

stairs and landings

Structures can be broadly grouped into five categories. These categories are based on how the structure transmits the imposed loads:

- Form Active Structures – transmit loads through the shape and geometry of their material (e.g. cable or arch).

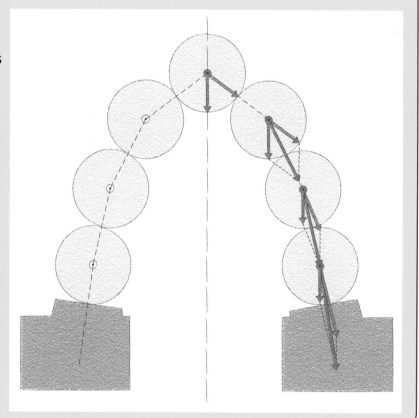

Form active structure (e.g. arch).

- Vector Active Structures – transmit loads through the composition of tension and compression members (e.g. roof truss).

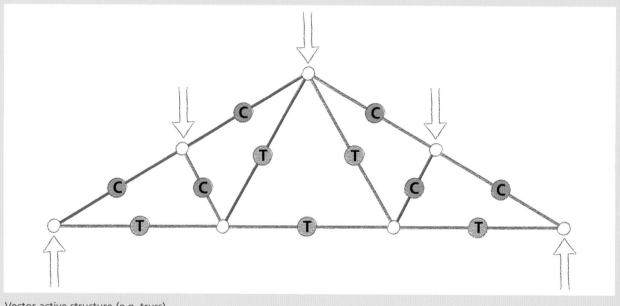

Vector active structure (e.g. truss).

- Bulk Active Structures – transmit loads through the bulk and continuity of their material (e.g. beam, column or portal frame).

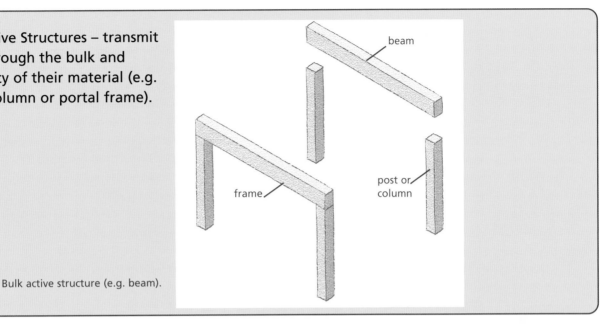

beam

post or column

frame

Bulk active structure (e.g. beam).

Plate structure: plate roof.

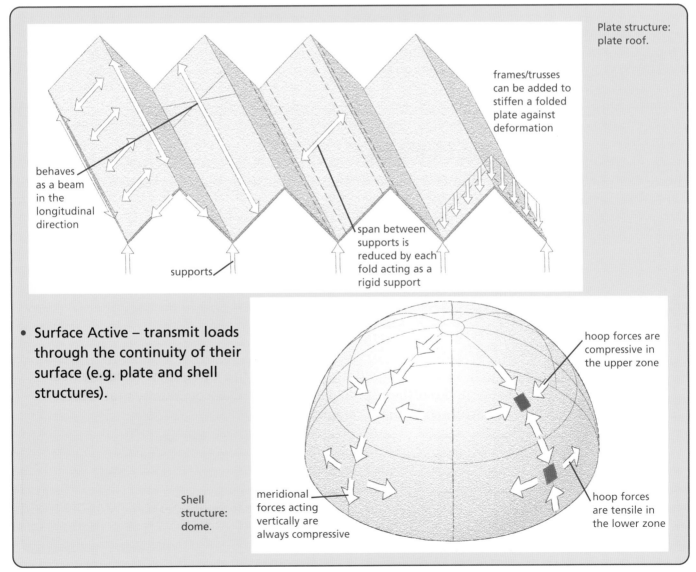

frames/trusses can be added to stiffen a folded plate against deformation

behaves as a beam in the longitudinal direction

supports

span between supports is reduced by each fold acting as a rigid support

- Surface Active – transmit loads through the continuity of their surface (e.g. plate and shell structures).

hoop forces are compressive in the upper zone

Shell structure: dome.

meridional forces acting vertically are always compressive

hoop forces are tensile in the lower zone

• Membrane Structures – transmit loads through air-supported or inflated fabric, or through pole-supported fabric (e.g. tent structures).

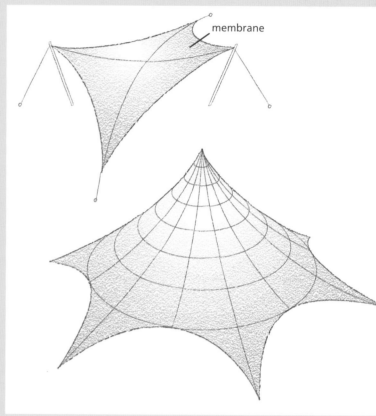

Membrane structures (e.g. tent).

Revision Exercises

1. Describe, with the aid of neat freehand sketches, the various loads experienced by a structure. Explain each briefly.
2. Describe, using neat annotated sketches, everyday examples of the following stresses:
   (a) tension,
   (b) compression,
   (c) shear,
   (d) torsion,
   (e) bending.
3. Structures can be broadly grouped into five categories. Generate neat annotated sketches of buildings in your local community that represent at least three of these categories.

Construction sites are not suitable places for children. A construction site is a hazardous place where very serious accidents can happen, especially if untrained people wander onto them. Every construction site has signage posted at its entrance advising of the hazards present. These signs also tell visitors and workers what special measures should be taken for their own safety when they enter a site, such as wearing hard hats, safety boots and high visibility vests.

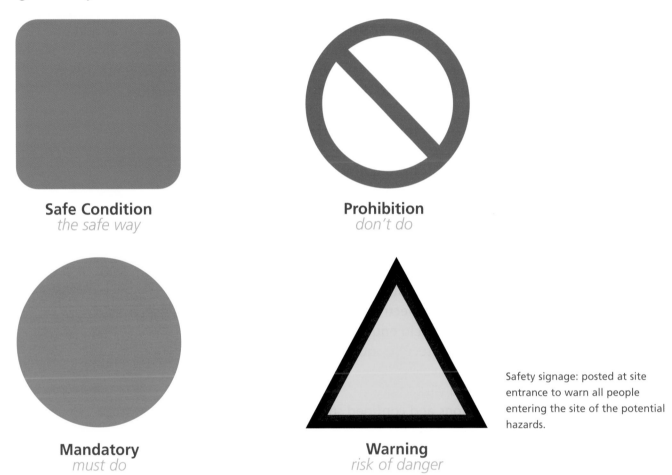

Safe Condition
the safe way

Prohibition
don't do

Mandatory
must do

Warning
risk of danger

Safety signage: posted at site entrance to warn all people entering the site of the potential hazards.

The safety of workers in the construction industry is of paramount importance. Construction is an inherently risky business: the use of electrical tools, heavy materials and large machinery in tight spaces under constant pressure to complete the job quickly can lead to dangerous situations. It is essential that these specialist tools and materials are used by trained workers only.

Since the late nineties, the rapid increase in the amount of building taking place in Ireland has been linked to a similar increase in the number of people getting seriously injured or killed on-site. Around twenty people die on construction sites in Ireland every year. In recent years various interest groups with links to the construction industry have made significant efforts to make construction sites safer places to work. One of these safety initiatives is the Safe Pass training programme.

The Safe Pass training programme is administered by FÁS, Ireland's national training and employment authority. In order to work on a construction site a worker must have completed the Safe Pass training programme. Professionals who regularly visit construction sites as part of their work (e.g. architects, quantity surveyors and engineers) should also have completed the Safe Pass training programme. FÁS maintain a register of qualified tutors and of everyone who has completed the training programme. The Safe Pass training programme must be completed every four years to ensure construction workers are kept up to date with changes in safety practices.

The programme consists of thirteen modules:

- The reasons for promoting safety.
- Health and safety at work legislation.
- Accident reporting and emergency procedures.
- Accident prevention.
- Health and hygiene.
- Manual handling.
- Working at heights.
- Working with electricity, underground and overhead services.
- Use of hand-held equipment and tools.
- Personal protective equipment.
- Safe use of vehicles.
- Noise and vibrations.
- Excavations and confined spaces.

Each of these modules addresses hazardous areas of construction that a worker is likely to encounter on-site.

Personal Protective Equipment

The best way to prevent injury is to remove the hazard completely. If this is not possible, the next best thing is to reduce the level of exposure as much as possible. The final step is to use personal protective equipment to try to avoid or limit the extent of the potential injury.

A number of pieces of personal protective equipment are considered absolutely essential for every worker on a construction site. These include:

- head protection (hard hat),
- high visibility vest,
- safety boots,
- safety goggles and glasses,
- gloves.

Every worker should have his/her own personal set of this equipment. The hard hat, vest and boots are worn at all times, while the glasses and gloves are used as necessary. It is common practice on-site nowadays for a worker to be refused entry to the site if s/he does not have the correct personal protective equipment. There is a very wide range of specialist personal protective equipment available for specific tasks. For example, respiratory protection and hearing protection are used when working in dusty or noisy environments.

Worker wearing personal protective equipment.

Specific Risk Areas

Scaffolding

Scaffolding is widely used in the construction industry for working at heights. It is important that scaffolding is erected by trained personnel to ensure that it is stable and secure when in use. The stability of scaffolding depends on a number of things:

- how it is used,
- secure footings,
- correct bracing,
- adequate ties to structure.

Before installing a system of scaffolding, consideration must be given to how it will be used. Examples of uses of scaffolding, that would increase the loads exerted on it, include:

- a loading platform for concrete blocks or roof tiles,
- wrapping it in netting to contain waste and protect the public from falling objects,
- attachments being fitted such as waste chutes or lifting appliances.

There are two main types of scaffolding: system scaffolds, and tube and fitting scaffolds. System scaffolding has become the most common type of scaffolding in use because it is easy to use and can be quickly assembled by a small number of people.

Each type of system scaffolding consists of a range of components such as standards, ledgers, transoms and base plates and has its own specific assembly requirements. Tube and fitting scaffolding is constructed from steel tubing and several types of couplers.

Waste chute attached to scaffolding: these chutes are a safe and effective way of removing waste. However, they increase the load carried by the scaffolding and should be properly installed using a supporting tower.

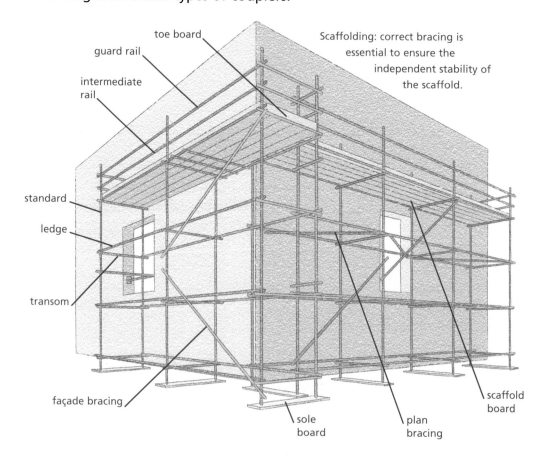

Scaffolding: correct bracing is essential to ensure the independent stability of the scaffold.

toe board
guard rail
intermediate rail
standard
ledge
transom
façade bracing
sole board
plan bracing
scaffold board

The correct design of system scaffolding depends on the following principles:

- safe transfer of the load to the ground (steel base plates resting on timber soleplate),
- diagonal bracing of vertical and horizontal members,
- tying the scaffolding to the building at regular intervals,
- correct use of component parts (no modification on-site),
- scaffolding should not be overloaded.

Bracing is used to stiffen the scaffold so that it does not sway. Swaying can cause instability, cracking of welds and overstressing of the standards. Bracing used on the front of the scaffold is called façade bracing. Usually this is the only bracing necessary, however where ties are not possible or where loading bays are required, plan bracing is also necessary.

There are four types of ties used to secure the scaffold to the building structure:

- Through ties – secured through a window or other opening and bearing against a solid face.
- Reveal ties – using a scaffold tube jacked tight against opposing faces of a window or door opening.
- Box ties – secured around a column.
- Ring bolts – drilled or cast in anchor.

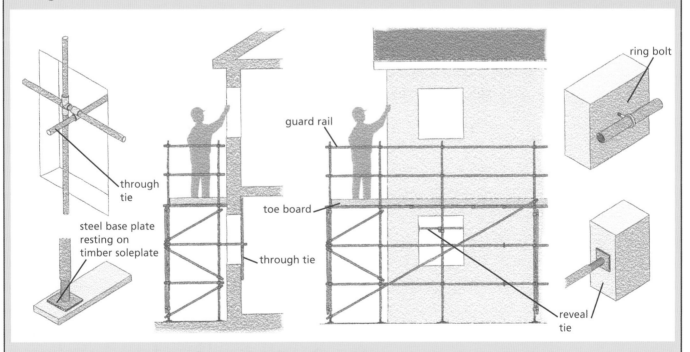

Scaffolding: tying the scaffolding to the structure greatly improves stability.

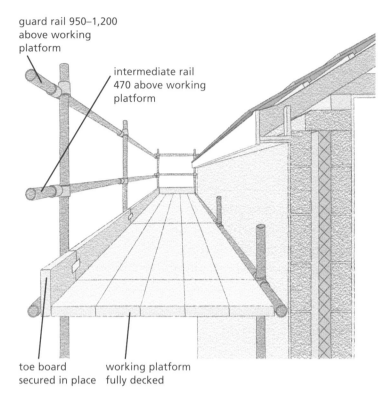

guard rail 950–1,200 above working platform

intermediate rail 470 above working platform

toe board secured in place

working platform fully decked

Scaffolding: working platform.

The working platform should be wide enough to allow safe passage of workers along the platform. The platform (scaffold boards) should also be capable of supporting the loads imposed upon it. The actual width required depends on the function. For example, if the platform is to accommodate both workers and materials it must have a minimum width of 800 mm.

For working on a scaffold above a height of two metres, it is essential that the working platform is fitted with a guard rail, intermediate rail and toe board to prevent workers from falling. The height of the guard rail should be 950–1,200 mm. The intermediate rail should also be installed no higher than 470 mm above the toe board.

Access to scaffolding is best achieved using ladder access towers. A ladder access tower is fixed to the outside of the scaffold and allows for safe access to the working platforms at each level. If an access tower is not used it is essential that the following guidelines for ladder use be followed:

- Ladder stiles should extend at least one metre above the platform.
- The top of the ladder stiles should be securely lashed to the scaffold.
- The slope of the ladder should not exceed four vertical to one horizontal.
- The stiles should be supported on a firm footing.

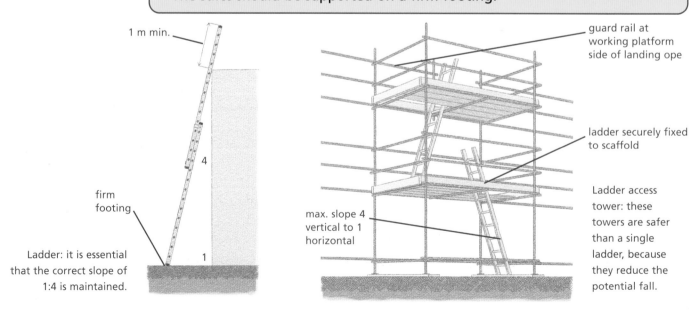

1 m min.

firm footing

Ladder: it is essential that the correct slope of 1:4 is maintained.

max. slope 4 vertical to 1 horizontal

guard rail at working platform side of landing ope

ladder securely fixed to scaffold

Ladder access tower: these towers are safer than a single ladder, because they reduce the potential fall.

Excavations

Excavation is the removal of soil or similar material from an area to facilitate the construction of a foundation or the laying of pipework. This work is usually completed using heavy machinery such as bulldozers and back hoes. The common hazards associated with excavation works are:

- contact with underground services (gas, electrical etc.),
- contact with overhead power lines,
- collapse of the excavation's sides,
- materials falling onto people working in the excavation,
- people and vehicles falling into the excavation,
- people being struck by machinery,
- undermining of nearby structures and buildings,
- accidents to members of the public.

Before commencing excavation work it is important to contact the service providers (e.g. ESB/Bord Gáis) for advice on possible underground services. Special scanners can be used to look for buried services. The area should also be checked for overhead power lines. Every potential hazard found at this stage should be marked. In the case of underground services this can be done by marking the ground with paint or flags. For overhead power lines, posts and bunting should be erected. This will remind excavator drivers of the danger overhead.

Locating and marking the position of underground services before excavation begins (i.e. electricity, gas, etc.).

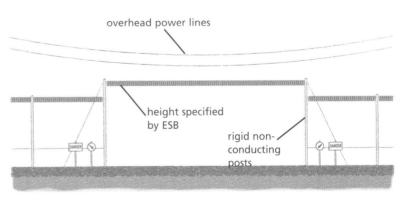

Overhead power lines: safe routes for driving of plant (e.g. diggers, telescopic lifts) on-site must be clearly marked.

When soil is removed from a trench there is always a risk that the sides of the trench may collapse inward. If a person is required to work in the trench (e.g. pipe laying) it is essential that steps are taken to ensure the trench is secure. Workers should never enter an unsupported trench greater than 1·25 m in depth. There are many types of trench boxes and hydraulic walls that can be

Excavation: trench sides can also be battered (sloped back) to prevent collapse of the trench wall.

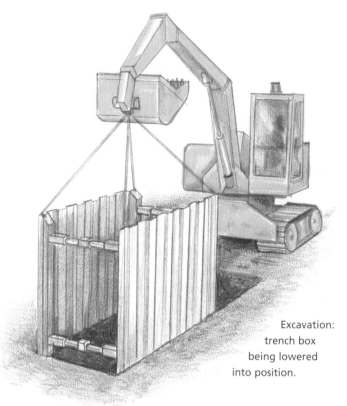

Excavation: trench box being lowered into position.

Excavation: trenches over 1.25 m in depth must be supported before a worker can enter them.

lowered into trenches to secure their sides without the need for a worker to enter the trench beforehand. Alternatively, the sides of the trench can be battered to remove potential slippage material.

It is also essential that workers do not enter a trench while an excavator is working above it. This might lead to the worker being struck by the machine or the collapse of the side of the trench due to the extra weight of the excavator.

Electrical Tools and Equipment

Electrical equipment is very familiar to us and is used on construction sites every day. While the use of battery-operated tools has reduced the risk of electrical injury considerably, there is still a need to be very cautious when using electrical tools. Where mains-operated tools are being used it is important to check regularly for damage (especially to cords and plugs). All electrical tools should operate on 110 V and should be double-insulated (have two independent layers of insulation over live conductors).

It is common practice on building sites to use isolation transformers with a secondary voltage of 110 V which has the centre point connected to earth. This arrangement ensures that while the full 110 V are available to power the tool, if the user touches a live wire the maximum voltage will be 55 V, which is considered to be relatively safe.

These tools should be connected to a supply which is protected by a residual current device (see chapter 23, Electricity). For some purposes, such as lighting, even lower voltages can be used. The case, cord and plug of mains-operated tools should be checked every day to ensure they are in good condition. Mains-powered electrical equipment should not be used outdoors in wet conditions.

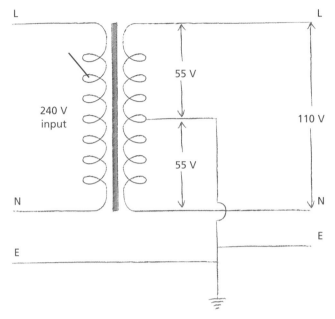

Electrical safety on site: isolation transformer with centre tapped to earth. This reduces the risk of electrocution on-site.

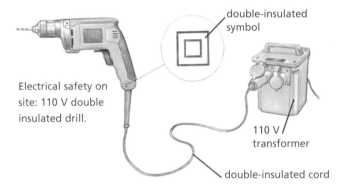

Electrical safety on site: 110 V double insulated drill.

double-insulated symbol

110 V transformer

double-insulated cord

Revision Exercises

1. Discuss briefly the importance of safety in the construction industry.
2. Why is it essential that construction workers have completed the Safe Pass programme before working on a building site?
3. Explain, using neat annotated sketches, the principles of safe system scaffolding design.
4. Explain, in your own words, the hazards associated with excavations and how the risk associated with at least two of these can be minimised.
5. Describe the personal protective equipment that must be worn at all times on-site by construction workers.
6. Ordinary Level, 2002, Question 6
    (a) The following are four safety signs that are commonly used in the construction industry.

State clearly what is meant by each of the above.

Two of the above signs are circular in shape and two are triangular. Briefly describe the meaning of these different shapes.

(b) Using notes and neat freehand sketches describe three safety precautions that should be observed when using scaffolding on a construction site and briefly explain the reason for each.

7. Ordinary Level, 2003, Question 9
List and explain two appropriate safety precautions that must be observed in each of the following situations:
(a) Excavating a foundation trench for a dwelling house.
(b) Using an extension ladder during the construction of a house.
(c) Visiting a construction site.
(d) Slating a pitched roof.

8. Ordinary Level, 2004, Question 5
(a) List two specific safety precautions to be observed in each of the following situations:
   • placing fibreglass insulation in the attic space of a dwelling house,
   • wiring a three-pin plug,
   • erecting a scaffold,
   • visiting a construction site when trenches are being excavated using machinery.
(b) Give one reason why each safety precaution listed in (a) should be observed.

9. Higher Level, 2006, Question 2
(a) Identify **two** possible risks to personal safety associated with each of the following:
   (i) Scaffolding;
   (ii) Deep Excavation;
   (iii) Use of electrical tools out-of-doors.
(b) Using *notes and freehand sketches as appropriate*, outline **two** specific safety precautions that demonstrate best practice in order to eliminate **each** risk identified at (a) above.
(c) Under the Safety, Health and Welfare at Work Regulations, it is compulsory for employers to have a safety statement. Discuss in detail **two** benefits of such a safety statement for employees in the construction industry.

Soils

Most people would agree that walking in sand is harder than walking on a concrete footpath. This is because your feet tend to sink into the sand when you try to move forward. A similar thing happens when a house is built – it sinks into the ground. Of course, the extent to which a house sinks into the soil is so small that it isn't noticeable. This is because investigations are carried out to ensure that the ground is strong enough to support the weight of the building before construction begins. This is done by checking what type of soil is below ground level. Some soils are good at supporting loads, others are not. By establishing the soil type and comparing it to known performance figures, the designer can be confident that the building will not sink when complete.

Soil Investigation

The purpose of carrying out a soil investigation can be summarised as follows:

- to establish what type of soil is below ground level and its characteristics,
- to determine the suitability of the site for the proposed building,
- to ensure that the correct foundation design is chosen for the soil and for the type of building planned,
- to ensure that the building is constructed safely, efficiently and economically.

Before visiting the site a desk study of existing records should be carried out. This involves examining Ordnance Survey maps, planning records, historical documents and any other available information. It is also a good idea to consult with the planning authority and local people, to gather as much information as possible about the land. Examples of the type of information required include:

- the geology of the area,
- topography,
- vegetation,
- drainage (locations of water drains, streams),
- previous uses of the land,
- typical foundation designs used in the area.

Other factors that should be considered include:

- Are there preservation orders in place, e.g. protected structures such as heritage buildings?
- Is the site in a special area of conservation, e.g. wetlands?
- Is the area prone to flooding, e.g. low-lying land?
- Are there overhead or underground services crossing the site, e.g. electricity power lines?

Once this information has been gathered the next step is a visit to the site. Walking the site is a very useful way to assess a number of features including:

- the lie of the land, the nature of vegetation and plant growth, the feel of the soil (e.g. waterlogged), existing buildings and natural features (e.g. rock outcrop).

The final stage in site investigation involves:

- collecting and analysing soil samples,
- determining the level of the water table.

Traditionally, the investigation of soil on-site has been carried out by digging trial holes. This involves digging a square pit in the ground which allows the builder to determine what type of soil is present by observation.

This approach is suitable for small-scale projects (i.e. foundation depth <3 m) where the soil on-site is firm and stable. Alternatively, the collection of soil samples can be done by drilling into the ground and removing samples using a large auger bit. The disadvantage of this approach is that the samples of soil are broken up or disturbed in the process, so an accurate picture of the soil strata is not gained.

Poorer quality soils or larger projects require more extensive investigation. In these cases, a core sample can

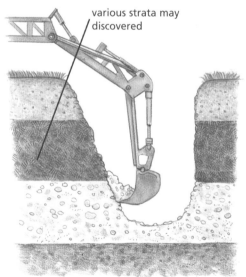

various strata may discovered

Soil investigation: digging a trial hole allows a visual inspection of the soil to be carried out and the level of the water table to be established.

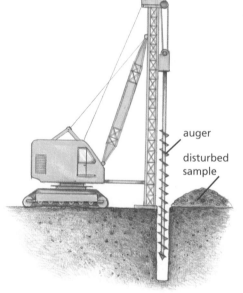

auger

disturbed sample

Soil investigation: removing soil samples using an auger.

be taken by driving a sample tube into the ground. The sample collected in this way is undisturbed and gives a very accurate picture of the subsoil conditions.

Based on the types, and arrangement, of the soils found an accurate measurement of the soil's bearing capacity (strength) can be determined.

Water Table

Water occurs naturally below ground – rainwater soaks into the ground and flows through the various layers of soils and rock until it eventually reaches the sea (see hydrological cycle, chapter 17, Water). The natural level of the water held in the soil is called the *water table*.

The level of the water table falls and rises throughout the year depending on the weather. This can lead to problems for the builder in two ways:

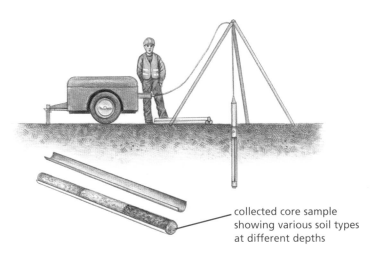

collected core sample showing various soil types at different depths

Soil investigation: removing undisturbed soil samples using a core sampling machine.

- subsoil water can cause problems during the excavation of foundations,
- a high water table can lead to flooding during wet periods.

The level of the water table should be noted when the soil investigation is being carried out. Usually, if the level of the water table in a particular area is quite high, it will be well known locally and the land will not usually be used for small-scale development (e.g. a single house).

Another important test that is usually carried out at this stage is a Percolation Test. This involves ascertaining the rate at which the soil can absorb moisture. This is very important if the house is going to have its own wastewater treatment system (e.g. septic tank). The Percolation Test is described in detail in chapter 19, Drainage.

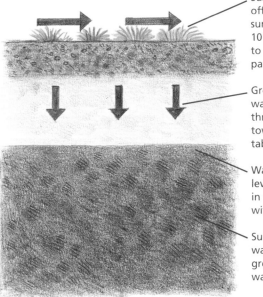

Surface water: run-off from the surface varies from 10% in green areas to 90% in hard-paved areas.

Ground water: water percolating through subsoil toward water table.

Water table: upper level of water held in the soil (varies with seasons).

Subsoil water: water held in the ground below the water table.

Soil Types

The wide variety of soil types in Ireland can be organised into five categories. Apart from rock, the five categories are gravels, sands, silts, clays and peats. These categories generally refer to the size of the soil particles. The size of the soil particles has an influence on the level of cohesion displayed by the ground. Smaller particles tend to be more cohesive (e.g. clays) whereas larger particles tend to be less cohesive (e.g. gravels). You can explore the cohesiveness of soil types by trying to crush them between your fingers. Cohesive soils will stretch and change shape, like modelling clay, while non-cohesive soils will crumble into small parts.

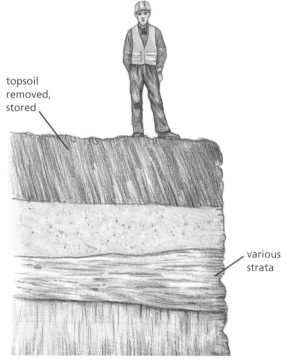

topsoil removed, stored

various strata

Soil investigation: it is important to know the type and depth of the soil on site before designing the foundation.

The ability of a soil to carry a load is referred to as its *bearing capacity*. The bearing capacity is measured in kN/m^2. It is essential to know the bearing capacity of the soil on a site before building because it will have a significant influence on the design of the house, especially the design of its foundations. It is possible to find more than one soil type on a site and various strata of different soils are sometimes revealed during excavations.

The uppermost layer of soil on most sites is called the topsoil. The topsoil is the upper 300–600 mm of soil in which the vegetation (grass, plants etc.) grows. This soil is very soft and would be easily compressed if a structure were built on it. For this reason the top 300–600 mm of soil is always removed where the house is to be built before foundations are marked out and excavated.

The bearing capacity of soils vary greatly so general guidelines are provided for the typical safe bearing capacity of soil types. While these figures are reliable, the only way to be certain of the bearing capacity of a soil is to test it in a laboratory. A concrete party wall of a typical two-storey semi-detached house will generate a load in the region of $75 \ kN/m^2$.

Soils and Site Investigation

The following table shows the typical safe bearing capacity of various soil types:

Group	Types of Soil	Bearing Capacity kN/m²
Rock	Igneous	10,000
	Limestone	4,000
	Slate	3,000
	Shale	2,000
Non-cohesive soils	Compact sand	greater than 600
	Compact gravel	greater than 300
	Loose gravel	less than 200
	Loose sand	less than 100
Cohesive soils	Very stiff	300–600
	Stiff clay	150–300
	Firm clay	75–150
	Soft clay/silt	75
	Very soft clay/silt	less than 75
	Peat/fill	unpredictable

In general, any soil with a bearing capacity above 300 kN/m² cannot be excavated by hand and requires the use of mechanical excavators. Soils with a bearing capacity below 100 kN/m² can be moulded in the hand like modelling clay.

Angle of Repose

The angle of repose is easiest to understand if you imagine yourself leaning back in a chair. If you lean back beyond a certain angle, you will fall. The critical angle, at which you go from being stable to falling, is the angle of repose.

Soils behave in a very similar manner. When a trench is excavated the sides of the trench may or may not collapse – depending on the angle of repose of the soil. If the soil is a stiff clay then the trench will most likely hold (provided the soil is not saturated). However, if the trench is excavated in sand or gravel the sides will most likely collapse. We've all experienced this, digging holes at the beach! This is because the sides of the trench are at an angle greater than the natural angle of repose of the soil.

Angle of respose.

Another everyday construction example of this is seen when aggregates are being delivered to a building site. When the truck tips up the load, the sand and gravel holds fast until a certain point at which the pile begins to collapse and pour out. The angle of repose in this case is overcome because the angle between the side of the sand and gravel is increasing (relative to the ground). Usually, this is quickly overcome if the angle continues to increase – the entire pile slides out. However, we quickly meet the angle of repose again because when bulk materials are poured onto the ground, a conical pile will form. The angle between the edge of the pile and the horizontal surface is the angle of repose. It is related to the density, surface area and coefficient of friction of the material. Material with a low angle of repose forms flatter piles than material with a high angle of repose. Dry sands and gravels can stand at slopes equal to their angle of repose no matter how deep the excavation.

The angle of repose is important when deciding on whether or not to provide supports (timbering, sheet piling) to the sides of the trench during construction. However, for the construction of houses it is usually more practical and economical to select a different foundation design (e.g. raft instead of strip) rather than to provide timbering.

Revision Exercises

1. Explain, in your own words, why it is necessary to investigate the quality of the soil on a site.
2. Outline the type of information that can be gathered by carrying out a desk study of a site.
3. Describe, with the aid of neat labelled freehand sketches, one method of investigating the quality of the subsoil on a site. State four reasons why this is done.
4. Explain, in your own words, how the level of the water table varies at different times of the year.
5. Describe, using neat sketches, how the level of the water table can affect construction work.
6. Explain the term *angle of repose* and outline its significance in relation to excavations.

Foundations

A foundation is the part of a structure that transfers the loads from the structure to the ground. It is essential that the loads are spread safely and evenly over the supporting ground to ensure the stability of the building.

In most cases we do not see the foundation of a building because it is below ground level. As well as supporting the weight of the building, the foundation provides a level bed on which to build. While we usually think in terms of the building's weight keeping it in place, in some cases, especially for taller buildings, the foundation actually anchors the building to the ground.

A properly designed foundation will limit settlement, i.e. the tendency for a new building to *sink* into the ground. It is normal for a building to experience some settlement. This is because most soils are a mixture of the soil, air and water. When the building load is exerted on the soil, the air and water are driven out and the soil consolidates. In non-cohesive soils this happens during the construction phase, although in cohesive soils (e.g. clay) it can occur over a period of years. Settlement is defined as downward movement of the soil, or any structure on it, as a result of soil consolidation, usually caused by the load applied by the structure.

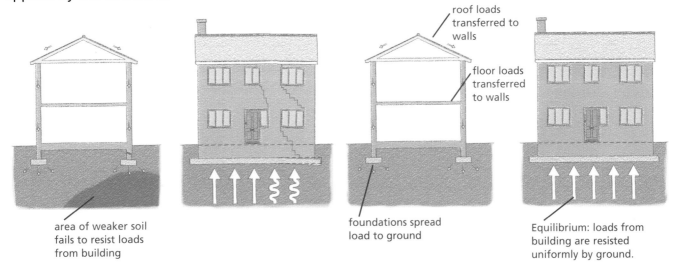

area of weaker soil fails to resist loads from building

roof loads transferred to walls

floor loads transferred to walls

foundations spread load to ground

Equilibrium: loads from building are resisted uniformly by ground.

Foundations: differential settlement will occur if an area of softer soil is undetected.

Foundations: a properly designed foundation will safely and evenly transmit the load over the supporting ground to ensure the stability of the building.

Differential settlement occurs when one area of the foundation settles at an increased rate. This can lead to cracking and ultimately failure of the foundation. Differential settlement is usually caused when the presence of an area of weaker soil below the foundation goes unnoticed. This weaker soil consolidates to a greater extent than the rest and the foundation is left unsupported.

Functions

The function of a foundation can be summarised as follows:

> • to transmit all building loads to the ground,
> • to limit settlement and prevent subsidence,
> • to provide a level bed on which to build,
> • to anchor the structure to the ground.

Design Features

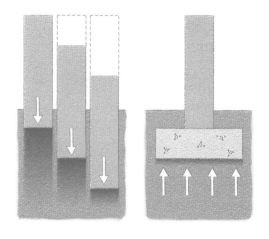

Foundations: spreading the load over a wider area prevents settlement.

A foundation is always wider than the element (e.g. wall or column) which it is supporting. This is so that the load is spread over a greater area. The bearing pressure exerted on the soil by a structure is a product of the force per unit area. For example, if the bearing pressure is 100 kN/m^2 this could be 10 kN exerted on 10 m^2 or 5 kN exerted on 20 m^2. Therefore, as the area increases the force exerted decreases. By increasing the area of a foundation under a house, the force exerted per square metre on the soil is decreased. For this reason, a traditional strip foundation (commonly used for house construction) is always three times wider than the overall width of the wall.

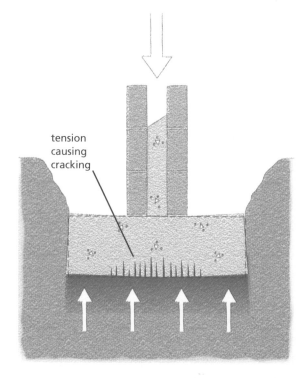

tension causing cracking

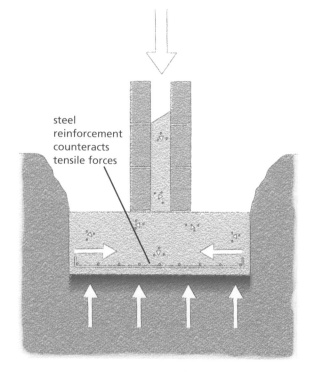

steel reinforcement counteracts tensile forces

Strip foundation: the lower portion of the strip is stretched (tension) as the foundation bends under loading. Placing steel reinforcement in the foundation counteracts this and ensures the foundation remains stable.

A foundation must be strong and rigid enough to ensure it does not fail (crack or break up) during the lifetime of the building. Foundations are often made from reinforced concrete for this reason. The reinforcement ensures that the foundation will not fail under any uneven stresses that may occur. A foundation experiences compressive and tensile stresses from the weight of building above and from the soil reacting to these forces from below.

These forces are added to by other factors, particularly in clay soils, including:

- Climatic Factors – variation in the levels of rainfall causes seasonal swelling and shrinkage of clay soils due to volume changes with varying moisture content. Also, if the water content of a soil is quite high the moist soil may freeze and expand upwards in severely cold weather, exerting a lifting force on the foundations known as *frost heave*.

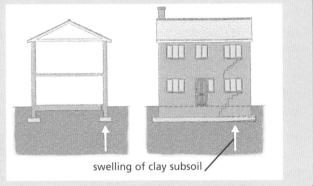

Foundations: movement of clay subsoils, caused by climactic factors, can cause failure in a foundation.

swelling of clay subsoil

- Effects of Vegetation – large trees absorb very large quantities of water from the soil every day. If a builder cuts down a large tree (growing on a site with a clay soil) and then builds close by (i.e. within 30 m), the foundations could be subjected to heave (lifting) from the excess water that builds up in the soil in the absence of the tree. The movement caused by the removal of mature trees can be as high as 100–150 mm and can continue for 10–20 years.

  Similarly, during a period of very hot weather a large tree will tend to soak up all the moisture in the soil causing the soil to contract. When this occurs near or under a foundation it is possible that the unsupported foundation could crack. The drying effect of a large tree (growing in a heavy clay soil) can extend to a depth of 3–5 m.

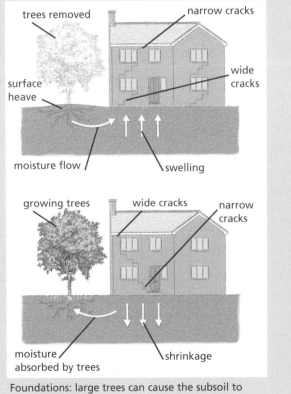

Foundations: large trees can cause the subsoil to shrink and expand, leading to structural damage.

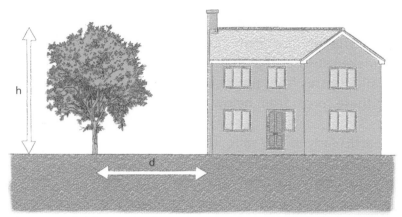

h = mature height   d = distance

Foundations: newly planted trees should be positioned at a distance equal to their mature height from the house.

In both cases the swelling and shrinkage effects can cause failure of the foundation and cracking in the walls of the building. Newly planted trees should be positioned at a distance equal to their mature height from the house.

In general, most buildings are constructed in areas where the soil close to the surface is suitable for construction. In a suitable area the ground will be uniform, stable and level allowing for the use of shallow economical foundations. The factors that affect the type of foundation to be used include:

- Bearing capacity of the ground – ability of the soil type to support structural loads.
- Depth of good strata – the depth to which the foundation must reach to find good support. Influenced by the bulb of pressure (see below).
- Composition of the ground – the uniformity of the soil. The reuse of sites (brownfield sites) sometimes means that the soil is made up of a variety of soil types and backfill, and therefore will not perform consistently.
- Ground level and gradients – building on slopes may require the use of stepped foundations or retaining walls.

The *bulb of pressure* is an important concept in foundation design. It tells us that the pressure exerted by a foundation on a soil decreases with increasing depth – the soil at the top is under most pressure, while soil further down is under less pressure. This is important because it guides the builder or structural engineer as to what depth the soil must be checked when carrying out a soil investigation. For strip foundations the bearing pressure exerted on the soil is only significant to a depth of approx. 3 times the width, whereas for a raft foundation the bearing pressure exerted on the soil is only significant to a depth of approx. 1.5 times the width of the raft. Pressure bulbs of less than twenty per cent of the original loading can be safely ignored for all foundation types.

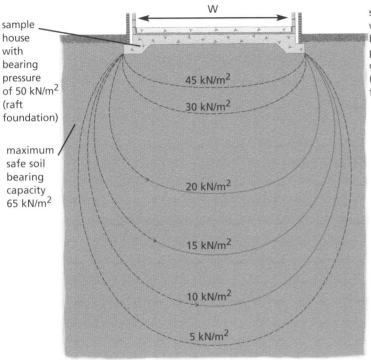

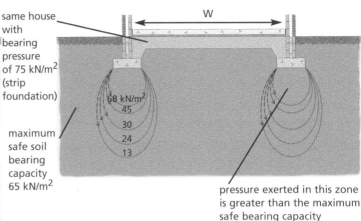

sample house with bearing pressure of 50 kN/m² (raft foundation)

maximum safe soil bearing capacity 65 kN/m²

45 kN/m²

30 kN/m²

20 kN/m²

15 kN/m²

10 kN/m²

5 kN/m²

Bulb of pressure: because the raft foundation is very wide the bulb of pressure extends quite far down into the soil. However, the level of pressure exerted on the soil is very low.

same house with bearing pressure of 75 kN/m² (strip foundation)

maximum safe soil bearing capacity 65 kN/m²

68 kN/m²
45
30
24
13

pressure exerted in this zone is greater than the maximum safe bearing capacity

Bulb of pressure: while the bulb of pressure for a strip foundation does not extend very far down into the soil, the level of pressure exerted on the soil is much higher.

Setting Out Foundations

The first step in the construction of a typical house involves clearing the area where the house is to be built. It is very important to retain as much of the existing landscape features as possible. Hedges, trees and other naturally occurring features should not be unnecessarily removed for convenience. As mentioned earlier, the topsoil (upper 300–600 mm of soil in which vegetation grows) is removed and stored for landscaping use later. The frontage line is then established and the outline of the building is marked out from this. Most construction requires the setting out of right-angled lines. This is usually achieved on-site using a right-angled triangle. This is based on a right angle whose sides are: 3 units × 4 units × 5 units. For example, a triangle made using 50 × 50 mm white deal measuring 900 × 1200 × 1,500 mm, alternatively a laser sight may be used.

In the case of strip foundations, profile boards are used to indicate

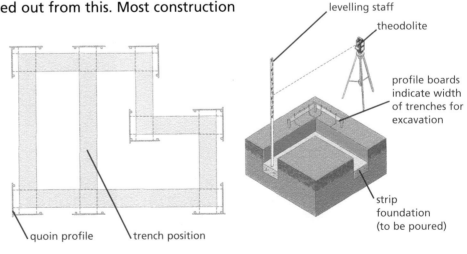

quoin profile   trench position

levelling staff

theodolite

profile boards indicate width of trenches for excavation

strip foundation (to be poured)

Setting out of a strip foundation using profile boards, a theodolite and a levelling staff. The use of laser levelling equipment has made this task much quicker and more accurate.

the position of the load-bearing walls. String can be attached to the profile boards to generate an accurate picture of where the foundation trenches are to be excavated and to facilitate the accurate marking of the foundation depth.

Types of Foundations

There are three main types of foundations used in the construction of houses: strip, raft and piled. Each of these foundations is suited to particular soil types and conditions.

Strip Foundation

Strip foundation: a continuous strip of concrete that supports all load-bearing walls.

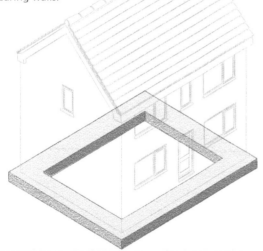

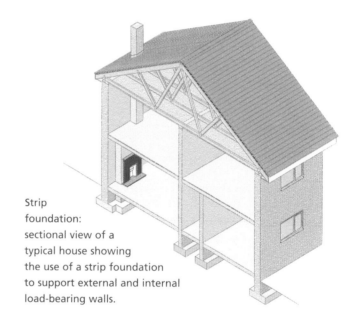

Strip foundation: sectional view of a typical house showing the use of a strip foundation to support external and internal load-bearing walls.

Strip foundation excavated in preparation for pouring of concrete.

Strip foundation after concrete has been poured.

Strip foundations are used where the soil is of good bearing capacity. The traditional strip foundation is the most commonly used foundation for the construction of houses for a number of reasons:

- The load from a typical house is carried evenly (at its perimeter) by the external walls.
- The load exerted by a house is relatively low.
- It is an economical way to support the load.

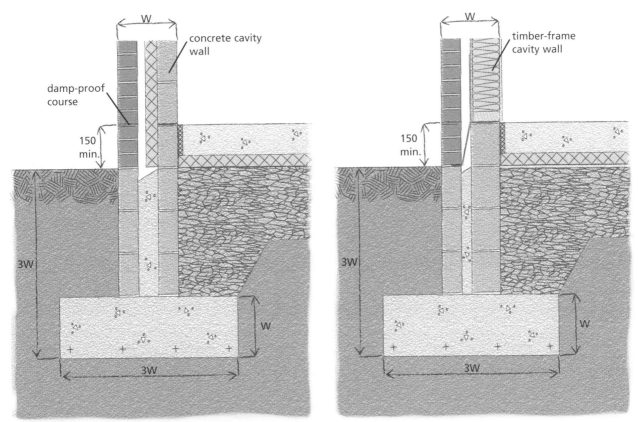

Strip foundation: the key sizes of a strip foundation for concrete cavity wall construction and timber frame cavity wall construction are similar. The size and position of the strip is directly related to the overall width of the wall.

Note: Relationship is between overall wall thickness and foundation width, irrespective of individual thicknesses of wall components.

The principle design features of a strip foundation are based on the fact that the load is transmitted at a 45° angle from the base of the wall to the soil. This means that the depth of the foundation must be equal to (or greater than) the overall width of the supported wall. The width of the foundation must be three times the width of the supported wall. As the load is resisted by the soil, it becomes clear that any foundation wider than three times the width of the wall would be redundant and may possibly suffer shear failure.

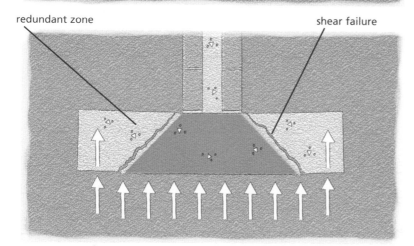

load distribution in concrete

W

W

45°

redundant zone

shear failure

Steel reinforcement may be added to the concrete to resist the tensile stresses and to ensure the entire strip performs as one unit.

On sloped sites it is sometimes necessary to step a strip foundation in order to avoid excessively deep excavations which are hazardous to workers. The use of stepped strip foundations also reduces unnecessary waste of concrete and blockwork. It is important that stepped foundations are properly designed to ensure strength and stability. For example, the strip is overlapped – by at least twice the step height, or the thickness of the foundation, or 300 m, whichever is greater – to improve strength. Steps are arranged to suit the coursing (height) of blockwork in the rising wall.

Strip foundation: for a plain concrete strip foundation (i.e. not reinforced with steel) the width of the foundation should be three times the width of the wall to avoid shear failure.

Stepped strip foundation: it is essential that the overlapped portion of the foundation is correctly designed.

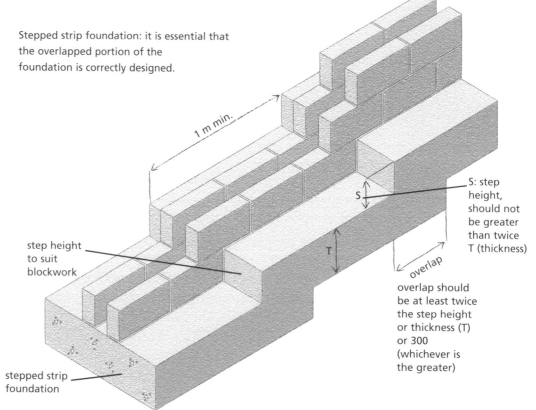

1 m min.

S

T

step height to suit blockwork

stepped strip foundation

S: step height, should not be greater than twice T (thickness)

overlap

overlap should be at least twice the step height or thickness (T) or 300 (whichever is the greater)

Raft Foundation

A raft foundation floats on the soil spreading the load exerted by the building over as wide an area as possible, resulting in a much lower level of pressure (force per unit area) being exerted on the soil (see bulb of pressure above). Raft foundations are used on sites where the soil is of low bearing capacity. One of the advantages of the raft foundation is that if settlement occurs, the building moves as a whole unit and differential settlement is avoided. Raft foundations are most suitable for light buildings such as houses.

A raft foundation consists of a continuous slab of reinforced concrete with a thickened edge. This edge thickening is an important design feature of raft foundations. It provides extra support to the area where load-bearing walls exert most load on the foundation. The raft is also thickened at intermediate points where internal load-bearing walls (if any) will be constructed.

Raft foundations are normally poured on-site. To facilitate this, a timber formwork is made up and steel reinforcement is placed and secured. The concrete is then poured and allowed to set for several hours before the surface is power floated to a smooth finish.

The edge of the raft is usually stepped to create a toe. The purpose of the toe is to allow the outer leaf to continue below

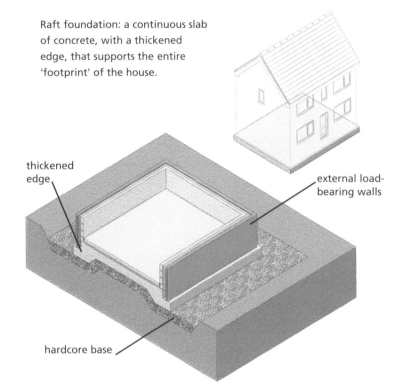

Raft foundation: a continuous slab of concrete, with a thickened edge, that supports the entire 'footprint' of the house.

thickened edge

external load-bearing walls

hardcore base

Raft foundation: this raft foundation has just been poured. This photo shows the formwork used to create the 'toe' at the edge of the foundation.

Raft foundation: the concrete raft is supported on a compacted hardcore base that extends at least one metre beyond the edge of the raft.

'toe'

hardcore

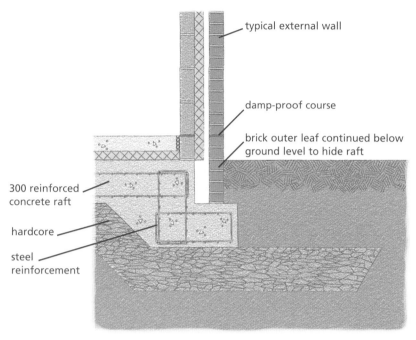

typical external wall

damp-proof course

brick outer leaf continued below ground level to hide raft

300 reinforced concrete raft

hardcore

steel reinforcement

ground level thereby masking the edge of the raft. This is particularly important for houses with a red brick external leaf.

Raft foundation: the edge of the raft is thickened to support the external load-bearing walls. The edge is usually stepped to create a 'toe' to mask the edge of the raft. Note also, the use of steel reinforcement to strengthen the raft.

Piled Foundation

A pile is a column of concrete that extends downward deep into the soil. Piled foundations are used in a number of situations:

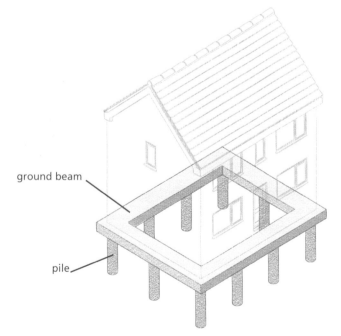

ground beam

pile

Piled foundation: consists of a number of piles connected by a ring of concrete, called a ground beam, which is similar to a strip foundation but not as wide.

Piled foundation: these friction piles have been driven into place on a site with a soil of low bearing capacity and will later be cut down to the correct height. Note the steel band around the top of the pile to protect it during driving.

- where the soil near the surface is of low load-bearing capacity, but the soil at greater depth is of sufficient load-bearing capacity,
- where the soil is prone to swelling or shrinkage due to seasonal changes,
- where the building is subject to an uplifting force and must be anchored to the ground,
- on brownfield (previously used) sites which have been filled.

Piled foundations work in one of two ways depending on the soil conditions. End-bearing piles reach below soft ground to firmer ground of good bearing capacity. Friction piles rely on adhesion or friction between the surface of the pile and the soil. This is similar to the way in which it becomes harder to hammer a stake into soft soil. As the friction between the surface of the stake and the surrounding soil increases, it is progressively harder to drive the stake.

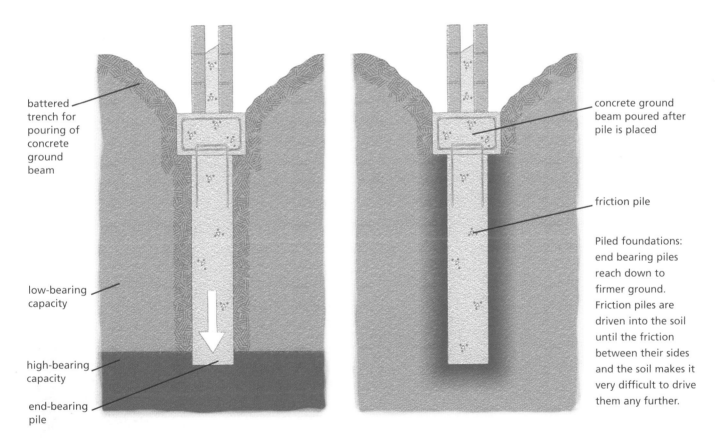

battered trench for pouring of concrete ground beam

low-bearing capacity

high-bearing capacity

end-bearing pile

concrete ground beam poured after pile is placed

friction pile

Piled foundations: end bearing piles reach down to firmer ground. Friction piles are driven into the soil until the friction between their sides and the soil makes it very difficult to drive them any further.

Piled foundations can be constructed in a number ways. The simplest method is to drive precast piles into the soil using specialised percussion drivers. This is a noisy operation and can lead to damage of the pile. A similar method involves driving a hollow steel casing into the ground. When the required depth is reached, steel reinforcement is placed into the casing and it is filled with fresh concrete.

The short bored pile is an alternative method. Most commonly used in the construction of houses, it involves boring a hole into the soil and filling it with concrete. Steel reinforcement is then pushed down into the fresh concrete. The relatively light loads generated by houses are readily accommodated by this type of pile. Steel reinforcement is often used in the upper portion of the pile only.

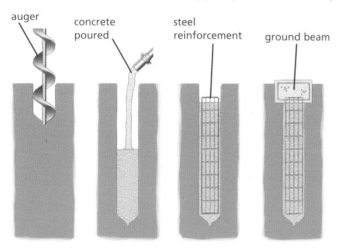

Piled foundation: after the soil has been removed, the concrete is poured and the steel reinforcement is added. The steel projects up to create a link with the ground beam, which is poured once the pile has cured.

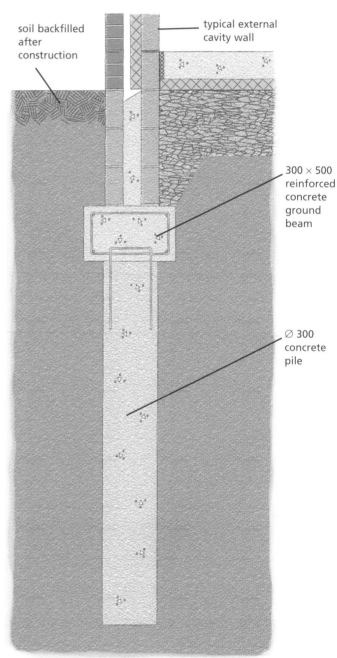

Piled foundation: cross-section of a pile foundation showing a typical concrete cavity wall and solid concrete ground floor.

Once the piles are complete a surface must be created on which to build the walls of the house. This surface is called a *ground beam*. Similar in appearance but smaller than a strip foundation, it is poured into the intended position along the line of the piles to create a continuous beam around the perimeter of the dwelling.

The size of the beam and the size, depth and spacing of the piles is dependent on the particular soil conditions and on the building design. In areas where extra loading occurs (e.g. under a two-storey chimney breast) piles can be grouped to provide greater support.

Revision Exercises

1. Explain the importance of any four considerations that should be taken in account when selecting a foundation type for a structure.
2. Define the term *settlement* and give examples of ways in which it can adversely affect a structure.
3. Describe the risks associated with mature trees when building on clay subsoils.
4. Explain how the *bulb of pressure* concept is used in foundation design.
5. A typical two-storey detached domestic dwelling is to be constructed on a site with a stiff clay subsoil. Suggest a suitable foundation type, give reasons for your answer and describe it with the aid of neat labelled freehand sketches.
6. A typical two-storey detached domestic dwelling is to be constructed on a steeply sloped site with a stiff clay subsoil. Suggest a suitable foundation type, give reasons for your answer and describe it with the aid of neat labelled freehand sketches.
7. A typical single-storey bungalow is to be constructed on a site in a bogland area. Suggest a suitable foundation type, give reasons for your answer and describe it with the aid of neat labelled freehand sketches.
8. A large two-storey domestic dwelling is to be constructed on a site which has a subsoil consisting of three-metre deep soft clay, with rock below. Suggest a suitable foundation type, give reasons for your answer and describe it with the aid of neat labelled freehand sketches.
9. Higher Level, 2002, Question 5
    Trial holes indicate that the site, on which a house is to be built, has a loose gravel subsoil.
    (a) Discuss in detail the considerations governing the choice of foundation for this house.
    (b) Describe, with the aid of notes and detailed sketches, two types of foundation that would be suitable for the house.
    (c) In the case of each type of foundation selected, state clearly two reasons why it is considered suitable.
10. Higher Level, 2006, Question 4.
    Investigations indicate that a site on which a house is to be built has a moderately firm clay subsoil. Consideration is being given to using either a traditional strip foundation or a raft foundation.
    (a) Show, with the aid of notes and freehand sketches, the design detailing for each type of foundation listed above. Indicate typical dimensions for each foundation.
    (b) Recommend one of the above foundation types for the house and give two reasons in support of your recommendation.
    (c) Identify two factors that could adversely affect the strength of concrete in a foundation.

Functions

The primary functions of the external walls of a house are to support the loads generated and to create a comfortable living space.

Design Features

Walls must be designed to meet a number of performance requirements including:

- Strength and Stability – The strength and stability of an external wall depends on:
  - strength of the components used (i.e. blocks, mortar),
  - Slenderness Ratio – tall, thin walls are less stable than short, thick walls,
  - eccentricity of the applied load – the force experienced by the wall from above must be centred.

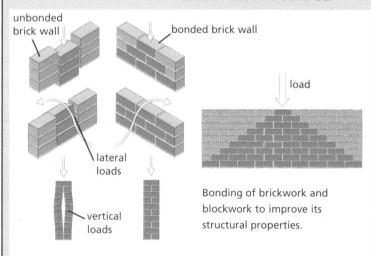

unbonded brick wall

bonded brick wall

lateral loads

vertical loads

load

Bonding of brickwork and blockwork to improve its structural properties.

Bonding is an essential design feature of masonry walls and greatly improves the strength of a wall by ensuring that it performs as a continual unit. Bonding is defined as the overlapping of blocks or bricks to ensure that the vertical and lateral loads are dispersed evenly throughout the wall. Various bonding patterns also used to improve the appearance of walls.

- Weather Resistance – Designing for weather resistance in the Irish context is mainly about preventing the penetration of wind-driven rain. This can be approached in a number of ways. The oldest and simplest approach is to make the external walls very thick. The porous solid masonry allows the moisture to penetrate its surface. It relies on the fact that the rain will eventually stop and the wind and sun will dry out the wall before the moisture reaches the inner surface. This method is still used in many parts of the world (e.g. Scotland) that experience extreme weather conditions.

A more modern approach is the impervious fabric method. This involves the use of weather-resistant materials such as aluminium or plastic cladding. This form of construction is common in larger buildings but is not commonly used in housing in Ireland.

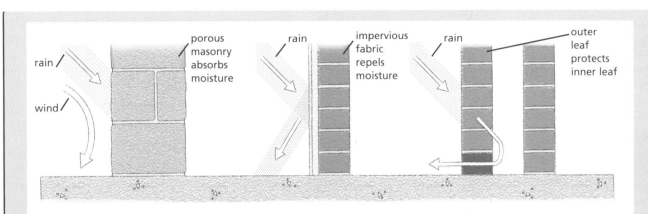

The most common method used for housing is the cavity construction. This involves building a wall that consists of two separate walls, an inner wall and an outer wall, with a small gap called a cavity between them. The cavity prevents the moisture absorbed by the outer wall reaching the inner wall. The inner and outer walls used in cavity construction are normally referred to as the *inner leaf* and the *outer leaf*. The function of the outer masonry leaf is to protect the inner leaf from the effects of weathering. The function of the inner leaf is to support the loads of the house.

Weather resistance of walls.

- Fire Resistance – The fire resistance of the wall is determined by the materials used in its construction. For example, concrete walls are naturally fire resistant because concrete is not a combustible material. The extent of the requirement for fire resistance depends on the height of the building, the floor area of the building (i.e. the distance from any point inside the building to a fire exit) and the function of the building (e.g. a hotel has a greater requirement than a warehouse). The minimum requirement for fire resistance of walls in conventional housing is 30 minutes. This means that the wall should remain stable for at least 30 minutes in the event of a fire. This allows sufficient time for the occupants to escape.

'open-plan' layout where stair rises from an internal room

Fire escape: the bottom of the stair should be within 4·5 m of an external door.

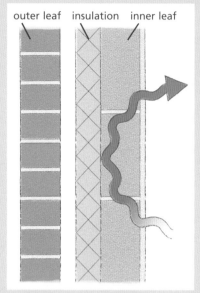

outer leaf   insulation   inner leaf

Thermal insulation: concrete cavity walls typically rely on insulation held against the inner leaf to retain heat.

- Thermal Insulation – The proper thermal insulation of walls is essential to ensure that heat is not lost from the home. Poor thermal insulation leads to greater fuel use which in turn is damaging to the environment and expensive for the home owner. Masonry cavity walls are usually insulated by placing a continuous layer of rigid expanded polystyrene foam insulation into the cavity. This insulation board is held tightly against the inner leaf. This means that when the house is heated the heat energy is absorbed by the inner leaf. For this reason it can take some time for a comfortable temperature to be achieved. However, once warm, concrete cavity-walled homes can remain so for several hours after the heating is turned off, as the concrete inner leaf acts as a heat store.

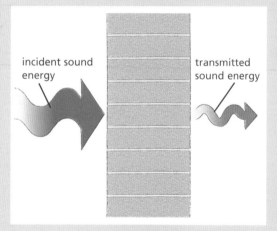

incident sound energy   transmitted sound energy

Acoustic insulation: a typical concrete party wall consists of blocks laid on flat. This creates a thick, heavy wall that prevents sound transmission.

- Acoustic Insulation – Acoustic insulation is generally not a problem in cavity wall construction, particularly in the case of masonry walls. Concrete cavity walls have three features that make them effective sound insulators. The mass of the concrete means that it readily absorbs sound energy. The cavity between the inner and outer leaves reduces the transmission of sound through the wall and the plaster and render applied to the inner and outer surfaces seals any gaps in the blockwork that may allow sound to penetrate. In the case of timber-frame walls, the outer masonry leaf also absorbs sound energy, the cavity functions in the same way as before and the quilted insulation placed into the timber inner leaf prevents excessive noise (see construction to timber-frame cavity walls). The plaster on the internal surface of the wall also prevents noise transmission. (See chapter 22, Sound and Acoustic Insulation.)

- Aesthetics – The visual appearance of the external walls contributes greatly to the overall aesthetic quality of a house. While much of this is open to taste it is essential that a high level of finish is achieved if a wall is to look well. This is particularly important in the case of brickwork, where the constructed wall will not be rendered. Various patterns of brickwork can be used to add visual interest to brickwork.

Alternatively, external walls can be rendered to a smooth or dashed finish (see chapter 24, Rendering and Plastering). In formal heritage buildings (e.g. courthouses) it is common to see ashlar work. *Ashlar* is the term used to describe masonry constructed using squared stone with very tight joints. This stone was dressed (shaped) to square using hand tools and secured using fine lime mortar joints. Various surface finishes can be applied.

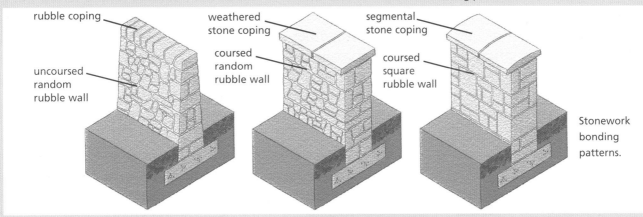

English bond

Flemish bond

English garden wall bond

Flemish garden wall bond

Brickwork bonding patterns.

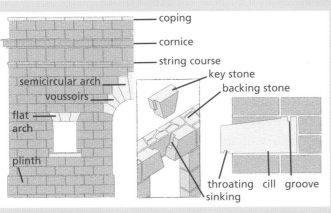

rubble coping

weathered stone coping

segmental stone coping

uncoursed random rubble wall

coursed random rubble wall

coursed square rubble wall

Stonework bonding patterns.

coping

cornice

string course

key stone

backing stone

semicircular arch

voussoirs

flat arch

plinth

throating   cill   groove

sinking

Ashlar stonework: usually used in formal and ecclesiastical buildings (e.g. courthouses, churches). This work is recognised by the square, clean stone and the fine mortar joints.

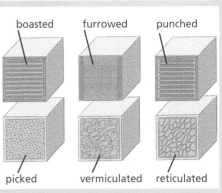

boasted   furrowed   punched

picked   vermiculated   reticulated

Surface dressing of ashlar stones.

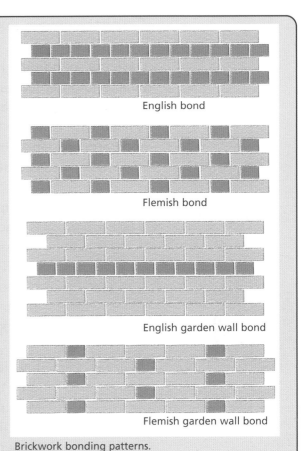

Construction of Concrete Cavity Walls

Once the foundation has cured, the external walls can be raised. Both the inner and outer leaves are built up to a level 150 mm above ground level. This portion of the wall is called the *rising wall*. The cavity in the rising wall is filled with concrete to resist lateral pressure exerted on it from the soil and hardcore. Once the rising walls are secure, the concrete ground floor can be poured.

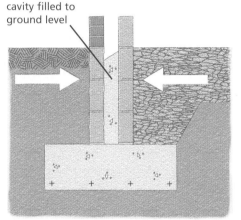

cavity filled to ground level

Concrete cavity walls: below ground level, the cavity is filled to resist lateral pressure exerted from either side by the hardcore and soil.

A strip of heavy duty plastic called a damp-proof course (DPC) is placed in the inner and outer leaf of the wall and the blockwork is continued. The DPC is always positioned so that it will be 150 mm above the external ground level (e.g. level of footpath and/or grass) when the construction is complete. The purpose of the DPC is to prevent moisture from rising up through the walls by capillary action.

Capillary action is usually described as the ability of a narrow tube to draw a liquid upwards against the force of gravity. Most people will have noticed how, when a straw is placed in a drink, the liquid in the straw curves upward and is often at a higher level than the rest of the liquid. In simple terms, this is because the molecules of the liquid are more attracted to the molecules of the straw than they are to each other (the adhesive forces between the molecules of the liquid and the molecules of the straw are stronger than the cohesive forces between the molecules of the liquid). It is also worth noting that a narrow tube will draw a liquid higher than a wide tube. For example, a glass tube with internal diameter 0·1 mm can raise a 300 mm column of water.

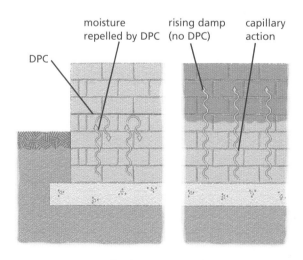

DPC

moisture repelled by DPC

rising damp (no DPC)

capillary action

This is very significant in construction because it means that there will always be a risk that moisture will be drawn up through a rising wall due to the small voids in a typical concrete block. Without a DPC the lower portion of the walls would be damp for most of the year causing paint or wallpaper to peel and mould to grow.

Where the site changes level, or is sloped, the DPC must be stepped with the slope to ensure it remains at least 150 mm above finished ground level.

Concrete cavity walls: the damp-proof course prevents moisture from soaking up through the masonry.

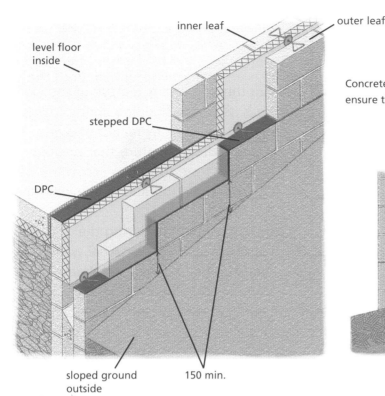

level floor inside

inner leaf

outer leaf

stepped DPC

DPC

sloped ground outside

150 min.

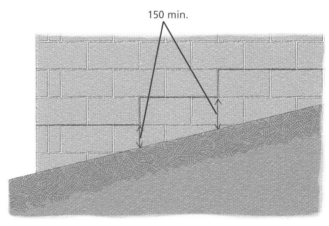

Concrete cavity walls: the damp-proof course is stepped to ensure the DPC remains at least 150 mm above ground level.

150 min.

Above DPC level the cavity contains the rigid insulation discussed earlier. This is held in place by wall ties. It is essential that the wall ties hold the insulation tightly against the inner leaf. This is to prevent interstitial condensation. Interstitial condensation is condensation that develops within the fabric of a building element (e.g. wall or roof). This can happen in the inner leaf of a cavity wall if the insulation is loose and the temperature within the blockwork drops to the dew point (temperature at which water condenses from vapour to liquid form).

Wall ties are stainless steel, galvanised steel or polypropylene links that are built into the horizontal joints of the inner and outer leaves to form a connection between them. This is very important as it ensures that the inner and outer leaves act as one unit, thereby improving the structural stability of the external wall. Wall ties are spaced 900 mm apart horizontally and 450 mm apart vertically.

The wall tie also has a component (e.g. plastic disc) that holds the insulation board firmly against the inner leaf. It is important that the insulation is placed correctly. This is achieved by building up one leaf to a height of one batt of insulation plus one course. The insulation is then securely fixed in place and the opposite course is raised to match the first.

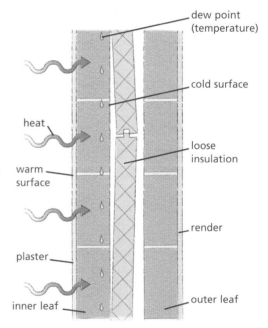

dew point (temperature)

cold surface

heat

loose insulation

warm surface

render

plaster

outer leaf

inner leaf

Concrete cavity walls: interstitial condensation will occur if the insulation is not properly secured against the inner leaf.

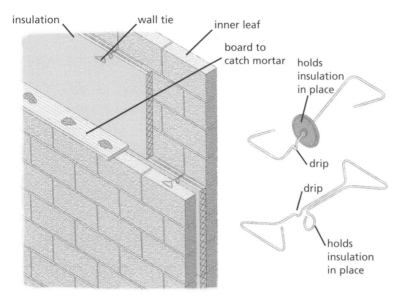

insulation   wall tie   inner leaf   board to catch mortar

holds insulation in place

drip

drip

holds insulation in place

Concrete cavity walls: installation of insulation using wall ties. Note how the mortar droppings are caught on a board to prevent them falling into the cavity.

Another important feature of the wall tie is its water-shedding ability. This is usually achieved by means of a drip. The drip is a design feature that prevents moisture from travelling across the wall tie from the outer leaf to the inner leaf.

It is essential that, during the construction of the external walls, mortar is not allowed to fall into the bottom of the cavity. If this is allowed to happen it will fill the cavity at the base of the wall and provide a path for moisture to cross the cavity. A board can be used to catch any mortar that falls toward the cavity.

Openings in Concrete Cavity Walls

An opening or ope must be created in the external walls to allow for doors and windows. This creates a number of challenges – the load exerted from above must be carried by the sides of the opening and the cavity must be sealed around opes to prevent heat loss and the spread of fire or smoke. Closing the cavity is problematic because it provides a pathway for moisture and cold to travel between the outer and inner leaf.

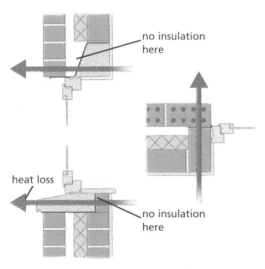

no insulation here

heat loss

no insulation here

Concrete cavity walls: closing the cavity at opes (e.g. window and door opes) can lead to cold bridging and dampness. To prevent this the passage of moisture and heat must be prevented.

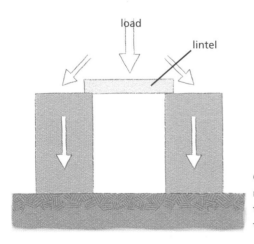

load

lintel

Concrete cavity walls: a reinforced concrete lintel transfers the load safely to either side of the ope.

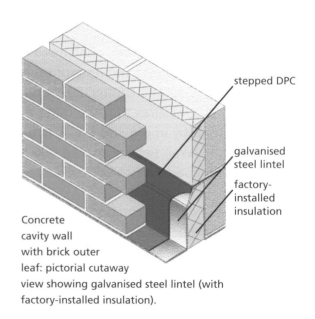

stepped DPC

galvanised steel lintel

factory-installed insulation

Concrete cavity wall with brick outer leaf: pictorial cutaway view showing galvanised steel lintel (with factory-installed insulation).

Door and window opes are usually spanned using a reinforced concrete beam called a lintel. A separate lintel is used for the outer and inner leaves. The lintel supports the load from above and transfers it safely to the wall at either side of the ope. The cavity is sealed above the window using rigid insulation board.

Where brickwork is used for the outer leaf, a galvanised steel lintel is used. This is because a concrete lintel would be visible from the outside. Steel lintels are shaped to increase their strength and prevent the transfer of moisture across the cavity. They also have insulation pre-installed to prevent thermal loss. A DPC is placed in the inner leaf above the lintel and is carried over the insulation and down to the outer leaf. This prevents the transfer of moisture across the cavity via the insulation.

At the sides (jambs) of the door or window the cavity is sealed using special L-shaped blocks. These blocks

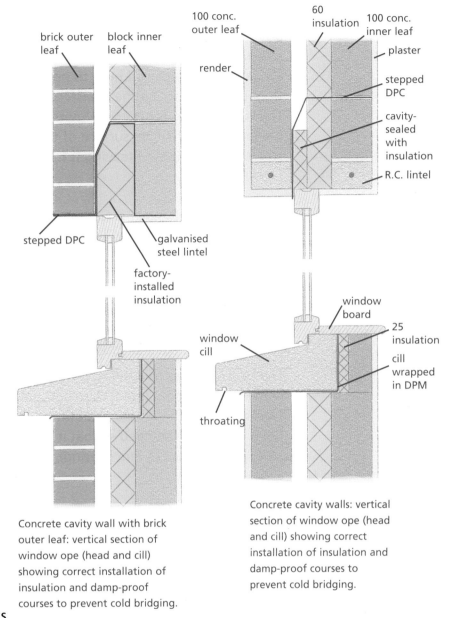

Concrete cavity wall with brick outer leaf: vertical section of window ope (head and cill) showing correct installation of insulation and damp-proof courses to prevent cold bridging.

Concrete cavity walls: vertical section of window ope (head and cill) showing correct installation of insulation and damp-proof courses to prevent cold bridging.

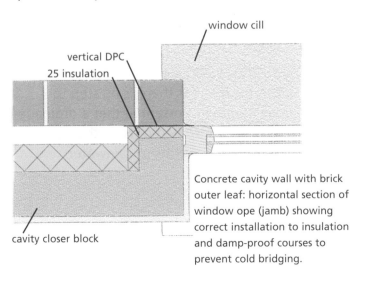

Concrete cavity wall with brick outer leaf: horizontal section of window ope (jamb) showing correct installation to insulation and damp-proof courses to prevent cold bridging.

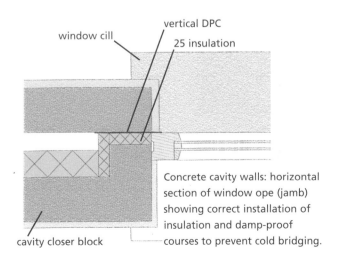

Concrete cavity walls: horizontal section of window ope (jamb) showing correct installation of insulation and damp-proof courses to prevent cold bridging.

also provide a fixing for the window frame. The problem of moisture transfer is overcome by placing a vertical DPC between the outer and inner leaf. Rigid insulation is also placed here to prevent cold bridging. *Cold bridging* is the term used to describe the transfer of cold from the outer to the inner leaf. This is a problem for home owners because when the warm air inside their home touches the cold surface of the wall (caused by cold bridging) it condenses on the wall leaving it moist. This moisture encourages mould growth.

The bottom of doors and windows are dealt with in different ways. Windows have a concrete cill below them in the outer leaf. The cill has a number of functions including:

> • to carry rainwater away from the wall,
> • to close the cavity,
> • to support the window.

The cill is wrapped in a DPC to ensure that the rainwater that soaks into it does not penetrate the wall. It is then placed on a bed of mortar on the outer leaf. It is positioned so that it just sits onto the inner leaf (approx. 25 mm bearing), thereby sealing the cavity.

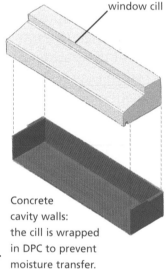

window cill

Concrete cavity walls: the cill is wrapped in DPC to prevent moisture transfer.

On the other hand, doors have a threshold rather than a cill. Door thresholds of newly constructed houses are universally designed to allow for convenient access by everybody (e.g. toddlers, old people, people on crutches, people who use a wheelchair). This means that there is no step to overcome. It is essential that a drainage channel is installed outside the door to prevent the ingress of rainwater.

25 edge insulation

transition units

threshold seal

DPC

cast in-situ concrete cill (max. slope 15°)

drainage channel

DPM

Door threshold designed for universal access. Note the design details that exclude rainwater: sloped cill, drainage channel, vertical DPC behind cill and threshold seal.

Construction of Timber-frame Cavity Walls

Timber-frame cavity walls are very similar to concrete cavity walls. The main difference is that in timber-frame houses the inner leaf is made from timber, the outer leaf remains masonry.

Timber-frame housing has a number of advantages over traditional concrete approaches, including:

- quick to construct,
- reduced site labour,
- quickly weatherproofed,
- recyclable,
- reduced waste during construction,
- energy efficient,
- low embodied energy.

The timber-frame inner leaf consists of a stud framework sheeted with plywood or oriented strand board (depending on the manufacturer). This sheathing provides structural stability as well as providing a surface to fix the breather membrane to.

The panels are made in a factory and delivered to the site. A crane is used to lift them into place. Each ground-floor panel is secured to the rising wall using galvanised steel straps. A DPC is fixed to the underside of the panel soleplate to prevent the timber absorbing any moisture from the concrete. The panels are fixed to each other and the internal wall panels.

Once the ground floor external and internal wall panels are installed, the upper floor is lowered onto them. The first-floor wall panels can then be installed. This type of timber-frame construction is called *platform construction*.

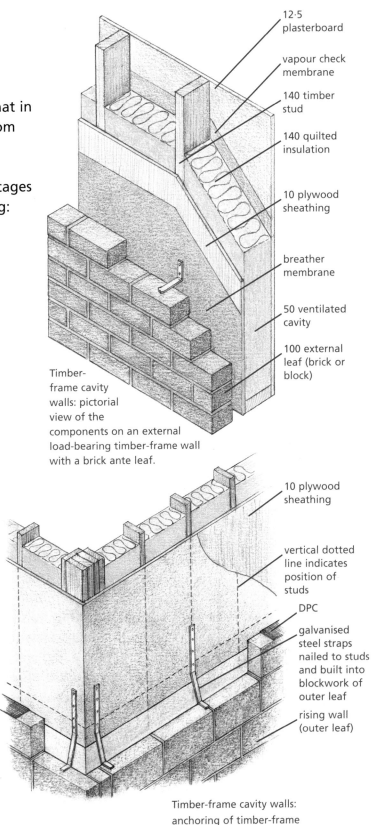

12·5 plasterboard

vapour check membrane

140 timber stud

140 quilted insulation

10 plywood sheathing

breather membrane

50 ventilated cavity

100 external leaf (brick or block)

Timber-frame cavity walls: pictorial view of the components on an external load-bearing timber-frame wall with a brick ante leaf.

10 plywood sheathing

vertical dotted line indicates position of studs

DPC

galvanised steel straps nailed to studs and built into blockwork of outer leaf

rising wall (outer leaf)

Timber-frame cavity walls: anchoring of timber-frame inner leaf.

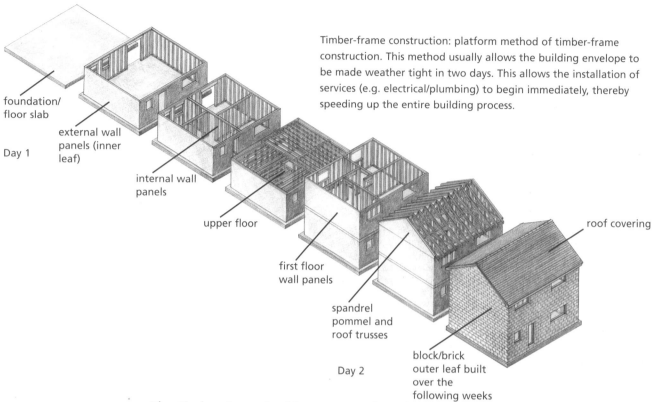

Timber-frame construction: platform method of timber-frame construction. This method usually allows the building envelope to be made weather tight in two days. This allows the installation of services (e.g. electrical/plumbing) to begin immediately, thereby speeding up the entire building process.

foundation/
floor slab

Day 1

external wall
panels (inner
leaf)

internal wall
panels

upper floor

first floor
wall panels

spandrel
pommel and
roof trusses

Day 2

roof covering

block/brick
outer leaf built
over the
following weeks

The timber inner leaf is connected to the outer masonry leaf using wall ties. In this case, the wall ties are nailed to the studs of the timber panel and pulled outward to meet the masonry – where they are built into the mortar joints.

Timber-frame wall panels achieve structural stability from the sheeting material applied to their outer surface. The sheeting prevents racking (sidewards collapse) in the panel when loading occurs.

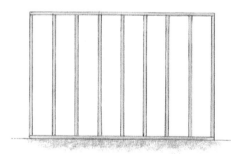

dynamic loading

Timber-frame cavity walls: the timber sheathing fixed to the timber-frame panels prevents racking by providing triangulation.

While the outer masonry leaf will protect the timber panel from weathering a special fabric or membrane is also used. This membrane, called a *breather membrane*, keeps out any rain that falls while the outer masonry leaf is being built. It is called a breather membrane because it has microscopic pores that allow water vapour to escape outward. This is important because it ensures that moisture does not get trapped within the timber leaf causing it to become damp.

Timber-frame wall panels are protected from fire by applying plasterboard to their inner surface. Plasterboard is non-combustible and will protect the timber panels for the required thirty minutes in the event of fire. Other measures taken to prevent the spread of fire include placing fire barrier strips within the cavity around opes, at eaves and verges and at the junction of party walls and external walls. A cavity fire barrier is designed to provide fire resistance for half an hour. A typical barrier is comprised of strips of non-combustible glass mineral wool compressed within a strong double-flanged polythene sleeve. This type of product is suitable for vertical and horizontal applications. It is held in place vertically by stapling both flanges to the timber sheeting and horizontally by stapling the upper flange only – the lower flange then acts as a drip.

The positioning of the thermal insulation is where timber-frame and masonry cavity walls differ most. In timber-frame houses the thermal insulation is placed within the timber panel inner leaf. Quilted insulation, similar to that used in most attics, is placed between the studs of every external panel. The low mass of timber-frame walls combined with the

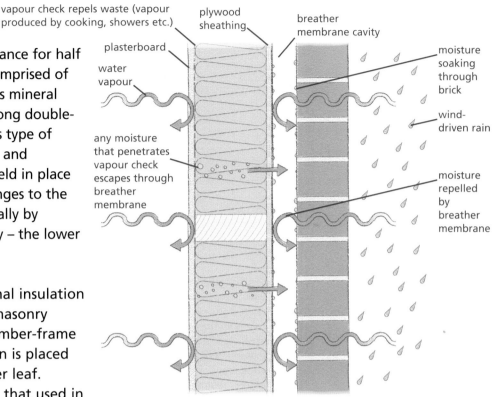

Timber-frame cavity walls: the combined protection provided by the vapour barrier (inside) and the breather membrane (outside) ensures the timber inner leaf remains dry at all times.

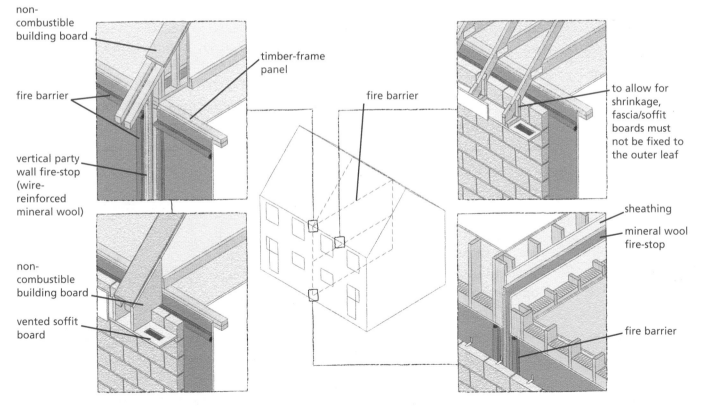

Timber-frame cavity walls: fire protection. To prevent the spread of fire with the wall, fire barrier strips are fixed to the inner leaf around opes, at eaves and verges and at the junction of party walls and external walls.

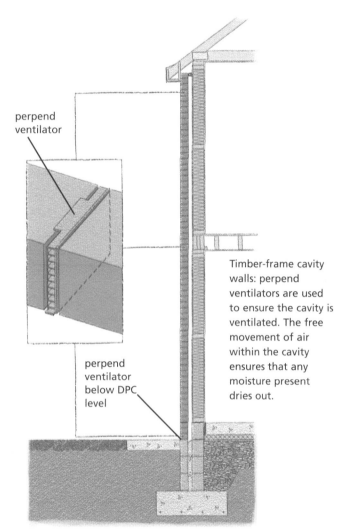

perpend
ventilator

perpend
ventilator
below DPC
level

Timber-frame cavity
walls: perpend
ventilators are used
to ensure the cavity is
ventilated. The free
movement of air
within the cavity
ensures that any
moisture present
dries out.

location of the insulation in this position means that timber-frame houses heat up quickly.

Another key difference between timber-frame and masonry houses is the ventilation of the cavity. In masonry cavity walls it is essential to seal the cavity in order to trap the air within it. This trapped air acts as an insulator for the wall. However, with timber-frame houses it is essential to ventilate the cavity so that any moisture that might build up within it can be dried out. To achieve this, special ventilators called *perpend ventilators* are placed in the perpend (vertical) joints between the blockwork in the outer leaf.

Another measure taken to prevent the build-up of moisture in the timber inner leaf is the use of a vapour check. A *vapour check* is a light sheet of clear plastic that is fixed to the inner surface of the timber panel after the insulation has been installed. This stops or checks the progress of moisture generated by cooking, showers and other steam-producing appliances.

In summary, we can say that moisture is prevented from penetrating the timber leaf both internally and externally and that any moisture that does manage to get into the timber leaf can evaporate outward through the breather membrane, where it will be dried off by the air ventilating the cavity.

Openings in Timber-frame Cavity Walls

Openings in timber-frame cavity walls are dealt with in a similar way to those in masonry walls. Timber lintels are used to span openings in the inner leaf. Where lintels are used in the inner leaf extra studs, called *cripple studs*, are used to carry the additional load.

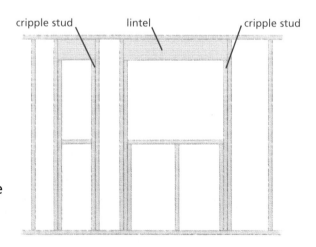

cripple stud   lintel   cripple stud

Timber-frame cavity walls: openings for windows and doors are spanned by timber lintels supported by cripple studs.

Instead of L-shaped blocks, a pressure-treated timber batten called a cavity barrier is fixed to the outer surface of the timber panel to seal the cavity around the ope and provide fixing for door and window frames.

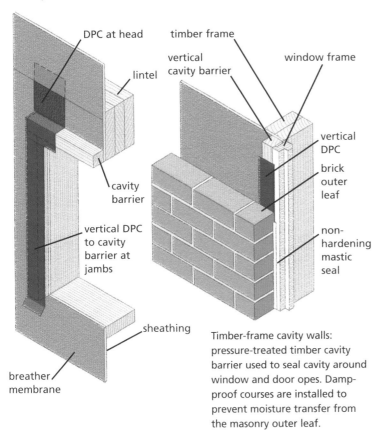

Timber-frame cavity walls: pressure-treated timber cavity barrier used to seal cavity around window and door opes. Damp-proof courses are installed to prevent moisture transfer from the masonry outer leaf.

Damp-proof courses are used in similar positions to those in masonry walls.

The window cills used in timber-frame construction are slightly smaller. The cill is supported by a timber cavity barrier and two courses of DPC are used to prevent the penetration of moisture from either the cill or the outer leaf.

When a timber-frame house is constructed a small amount of shrinkage in the timbers is common. To accommodate this shrinkage, gaps must be left to allow for the downward movement of the timber structure. For example, at the eaves (gutter) level the rafters must allow sufficient clearance so that they do not come into contact with the outer masonry leaf when shrinkage occurs.

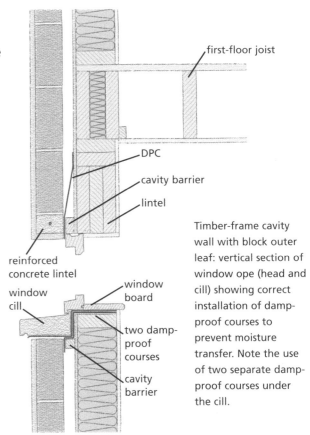

Timber-frame cavity wall with block outer leaf: vertical section of window ope (head and cill) showing correct installation of damp-proof courses to prevent moisture transfer. Note the use of two separate damp-proof courses under the cill.

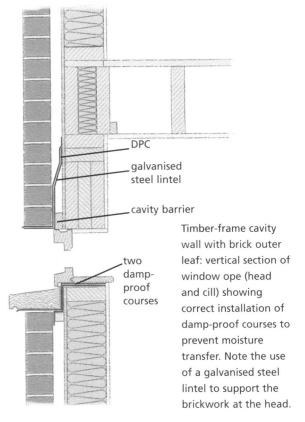

Timber-frame cavity wall with brick outer leaf: vertical section of window ope (head and cill) showing correct installation of damp-proof courses to prevent moisture transfer. Note the use of a galvanised steel lintel to support the brickwork at the head.

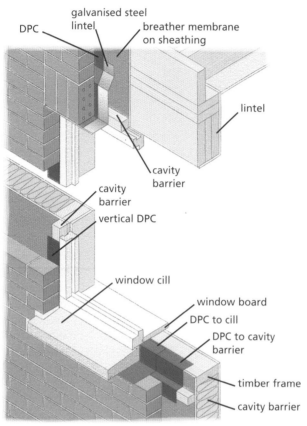

Timber-frame cavity wall with brick outer leaf: pictorial section (to aid visualisation) showing correct installation of components.

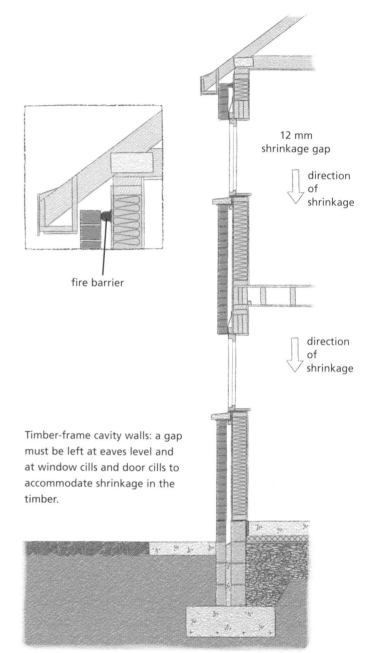

Timber-frame cavity walls: a gap must be left at eaves level and at window cills and door cills to accommodate shrinkage in the timber.

Revision Exercises

1. Explain, with the aid of neat labelled freehand sketches, the primary function of external load-bearing walls.

2. Describe, using neat annotated sketches, how the rendered, rubble, single-leaf walls of heritage buildings exclude wind-driven rain. Refer in your answer to the properties of the construction materials used.

3. Explain, with the aid of neat freehand sketches, the functions of wall ties in cavity walls.

4. Explain, with the aid of neat freehand sketches, the following terms: bonding, garden wall bond, coursed random rubble and ashlar.

5. Describe the correct method for the installation of expanded polystyrene insulation in a concrete cavity wall. Explain why it is important to follow this procedure.

6. Generate a neat freehand cross-section sketch of the cill of a window opening in an external load-bearing concrete cavity wall. The outer leaf is of block construction. The drawing should show the correct detail to prevent the loss of heat and the transfer of dampness.

7. Generate a neat freehand cross-section sketch of the head of a window opening in an external load-bearing timber-frame cavity wall. The outer leaf is of red brick construction. The drawing should show the correct detail to prevent the loss of heat and the transfer of dampness.

8. Higher Level, 2006, Question 1
    The main entrance to a dwelling house provides access for a person in a wheelchair. The door opening is located in a 300 mm external block wall with an insulated cavity and the door is a solid wooden door. The house has a solid concrete ground floor with a 20 mm quarry tile finish.
    Note: Make a neat freehand sketch of the solution.
    (a) To a scale of 1:5, draw a vertical section through the external wall and door, showing clearly the threshold and the door. The section should show all the constructional details from the bottom of the foundation to 300 mm above finished floor level.
    (b) Indicate on the drawing the specific design detailing that ensures that rainwater is removed from the threshold area and does not penetrate to the inner surfaces surrounding the door.

9. Higher Level, 2001, Question 1
    Note: Make a neat freehand sketch of the solution.
    A single-storey dwelling house has a 300 mm external concrete block wall with an insulated cavity, which includes an opening for a window. The dwelling has a suspended timber floor. To a scale of 1:5, draw a vertical section through the wall and floor showing all the details from the bottom of the foundation to the top of the concrete cill.

10. Higher Level, 2002, Question 2
    Poor design detailing can result in condensation occurring on the inner surfaces of external cavity walls, particularly at (i) wall plate level, and (ii) the wall surrounding window and door openings.
    (a) Discuss how condensation might occur at locations listed above and using notes and freehand sketches show the correct design details that would prevent the condensation occurring at (i) and (ii) above.
    (b) Condensation may also occur on the internal surfaces of the walls of an old house. Discuss two possible reasons for its occurrence and using notes and freehand sketches, show two means by which its occurrence might be eliminated.

11. Higher Level, 2004, Question 7
    A dwelling house has a standard 300 mm concrete block external wall with insulated cavity. Poor design detailing can result in the penetration of moisture to the inner leaf of the wall.

(a) Using notes and neat freehand sketches, show three locations where moisture may penetrate.

(b) For each location selected show, using notes and neat freehand sketches, the correct design detailing that would ensure that moisture does not reach the inner leaf.

(c) List two materials used to prevent the penetration of dampness in buildings. In the case of each material listed, state a location where it may be used and explain why the material is particularly suited for use in the location outlined.

12. Higher Level, 2005, Question 9

Timber-frame construction is now widely used for domestic dwellings in Ireland.

(a) To a scale of 1:10, draw a vertical section through the external wall and ground floor of a house of timber-frame construction. The top of a window cill is positioned 900 mm above floor level, the external leaf is of standard concrete block construction with a rendered finish and the ground floor is a solid concrete floor with 20 mm quarry-tile finish. Show all the constructional details from the bottom of the foundation to the top of the concrete cill.

(b) Discuss in detail two advantages of timber-frame construction and two advantages of standard concrete block-wall construction and recommend a preferred wall type for a new house.

13. Higher Level, 2006, Question 7

Note: Make a neat freehand sketch of the solution.

(a) To a scale of 1:10, draw a vertical section through the window, the external wall and the roof of a timber-framed house. The external leaf is of concrete block construction with a rendered finish. The roof has prefabricated trussed rafters, is slated and has a pitch of 45 degrees. Show all the constructional details from 300 mm below the window head, through the eaves and include three courses of slate.

(b) On the drawing, label and indicate the typical dimensions of four main structural members.

Functions

The primary functions of a roof are to protect a building from the weather and to retain the heat generated inside.

Design Features

Roofs must be designed to meet a number of performance requirements, including:

- **Strength and Stability** – The structural stability of a roof is tested every day. The loads exerted by the weight of roof tiles, the wind blowing against it and the additional weight of rain or snow, are considerable. If these forces are to be safely carried the roof must be well designed and constructed and the materials used must meet the required Irish standards. For example all roof trusses should bear the National Standards Authority of Ireland (NSAI) stamp indicating that the manufacturer is an approved roof truss manufacturer. Also, in order to avoid decay, all structural timber must have a moisture content of less than twenty per cent.
- **Weather Resistance** – Every roof in the world is sloped to some extent. Even so-called flat roofs are slightly sloped. The reason we slope roofs is to dispel rainwater. The roof also acts to reduce overheating of a building by absorbing heat energy from the sun. It is important that the materials used to construct a roof are able to respond to the extreme changes in temperature experienced each year.
- **Thermal Insulation** – To prevent heat loss the attic space of most roofs is insulated with blanket (quilted) insulation. This insulation is placed in two layers. The first layer is placed between the joists and the second layer is placed perpendicularly across the first layer. A typical glass-fibre wool (which has a conductivity value of 0·040 W/mK) should be laid to an overall depth of 250 mm to meet current regulations. This ensures that the heat generated inside is not lost. When air is heated it rises – this means that there is a natural tendency for much of the heat generated in a home to be lost through the roof.
- **Aesthetics** – Aesthetically, a roof has a significant visual impact. Features that influence the appearance of a roof include:
  - Roof Tile – The colour and texture of the chosen tile or slate, correctly installed so that each tile or slate sits down properly.
  - Slope of the Roof – For example, a roof that is too steeply sloped (or overhangs too far) will look out of character and *dominate* the house thereby ruining the visual balance of the house
  - Shape and Form – A simple uncluttered design is generally best. For example, asymmetrical roof designs, too many dormers, or dormers that are out of scale to the rest, do not look good.

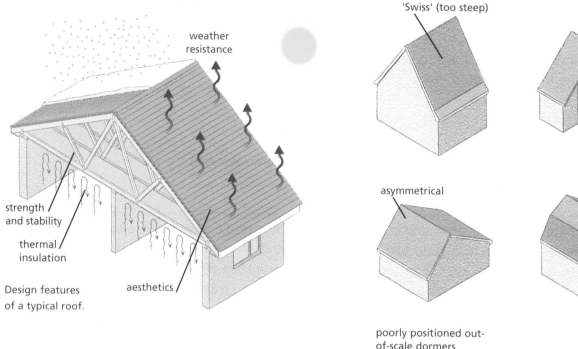

weather
resistance

strength
and stability

thermal
insulation

aesthetics

Design features
of a typical roof.

'Swiss' (too steep)

eroded

asymmetrical

Dutch/mansard

poorly positioned out-
of-scale dormers

too many
dormers

Roofs: the shape and form of a roof can
have a significant impact on the overall
appearance of a house. Each of these is
an example of inappropriate roof design.

Roof Forms

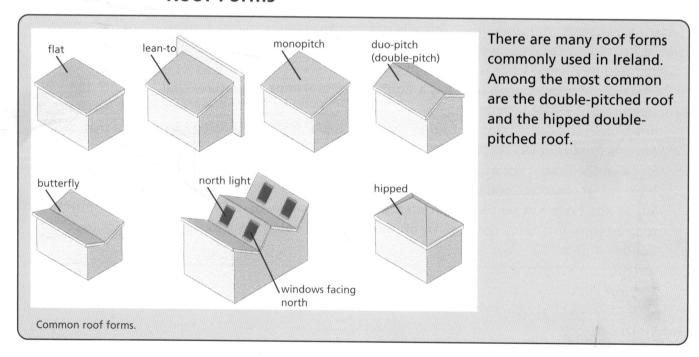

flat

lean-to

monopitch

duo-pitch
(double-pitch)

There are many roof forms
commonly used in Ireland.
Among the most common
are the double-pitched roof
and the hipped double-
pitched roof.

butterfly

north light

hipped

windows facing
north

Common roof forms.

Roof Terminology

Construction of Roofs

There are two approaches to the construction of roofs commonly used in Ireland. The traditional cut roof is used particularly for one-off houses in rural settings, while trussed roof construction is used for housing schemes where many houses of similar type are being constructed.

Construction of Traditional Cut Roofs

Each element of the traditional cut roof is individually cut and assembled on-site using hand and power tools. This is a slow method but can be economical especially for remote rural sites where the hiring of cranes is expensive and access to the site may be limited. Usually, it is uneconomical to get a trussed roof made for a single house. An advantage of the traditional cut roof is that the attic space is generally more open and accessible than in trussed roofs.

The traditional cut roof consists of sloped rafters which meet at the ridge board. The size of the timbers used depends on the span of the roof, the strength class of the timber and the spacing of the rafters. For example a rafter (with intermediate purlin support) for a 4 m span at 300 mm spacing, could be 44 × 125 mm (strength class C18) or 44 x 150 mm (strength class C16).

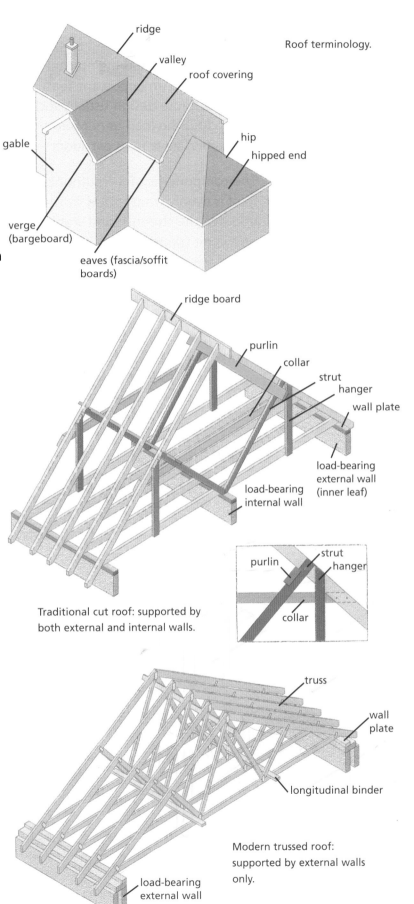

Roof terminology.

Traditional cut roof: supported by both external and internal walls.

Modern trussed roof: supported by external walls only.

The rafters are supported by purlins which are in turn supported by struts. The struts are supported by internal load-bearing walls (see chapter 12, Internal Walls). This is an important feature of traditional cut roofs because it means that the house design usually includes internal walls that are capable of supporting the load generated by the roof. This is why many traditional bungalows have long corridors – the corridor walls support the roof.

The components of a traditional cut roof are constructed to ensure that the load is evenly transmitted to the walls below. Collars and hangers are used to improve the strength and stability of the roof. The overall effect is to reduce one large triangle into a series of smaller ones. The triangulation of the roof in this way improves its structural stability.

The roof is secured to the walls using wall plates. A wall plate is a piece of timber that runs along the top of the inner leaf of the external wall and along internal load-bearing walls. The wall plate is secured to the inner leaf using galvanised steel straps. The wall plate must be treated with a preservative to protect it from decay.

The roof is also secured to the gables using galvanised steel straps.

Construction of Trussed Roofs

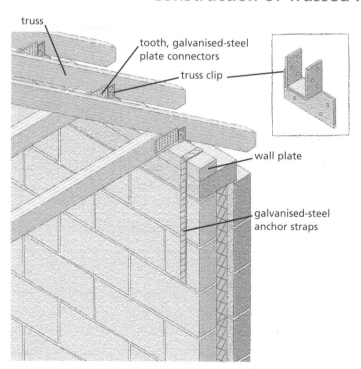

truss

tooth, galvanised-steel plate connectors

truss clip

wall plate

galvanised-steel anchor straps

Trussed roof construction: anchoring of roof to external walls. The wall plate is secured to the wall using a galvanised steel strap. The trusses are secured using galvanised steel truss clips.

Trussed roofs are constructed in factories and assembled on-site. Each truss is delivered to the site ready for installation. Trusses are designed to take account of the roof span, the load to be carried and the spacing between them. There is a variety of patterns for trusses. Commonly used trusses include fan, howe and double W trusses.

The trusses are butt-jointed with toothed galvanised-steel plate connectors. Trusses are usually delivered to site as required, but if they are stored on-site it is important that they are stored laid down on level bearers and are protected from the weather. Prolonged exposure to the elements could lead to rot of the timbers or corrosion of the connector plates.

Each truss is lifted into position using a crane and secured to the wall plate using galvanised steel straps.

The trusses are then braced to hold them in an upright position. The bracing is placed lengthways and diagonally, on the underside of the rafters, to ensure stability.

A gable ladder is used where an overhang is required at a gable end. The gable ladder must be supported by the gable blockwork.

If a hipped end is required this can either be cut on-site or mono trusses designed for the roof can be used. It is essential to ensure that measures are taken to prevent spreading of the roof at the corners. This can be done by using a half-lap joint to join the wall plate and incorporating an angle tie or a steel strap.

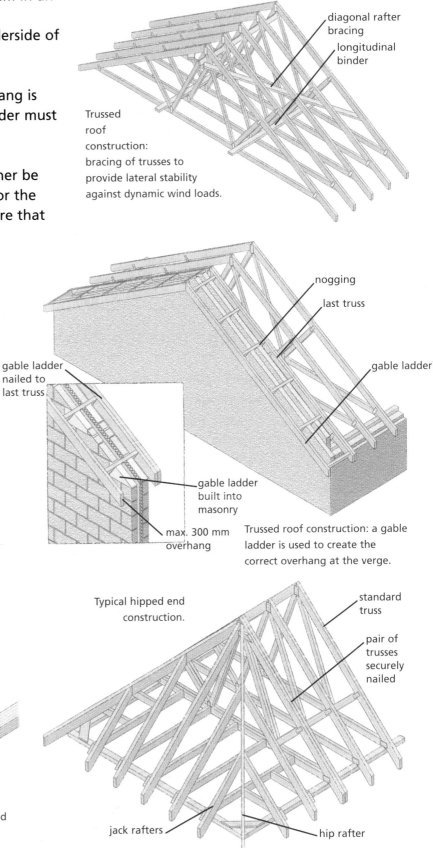

Trussed roof construction: bracing of trusses to provide lateral stability against dynamic wind loads.

Trussed roof construction: a gable ladder is used to create the correct overhang at the verge.

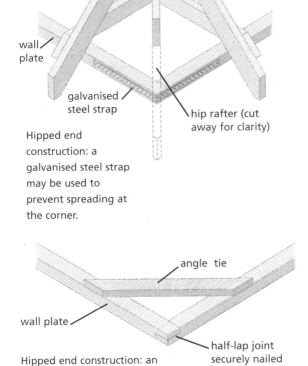

Hipped end construction: a galvanised steel strap may be used to prevent spreading at the corner.

Hipped end construction: an angle tie may also be used to strengthen the corner.

Typical hipped end construction.

Lean-to Roof Construction

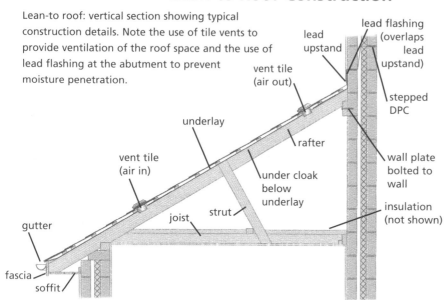

Lean-to roof: vertical section showing typical construction details. Note the use of tile vents to provide ventilation of the roof space and the use of lead flashing at the abutment to prevent moisture penetration.

Labels on diagram:
lead flashing (overlaps lead upstand)
lead upstand
vent tile (air out)
stepped DPC
underlay
rafter
vent tile (air in)
wall plate bolted to wall
under cloak below underlay
insulation (not shown)
joist
strut
gutter
fascia
soffit

Lean-to roofs are commonly used for house extensions and on a smaller scale for porticos and porches. There are two design details where lean-to roof construction differs from standard roof construction: the abutment of the roof to the house and ventilation. It is essential that the abutment is properly sealed to prevent moisture penetration. Adequate ventilation must also be provided to prevent decay of the timbers.

Living Spaces in Roofs

There is a long tradition in Ireland of using the roof (attic) space to increase the amount of living space in a house. The arrangement of the structural members in traditional cut roofs lends these roofs to conversion for use as living space. It is important to respect the character of traditional houses when converting a roof. There are a number of ways the character of an existing house can be retained when reconstructing the roof to create living space, including:

- pitch of new roof to be the same as the original in order to retain proportion,
- reuse original slates and place the best of the original slates on the principal elevation (e.g. front),
- maintain the original lap,
- source similar slates for the rear or if possible purchase new slates from the original quarry,
- retain and reuse original rainwater goods,
- retain the original fascia and soffit boards – new boards to have the same profile as the originals.

When converting the roof space of vernacular houses, the use of roof-lights as opposed to dormer windows is less obtrusive and will have less visual impact. The roof-lights should be at least ten per cent of the floor area of the living space. The construction of roof-lights is examined in detail in chapter 14, Windows (see pivot windows).

When use of the roof space is planned for a new house, structural steel is usually incorporated to ensure safe transmission of the loads. These roofs are commonly referred to as *dormer roofs* because of the use of dormer windows. Dormer windows should be in keeping with the scale and style of the house and should respect local traditions when used in houses in a rural setting.

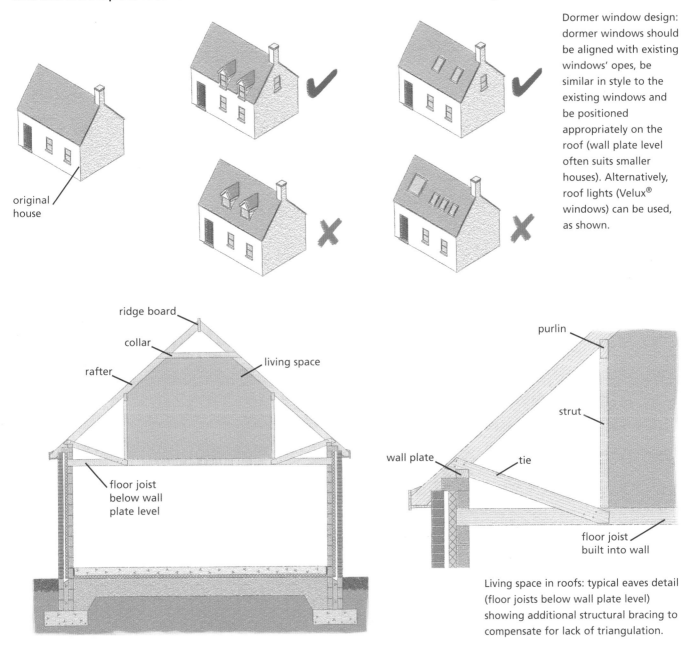

original house

Dormer window design: dormer windows should be aligned with existing windows' opes, be similar in style to the existing windows and be positioned appropriately on the roof (wall plate level often suits smaller houses). Alternatively, roof lights (Velux® windows) can be used, as shown.

ridge board

collar

rafter

living space

floor joist below wall plate level

Living space in roofs: vertical section of roof showing typical construction details when the floor joists are below wall plate level.

purlin

strut

wall plate

tie

floor joist built into wall

Living space in roofs: typical eaves detail (floor joists below wall plate level) showing additional structural bracing to compensate for lack of triangulation.

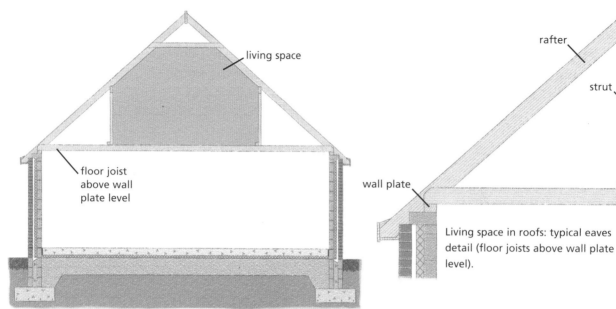

living space

floor joist
above wall
plate level

rafter

strut

wall plate

joist

Living space in roofs: typical eaves
detail (floor joists above wall plate
level).

Living space in roofs: vertical section of roof showing typical
construction details when the floor joists are above wall plate level.

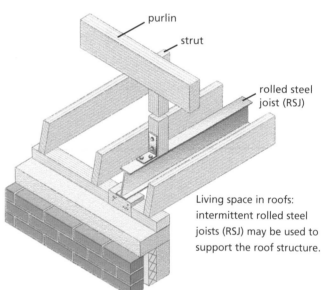

purlin

strut

rolled steel
joist (RSJ)

Living space in roofs:
intermittent rolled steel
joists (RSJ) may be used to
support the roof structure.

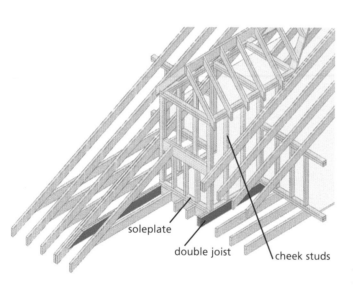

soleplate

double joist

cheek studs

Dormer window construction: pictorial
view of typical components. The joists
under the dormer cheek studs have
been doubled to support the extra
load. (Rafters and joists to the right of
the window have been shortened for
clarity.)

Living space in roofs:
alternatively, engineered
timber joists may also be
used to support the roof
structure.

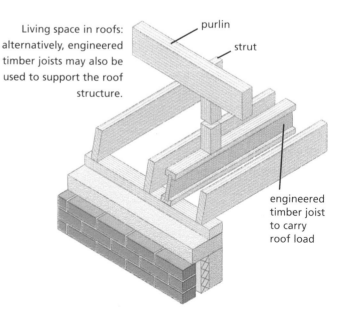

purlin

strut

engineered
timber joist
to carry
roof load

Roof Coverings

Once the roof structure is
complete the covering can
be fixed. The most
commonly used roof
coverings are tiles and
slates. Tiles are generally
made from concrete or
clay and come in a variety
of patterns, textures and
colours. Natural slate is
widely used in rural areas
where it blends well with
the landscape and other
houses. Fibre cement
slates are thin and smooth
and are available in a
variety of colours. The
material chosen will
depend on a number of factors including, location, cost and aesthetics.

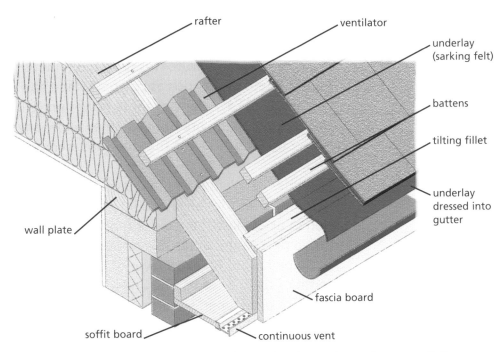

Roof coverings: pictorial view of eaves detail of a slated roof. Note the continuous ventilation strip in the soffit and the ventilator installed to ensure the air can flow past the insulation.

The roof covering must be securely installed to prevent the loss of tiles or
slates during windy weather. This happens because as the wind blows up the
slope of the roof it creates lift (just like the wing of a plane). This lifting force
will quickly dislodge any tiles that are not securely held.

Eaves
Before the roof covering can be installed the eaves should be completed. This
is done by securing a fascia board to the rafter ends and a soffit board to the
underside. A tilting fillet or continuous underlay support (plywood) is then
installed to ensure that rainwater does not cause the underlay to sag behind
the soffit board.

The gutter should then be secured to the fascia and the underlay dressed 50
mm into the gutter. Traditionally the fascia and soffit were constructed in
timber and gutters were of cast iron, although aluminium and uPVC products
are now widely used.

Underlay
The first step in roof covering is the installation of the roofing underlay. The
primary role of the underlay is to exclude any rainwater that penetrates the

roof tiles or slates. Bituminous sarking felt is traditionally the most commonly used roofing underlay. It consists of woven fibres (that make it durable and strong) covered in a water-resisting bituminous material. Lighter, more user-friendly alternatives, such as polypropylene and polyethylene fabrics, are now in widespread use. Many of these newer underlay products have the advantage of being breathable – they allow any moisture in the roof space to escape, thereby ensuring that the roof members remain dry.

The underlay is laid along the eaves of the roof first with the lower edge extending over the fascia and into the gutter. Subsequent rows should overlap by at least 150 mm for most roofs. At ridge level the underlay is carried over to the other side to overlap with the underlay laid on that side. For hipped roofs, an extra layer of underlay should be installed along the length of the hip rafter. The underlay should be positioned to ensure that all joints are overlapped to carry water safely into the gutter. The underlay is tautly secured to the rafters with galvanised nails so that it does not droop down between the rafters.

Battens

Once the underlay is secure, the battens can be installed. The function of the battens is to support the tiles or slates. The size and spacing of the battens will vary depending on the type of tiles or slates being used. It is essential that the battens are evenly spaced and installed according to the tile manufacturer's recommendations. The distance between the battens is called the *gauge*. Where joints are required they should be made over rafters. Battens should not be cantilevered or spliced between rafters. Battens vary in size according to the tiles and slates used, for example, typical concrete tiles require a 36 × 50 mm batten.

Tiles and Slates

Tiles and slates are made from a variety of materials, including:

- concrete,
- clay,
- fibre cement,
- natural slate.

Most tiles and slates can be placed into one of four commonly used categories:

- Clay and concrete tiles – can be either profiled (e.g. curvy) or flat (plain). Profiles vary widely but there are a number of common profiles that most manufacturers produce (e.g. Roman, Spanish roll).
- Fibre cement slates – flat and very smooth and are manufactured from a mixture of Portland cement, synthetic fibres, pigments and additives.
- Concrete slates – flat textured-surface finish (and colour) that mimic the character of natural slate.
- Natural slates – these are quarried from natural sources in countries such as Spain, South America and China.

Profiled concrete tiles.

Plain concrete tiles.

Fibre cement slates.

Concrete slates.

Natural slates.

Installation of Concrete or Clay Tiles

The procedure for installing concrete and clay tiles is very similar. The tiles are installed in courses (rows) from the eaves upwards. Tiles usually have two nibs that hook over the batten. Tiles are held securely in place by nailing through the holes provided or by using special clips provided by the manufacturer. In normal conditions, only every second course and the perimeter tiles are nailed to the battens. In windy areas every tile is nailed or clipped according to the manufacturer's instructions.

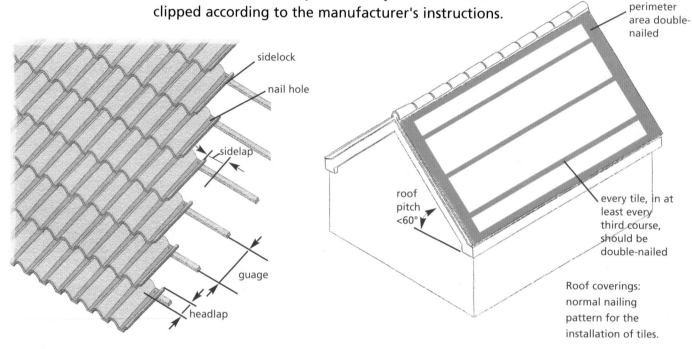

sidelock

nail hole

sidelap

guage

headlap

Roof coverings: installation of tiles.

perimeter area double-nailed

roof pitch <60°

every tile, in at least every third course, should be double-nailed

Roof coverings: normal nailing pattern for the installation of tiles.

A typical tile has an interlocking edge commonly called a *sidelock*. This interlocking feature ensures that all the tiles act as one unit, providing a secure stable roof covering. As each course of tiles is laid a vertical overlap, known as the *headlap*, is created. This is to ensure the rain is carried from one tile onto the next

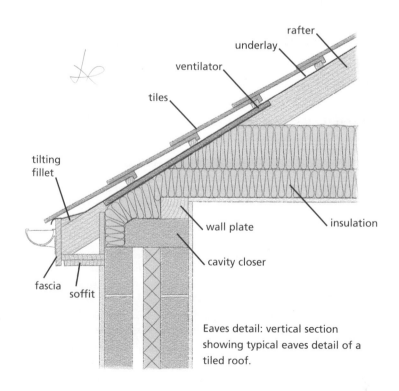

rafter

underlay

ventilator

tiles

tilting fillet

fascia

soffit

wall plate

cavity closer

insulation

Eaves detail: vertical section showing typical eaves detail of a tiled roof.

and safely into the gutter. The amount of headlap required varies according to the pitch of the roof and the exposure (wind) conditions, but for most roofs it is around 100 mm.

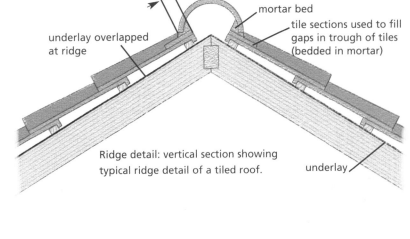

headlap 75 min.

The ridge tile of a tiled roof is usually bedded in mortar. Various shapes of tiles are used, although the half-round is very common. The mortar used to secure the ridge tiles is dyed to match the colour of the tiles. The ridge tile should have a headlap of at least 75 mm.

underlay overlapped at ridge

mortar bed

tile sections used to fill gaps in trough of tiles (bedded in mortar)

Ridge detail: vertical section showing typical ridge detail of a tiled roof.

underlay

The verge is an important area of the roof because it is from here that tiles are often dislodged during windy weather. Verge tiles are held in place using mortar in a similar way to the ridge tiles. It is essential that the mortar is supported until it has cured. This is achieved by providing an undercloak. A plain tile or slate 100–150 mm wide can be used to form the undercloak. The practice of nailing a batten to the bargeboard until the pointing has set is not

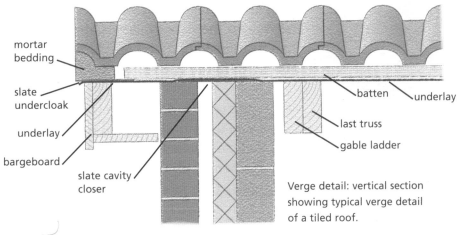

mortar bedding

slate undercloak

underlay

bargeboard

slate cavity closer

batten   underlay

last truss

gable ladder

Verge detail: vertical section showing typical verge detail of a tiled roof.

recommended. Some manufacturers provide dry verge (clipping) systems.

If the roof shape includes a valley it is essential that the penetration of water is prevented along the length of the valley. To achieve this a continuous polythene DPC must be installed along the length of the valley. This is then covered by a

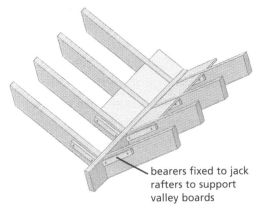

bearers fixed to jack rafters to support valley boards

Valley detail: showing installation of continuous valley support board in a trussed roof. The rafters of trusses should not be cut to accommodate valley support boards.

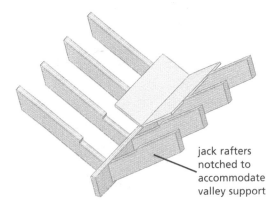

jack rafters notched to accommodate valley support

Valley detail: showing installation of continuous valley support board in a traditional cut roof. The jack rafters should be oversized to compensate for the notching.

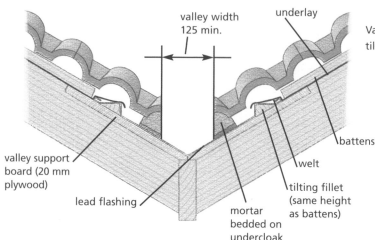

valley width 125 min.

underlay

Valley detail: vertical section showing typical valley details of a tiled roof.

battens

welt

tilting fillet (same height as battens)

valley support board (20 mm plywood)

lead flashing

mortar bedded on undercloak

lead (or plastic composite) lining which is overlapped by the tiles. A valley support board (plywood) must be installed to support the DPC and lead lining.

Installation of Fibre Cement, Concrete and Natural Slates

The procedure for installing fibre cement, concrete and natural slates is quite similar. Slates do not have nibs to hook onto the battens. Instead, they are installed by nailing each slate through two holes provided approximately halfway up the slate. Nailing slates at their centre holds them more securely than head nailing, thereby reducing the likelihood of breakage (due to wind vibration) during severe weather.

While fibre cement and concrete slates are supplied with nail holes, nail holes are created in natural slates on-site. The holes are always made on the bed (underside) of the slate and positioned so that the thinner end of the slate is at the top when fixed. The holes are punched out so that a small countersunk depression in the face of the slate is created – this accommodates the head of the fixing nail. Two holes are created in each slate. The position of the holes is approximately 20–25 mm in from the sides of the slate. The gauge distance varies according to the batten gauge and lap used. Some natural slates are supplied pre-holed to achieve a particular headlap. Natural slates are usually secured using copper nails, although some suppliers use a hook system where each slate is fixed with a stainless steel hook. This fixing system is more economical than the traditional method because of the greater speed and ease of application. It is used in conjunction with nailing in exposed areas.

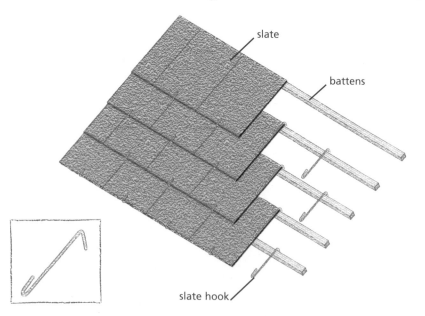

slate

battens

slate hook

Roof coverings: hook system used for the installation of slates.

Natural slate is supplied in a variety of thicknesses, from 5 mm up to approximately 20 mm. When the slates are delivered to the site they must be sorted through prior to fixing, to ensure that the slates on each course are of the same thickness and to prevent *kicking* slates or unsightly gaps between subsequent courses. Slates are removed from pallets and sorted into three or four grades of thickness. The thickest grade is used for the courses nearest the eaves and the thinnest grade at the ridge. On large roofs, the quantity of slates required for any one slope is sorted before the fixing starts on that slope.

Slates are laid in courses beginning at the eaves. One (or two) courses of undereaves slates are laid first. An undereaves slate is a shortened slate laid to provide a lift to the tail of the first course. This is done to ensure that the slates sit at the correct angle and do not droop downward toward the gutter. The first course of slates is usually raised slightly (10–25 mm depending on slate used) above the plane of the battens.

Slates usually have a 100 mm double headlap. To achieve this, slates are longer than tiles (e.g. typical tile length 300–400 mm, typical slate length 500–600 mm). The sidelap should occur at the midpoint of the slate immediately below.

The ridge of a natural slate roof is commonly finished using a clay angular ridge slate bedded in mortar. Alternatively, a lead ridge can be installed. This is done by nailing a timber round to the upper edge of the ridgeboard and dressing a 450 mm wide lead sheet over the round and onto the slates at both sides. The lead is held in place using lead tacks to prevent lifting during windy weather. These are 50 mm wide strips of lead that are nailed to the timber roll underneath the ridge sheet and turned up over the ridge sheet to anchor it.

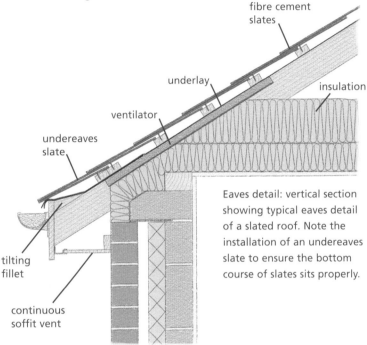

Eaves detail: vertical section showing typical eaves detail of a slated roof. Note the installation of an undereaves slate to ensure the bottom course of slates sits properly.

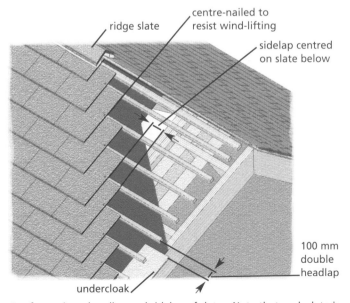

Roof coverings: headlap and sidelap of slates. Note that each slate is centre-nailed to reduce the movement/ rattling caused by strong winds. Only ridge and undereaves slates are head-nailed.

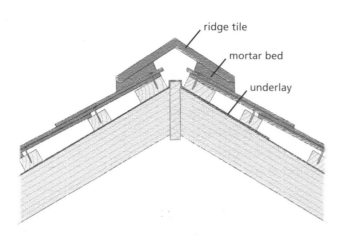

Ridge detail: vertical section showing typical ridge detail of a slated roof when a proprietary ridge tile is used.

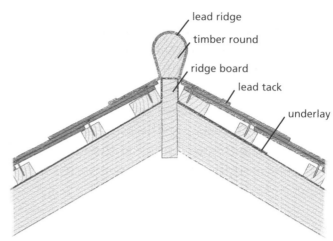

Ridge detail: vertical section showing typical ridge detail of a slated roof when a lead ridge is installed.

The ridge of fibre cement and concrete slates is normally finished using matching angular ridge tiles supplied by the manufacturer. These are often fixed in place using screws, washers and caps provided by the

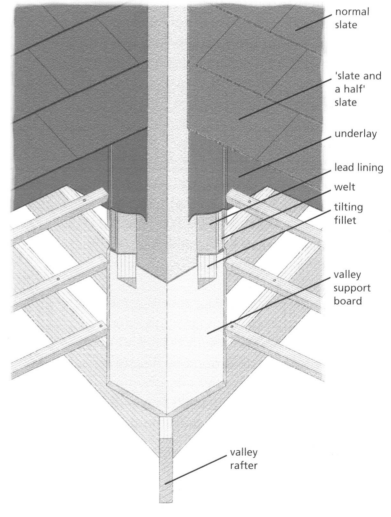

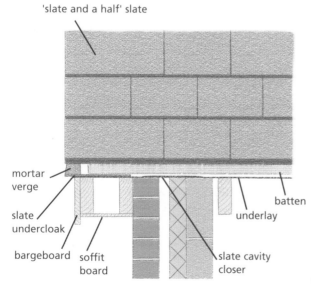

Verge detail: vertical section showing typical bedded verge detail of a slated roof.

Valley detail: pictorial view showing typical valley details of the valley of a slated roof. The lead lining is dressed over the tilting fillet and bent back to form a welt – this prevents the penetration of rainwater.

manufacturer to ensure a watertight finish. Alternatively they can be bedded in mortar.

The verge of a slated roof is finished in a similar way to the verge of a tiled roof. An undercloak is provided to support the pointing. The slates are allowed to overhang by a maximum of 50 mm.

A valley is constructed in a very similar way to that described above.

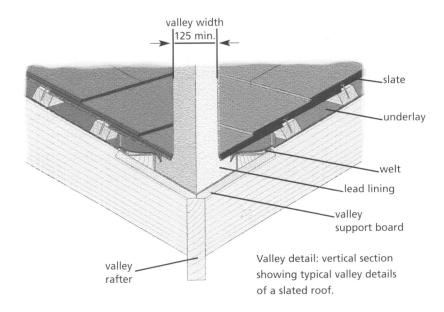

Valley detail: vertical section showing typical valley details of a slated roof.

Roof Ventilation

It is essential that the attic space of every roof is properly ventilated. This is to ensure that roof timbers remain dry and free from rot. Traditionally, the method used to ventilate a roof was to provide a continuous gap at eaves level. Where aluminium or uPVC products are used, vents are provided at regular intervals. In either case the ventilation provided for a typical house should be the equivalent of a continuous 10 mm gap between the fascia and soffit boards.

Roof Ventilation: eaves section showing the use of a soffit vent or a vent tile to allow air circulation.

An alternative method is to use a vent tile or slate. This is a special product designed to match the appearance of a particular tile or slate. It is mainly used where there is insufficient overhang available to ventilate at eaves level.

It is very important that the air can move freely past the blanket insulation in the attic. To facilitate this an eaves ventilator is installed.

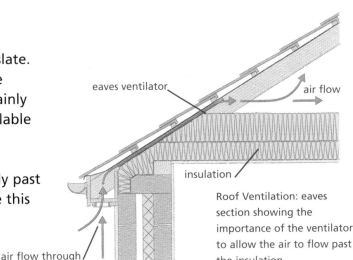

Roof Ventilation: eaves section showing the importance of the ventilator to allow the air to flow past the insulation.

Flat Roof Construction

Flat roofs: damage due to uneven movement of the covering caused by ponding coupled with changes in temperature.

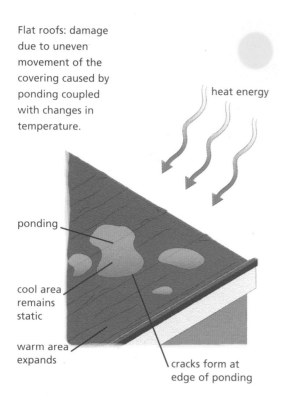

heat energy

ponding

cool area remains static

warm area expands

cracks form at edge of ponding

Flat roofs perform poorly in the Irish climate and should not be used if at all possible. Flat roofs tend to suffer from ponding and cracking of the surface. If the roof is not laid to a sufficient fall, water will gather on it forming small *puddles*. This is usually referred to as *ponding* and can lead to leaks developing in the roof.

Leaks are usually caused by the following mechanism. When the weather improves the sun will warm up the roof. The roof covering, usually a bituminous product, softens and expands when heated. However, the area of the roof under a pool of water is kept cool by the accumulated water (ponding) and does not heat up at the same rate as the exposed areas of the roof. This leads to differential expansion and contraction of the roof covering and hence to cracking. The water which has been sitting on the roof is now likely to seep through any tiny cracks that might have developed. This mechanism can take a number of years to cause damage but eventually it will.

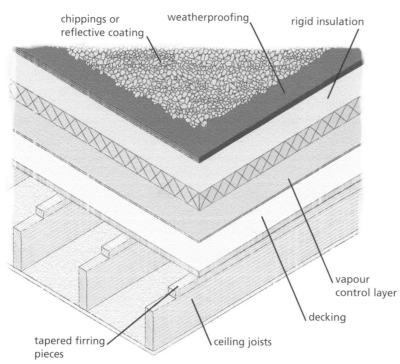

chippings or reflective coating

weatherproofing

rigid insulation

tapered firring pieces

ceiling joists

decking

vapour control layer

Flat roof construction: warm deck flat roof. The weather-proofing is installed above the rigid insulation and is protected by the reflective coating and chippings.

Also, flat roofs require a high level of specialist skill to install. Most weatherproofing used on flat roofs is of the *torch-on* variety. This involves heating the underside of the product and sticking it to the surface of the roof. Installing these products correctly is difficult and poor workmanship can lead to bubbling of the surface and to poor seals at joints. An overlap of at least 150 mm is normally required at joints. Approved weatherproofing systems are usually a polyester-reinforced bitumen product consisting of a number of layers. The top surface of some products is prefinished with chippings.

If a flat roof is to be constructed, the *warm deck* type should be chosen. The warm deck, flat roof design ensures the decking material is on the warm side of the insulation (underneath) thereby reducing the exposure of the decking to the expansion and contraction that is associated with changes of temperature. Positioning the decking in the warm zone means that it will have a better chance of staying dry in the event of a leak. The decking should be a suitable grade of plywood or oriented strand board. Particleboard (chipboard) is not a suitable decking material. Warm deck, flat roofs do not have to be ventilated. Flat roofs should be laid at a fall of 1 in 40 – this is achieved by installing a tapered firring piece along the top edge of the joists.

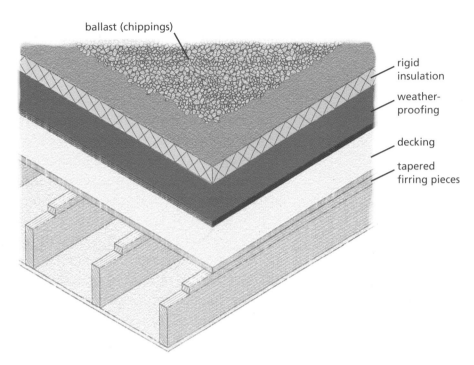

ballast (chippings)

rigid insulation

weather-proofing

decking

tapered firring pieces

Flat roof construction: warm deck, flat roof construction. The weather-proofing is installed below the rigid insulation. The insulation is held in place and protected by the chippings. This method offers maximum protection to the weatherproofing by keeping it at a consistent temperature.

Revision Exercises

1. Explain the functions of the roof of a domestic dwelling.
2. Describe, using a neat annotated sketch, the main components of a hipped roof.
3. Describe, with the aid of neat labelled freehand sketches, the main components of a traditional cut roof. Label each part and state its function.
4. Describe, with the aid of neat labelled freehand sketches, how a roof structure is secured to the walls.
5. Generate a neat labelled cross-section sketch of the eaves detail of a typical natural slate pitched roof.
6. Generate a neat cross-section sketch of the ridge detail of a typical concrete tile pitched roof.
7. Generate a neat cross-section sketch of the verge detail of a fibre cement slate pitched roof.
8. Generate a neat cross-section sketch of the valley detail of a concrete tiled pitched roof.

9. Indicate how ventilation of the attic space of a double-pitch roof is achieved. State three reasons why it is important that the attic space is ventilated.

10. Outline the difficulties often associated with flat roof construction.

11. Generate a neat labelled cross-section sketch of a warm deck, flat roof.

12. Higher Level, 2002, Question 8

    It is proposed to extend the kitchen area of an existing two-storey house. This requires the construction of a timber flat roof to the extension. The external wall of the house is a 300 mm insulated cavity wall.

    (a) To a scale of 1:5, show the design details of the roof construction at:
        (i)   eaves level, showing how the rainwater is to be removed.
        (ii) the abutment of the flat roof with the wall of the existing house.

    (b) Using notes and sketches show two design considerations in the roof construction which prevent the occurrence of:
        (i)   Condensation within the roof structure.
        (ii) Decay of the roof timbers.

13. Higher Level, 2004, Question 1

    A small porch, which projects 1·7m from a house, is shown in the accompanying sketch. The lean-to roof is slated and has a pitch of 30 degrees. The house and porch are constructed of standard 300 mm concrete block walls with an insulated cavity. The porch has a level plasterboard ceiling.

    (a) To a scale of 1:5 draw a vertical section through the porch showing the roof and wall of the house. The section should show all the construction details from 400 mm below the bottom of the ceiling joists to 300 mm above the abutment of the roof and wall of the house.

    (b) Indicate on your drawing two design details that ensure moisture does not penetrate at the abutment of the roof and wall of the house.

14. Higher Level, 2004, Question 8

The sketch shows a new dwelling house with a slated roof pitched at 45 degrees. The roof is a traditional cut roof and is designed to incorporate bedroom accommodation within the attic space. The house has an internal width of 7·0 metres. The external walls supporting the flooring joists are standard 300 mm concrete block walls with insulated cavity. The joists are also supported internally on a centrally located load-bearing concrete block wall.

(a) To a scale of 1:20 draw a vertical section through the roof structure. Show the constructional details from the bottom of the wall plates to the top of the ridge board. (It is not necessary to show slating or window details.)

(b) To provide natural light to the bedrooms in the attic space, a choice must be made to fit either pitched dormer windows or roof-light windows.

State two arguments in favour of fitting dormer windows and two arguments in favour of fitting roof-light windows.

15. Higher Level, 2005, Question 3

A small rural dwelling house in the vernacular tradition, built in the 1950s, is shown in the accompanying sketch. As part of a general restoration of the house it has been decided to renew the roof of the house and to incorporate bedroom accommodation within the attic space. A survey of the house reveals:

• traditional cut roof with original natural slate,
• softwood fascia and soffit,
• external uninsulated cavity walls of concrete block construction,
• solid block internal walls.

(a) Using notes and detailed freehand sketches, show the construction details of the roof structure to facilitate bedroom accommodation within the attic space. Indicate sizes for all roofing components. Show details of the insulation requirements for both walls and roof.

(b) Using notes and freehand sketches, outline one method of providing natural light to the bedrooms in the attic space in a manner that will respect the character of the original house.

Discuss two advantages of your preferred method of providing natural light to the bedrooms.

16. Higher Level, 2006, Question 3

It is proposed to provide bedroom accommodation in the attic space of a new house. The house has an internal width of 6.5 metres and the flooring joists are supported internally on a centrally located load-bearing wall. The roof is a traditional cut roof, is slated and has a pitch of 45°.

(a) Using *notes and detailed freehand sketches*, show the constructional details of the roof structure to facilitate bedroom accommodation in the attic space. Indicate clearly the ventilation *and* insulation detailing of the roof structure.

(b) The accompanying sketch shows a terrace of town houses. The dormer windows have been developed in an uncoordinated manner over a number of years. Using *notes and freehand sketches*, suggest a revised design for the dormer windows that would improve the visual appearance of the houses and enhance the character of the terrace.

Functions

The primary functions of a floor are to support the imposed loads and to provide a level surface for the activities that are carried out in the home.

Design Features

Floors must be designed to meet a number of performance requirements including:

- Strength and Stability – Floors experience both dead and live loading throughout the lifetime of the building. The live loads can vary significantly depending on the number of people living in a home and the activities occurring there (e.g. a simple activity, such as taking a bath can increase the load on the bathroom floor by up to 150 kg).
- Durability – The floors of every house must be hard-wearing if they are to withstand the demands of everyday use.
- Moisture Resistance – Ground floors must be designed to withstand moisture rising from the ground below. If moisture rises up through the floor, timber floorboards will cup and natural carpets will rot.
- Thermal Insulation – Ground floors must protect against the loss of heat through the floor. Generally, we think of heat as rising, however, all you have to do is to lie down on the ground during cold weather to realise how it draws heat out of your body. Uninsulated or poorly insulated floors will lead to cold homes and higher heating costs.

Ground Floors

Solid Floating Concrete Ground Floor

The solid floating concrete ground floor is the most common type of ground floor used in the construction of houses today. This floor consists of a reinforced concrete slab resting (floating) on rigid insulation which is in turn supported by compacted fill material. This floor relies on the properties of the reinforced concrete for strength and durability (see chapter 4, Construction Materials). A damp-proof membrane is used to resist the penetration of moisture from below – a layer of rigid insulation provides thermal insulation.

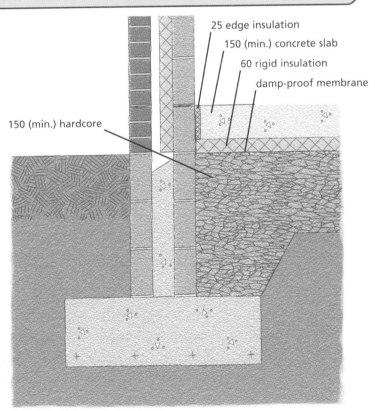

25 edge insulation
150 (min.) concrete slab
60 rigid insulation
damp-proof membrane
150 (min.) hardcore

Solid floating concrete ground floor: vertical section of floor, including strip foundation and typical concrete cavity wall with brick outer leaf.

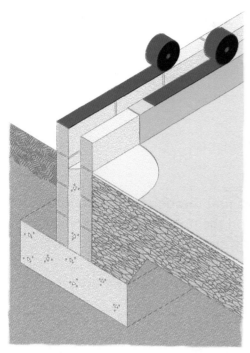

Solid floating concrete ground floor: Damp-proof membrane. It is essential that the membrane is carried up into the blockwork to form a water tight seal over the entire floor area.

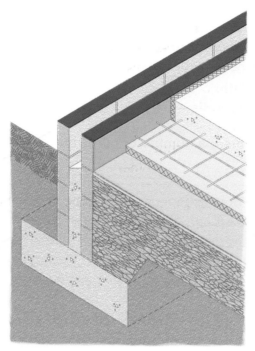

Solid floating concrete ground floor: Cutaway view of floor components. The steel reinforced concrete floats on rigid insulation.

This type of floor is constructed after the rising walls have been built. The area contained by the rising walls is filled with hardcore (see chapter 4, Construction Materials). The hardcore is well compacted in layers not more than 225 mm in depth. The overall depth of hardcore should not exceed 900 mm.

A thin layer of sand blinding is then spread on the hardcore. This is done to protect the damp-proof membrane against puncture. The damp-proof membrane, a 1,200 gauge plastic sheet, is spread out over the entire floor area and is carried up over the inner leaf of the rising wall to underlap the DPC.

The insulation is then laid on the DPM. The thickness of insulation required will depend on the product used, however, the overall elemental U-value of the floor must not exceed 0·25 W/m²K (see chapter 21, Heat and Thermal Insulation). The final step before pouring the concrete is to position the steel reinforcement. A typical steel mesh used consists of 10 mm steel bars at 200–450 mm centres. The steel reinforcement is held at least 50 mm above the surface of the insulation using special plastic supports.

The concrete is poured and compacted to drive out any air bubbles and ensure good contact with the steel reinforcement. It is then allowed to set for a few hours. Once the concrete has begun to harden it is smoothed using a power float.

Slab and Screed Floor

A variation on this type of floor is slab and screed floor. This floor was traditionally used in the construction of houses in Ireland before the floating floor became popular. This floor is constructed in two stages. The first stage involves the placing of the hardcore and the pouring of a concrete slab.

The second stage occurs much later in the construction of the house. This stage involves installing the DPM and insulation followed by the pouring of a floating screed. A screed is simply a cement and sand mix in the ratio of one part cement to three parts dry sand. A floating screed involves placing a layer of insulation between the screed and the slab. A floating screed should be at least 65 mm thick and reinforced with steel mesh.

This slab and screed approach allows for the installation of services (e.g. plumbing) before the screed is poured, saves the finished floor from damage during construction and will smoothen out any irregularities in the slab. However, it is slow, labour intensive and does not allow for an overlap of the DPM in the floor and the DPC in the wall to be achieved.

65 reinforced screed

60 rigid insulation

damp-proof membrane

150 concrete slab

150 (min.) hardcore

Slab and screed floor: vertical section of floor, including strip foundation and typical concrete cavity wall.

Suspended Timber Ground Floor

The suspended timber ground floor is used where the depth of fill (hardcore) exceeds 900 mm. A suspended timber ground floor is used in this situation because concrete slab floors have a tendency to settle or collapse where the fill is deep or poorly compacted. Suspended timber ground floors have some advantages, including:

- they make the installation of services much easier,
- they are aesthetically pleasing when finished, e.g. tongue and groove flooring boards.

However, they are more expensive to install in both material and labour, they require a high level of workmanship if they are to avoid being draughty and they tend to carry noise between rooms (see flanking transmission, chapter 22, Sound and Acoustic Insulation).

A suspended timber ground floor is raised above the concrete slab on small supporting walls called *tassel walls* or *sleeper walls*. The flooring joists rest on a timber wall plate which is strapped to the top of the tassel walls. A wall plate is typically 75 × 100 mm. The wall plate rests on a damp-proof course and is treated with a preservative to

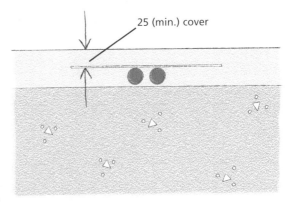

25 (min.) cover

Slab and screed floor: accommodation of services (plumbing/electrical) in the screed. A light steel mesh is placed above the services to reinforce the screed.

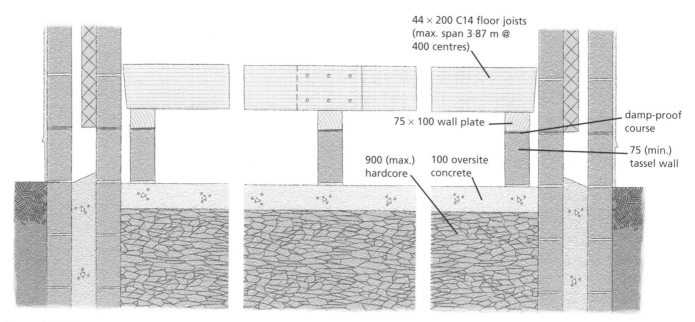

44 × 200 C14 floor joists
(max. span 3·87 m @
400 centres)

75 × 100 wall plate

damp-proof
course

75 (min.)
tassel wall

900 (max.)
hardcore

100 oversite
concrete

Suspended timber ground floor: vertical section of floor supported by tassel walls, including typical concrete cavity external walls. In this example, the floor joists are overlapped and nailed together above a supportive tassel wall. The ends of the joists are cut at a slope to prevent absorption of moisture from the wall. (Insulation between joists omitted for clarity.)

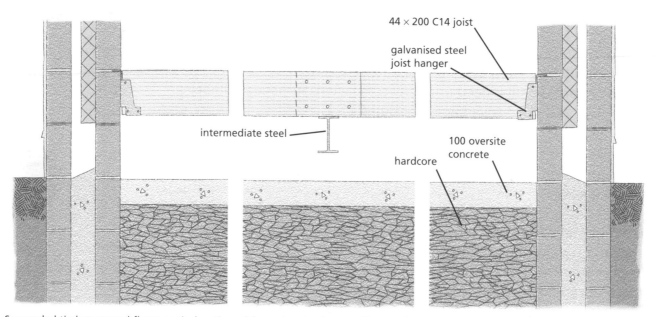

44 × 200 C14 joist

galvanised steel
joist hanger

intermediate steel

hardcore

100 oversite
concrete

Suspended timber ground floor: vertical section of floor, showing intermediate steel joist and galvanised steel joist hangers used to support timber joists when depth of hardcore exceeds 900 mm. (Insulation between joists omitted for clarity.)

prevent decay. The ends of the joists are also treated with a preservative. The size and strength class of the joists used depends on the span of the floor. Alternatively, the floor may be supported using galvanised joist hangers (built into the blockwork of the external walls) and an intermediate steel joist.

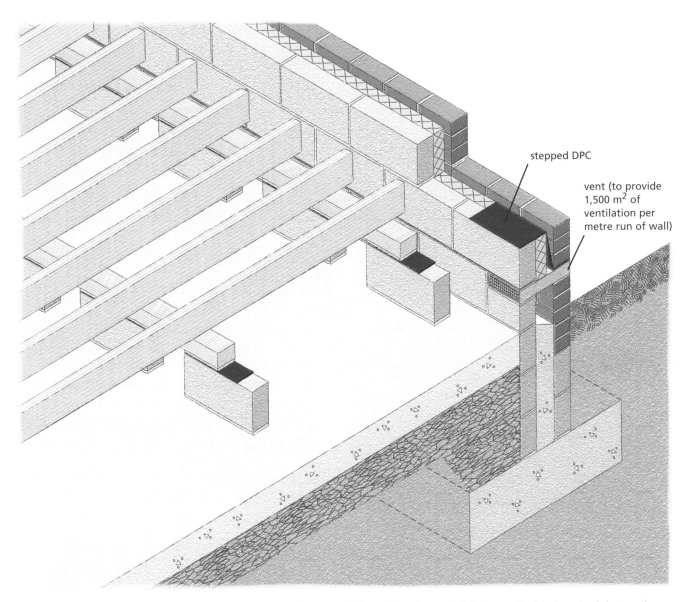

stepped DPC

vent (to provide 1,500 m² of ventilation per metre run of wall)

Suspended timber ground floor: pictorial view of tassel wall supported floor. Note the gap left between the blockwork of the tassel walls to aid air circulation. Note also the stepped damp-proof membrane, installed above the vent to prevent moisture penetration.

Vents are placed in the external walls and the tassel walls are built to allow for the circulation of air below the floor (gaps left in blockwork). This is essential to ensure the circulation of fresh air – this keeps the timbers dry and prevents decay.

Thermal insulation of the floor is achieved by installing either rigid or blanket insulation between the floor joists. The thickness of insulation required will depend on the product used, however, the overall elemental U-value of the floor must not exceed 0·25 W/m²K.

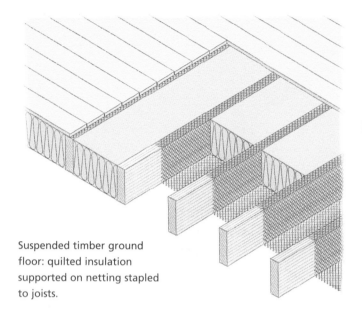

Suspended timber ground
floor: quilted insulation
supported on netting stapled
to joists.

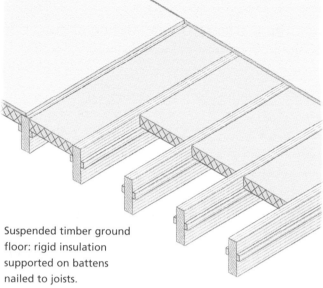

Suspended timber ground
floor: rigid insulation
supported on battens
nailed to joists.

Due to the combustible nature of timber, space for the fireplace must be left when a suspended timber ground floor is constructed. This involves *trimming* the joists around the fireplace.

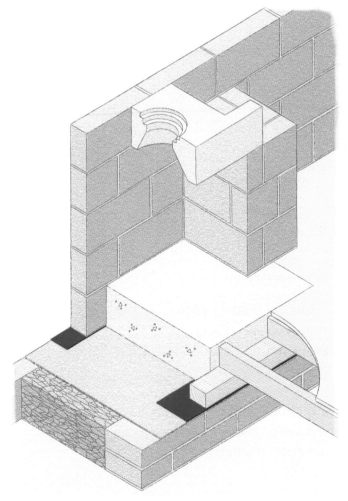

Suspended timber ground floor: cutaway view of floor and fireplace.
Timber joists are trimmed back to create space for the concrete hearth.

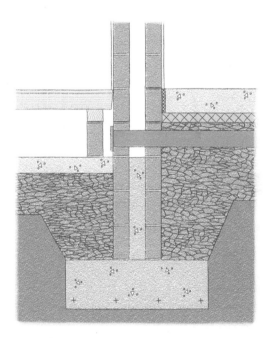

Suspended timber ground floor and solid floating
concrete ground floor: vertical section through the
interface of both floors showing the installation of air
ducting in the solid floor to allow air circulation under
the suspended floor to continue.

When an extension is added to an older house it is essential that steps are taken to ensure that air can continue to circulate freely under the suspended floor. This is achieved by connecting plastic air ducting pipe to the vent positions in the original wall.

Suspended Concrete Ground Floor

Suspended concrete ground floors are being widely used as an alternative to suspended timber ground floors. These floors are usually designed by a qualified engineer. There are two main types of suspended concrete ground floors:

> • Cast on-site – in this floor the slab is supported by the rising walls and does not rely solely on the hardcore for support. The benefit of this is that if any settlement occurs in the hardcore, the floor will not collapse.

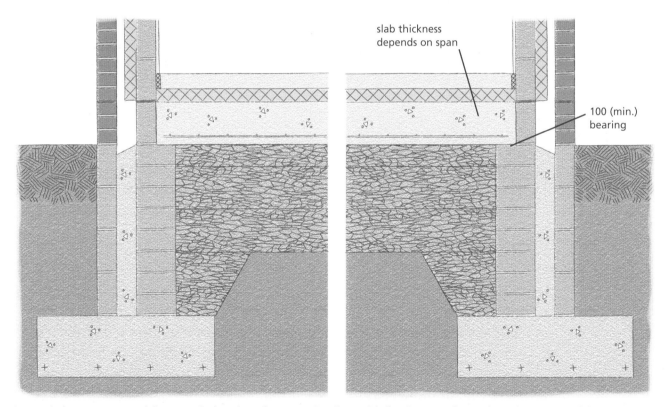

slab thickness depends on span

100 (min.) bearing

Suspended concrete ground floor: vertical section of 'cast on-site' floor with floating screed. Note how the concrete slab is suspended on the rising walls.

> • Precast – there are a number of precast products available. The beam and block type are commonly used in house construction. These usually consist of prestressed concrete beams with infill blocks. The beams are supported by the external rising walls.

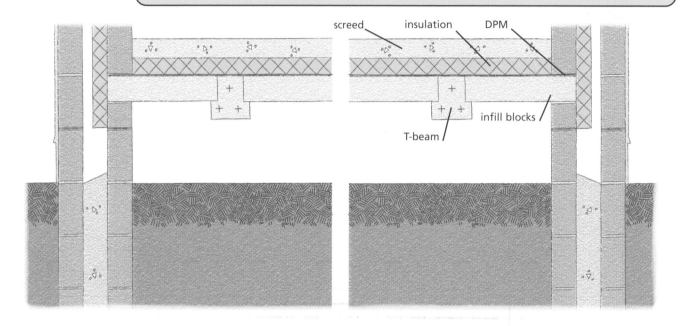

Suspended concrete ground floor: vertical section of typical precast block and beam floor.

Suspended concrete ground floors overcome the disadvantages normally associated with suspended timber floors – they do not transmit noise, they are not draughty and they do not suffer from decay. Precast floors are quick to install but are usually more expensive.

Radon

Radon is a naturally occurring radioactive gas found in soils and rocks all over Ireland. It is colourless, tasteless, odourless and can only be detected using specialist equipment. Radon gas seeps upward through the soil until it reaches the surface. Where this occurs outdoors it is quickly diluted by fresh air and is harmless. However, if it emerges into an enclosed space, such as a house, it can build up to dangerous levels. If the radioactive particles given off by radon gas are inhaled over a prolonged period of time they may increase the risk of lung cancer. The Radiological Protection Institute of Ireland estimates that approximately 150–200 people die from radon-related lung cancer every year.

Radon is measured in becquerels per cubic metre (Bq/m³). A presence of 1 Bq/m³ means that the concentration of radon gas present in a cubic metre of air emits one particle of radiation per second. The national reference level for radon in Ireland is 200 Bq/m³. Radon levels are indicated by the percentage of houses which are above this level in a given area. A high radon area is defined as an area where ten per cent of houses in the area have measured radon levels above 200 Bq/m³. Up-to-date maps showing radon levels are kept by the RPII (Radiological Protection Institute of Ireland).

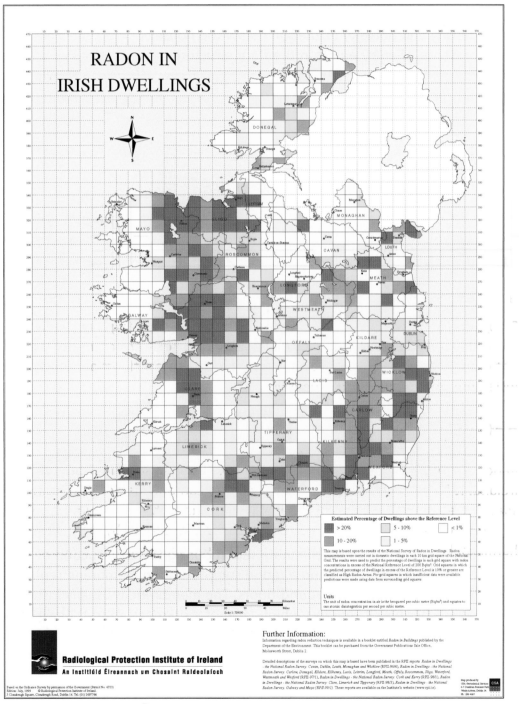

Radon can enter a building through small cracks in floors and through gaps around pipes or cables. Radon tends to be sucked from the ground into a building because the indoor air pressure is usually slightly lower than the outdoor air pressure. This pressure difference occurs because indoor air is less dense than outdoor air (it is warmer). The amount of radon entering a house depends on:

- the concentration of radon in the soil below the house,
- the density and porosity of the soil below the house,
- the presence of entry points in the floor structure.

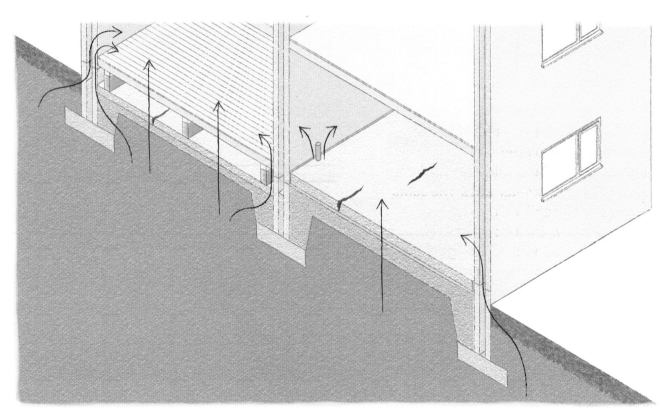

Radon: radon gas can enter a house through cracks or gaps in the substructure.

There are two steps taken to eliminate radon during the construction of new houses – a radon membrane and a radon sump. A radon membrane replaces the DPM usually installed in ground floors. The membrane is usually installed by specialists because it requires a high level of workmanship to achieve airtight seals around service pipes and stacks. The radon membrane must be continuous and step across the cavity in the external walls. A DPC is still used in the inner and outer leaf of external walls.

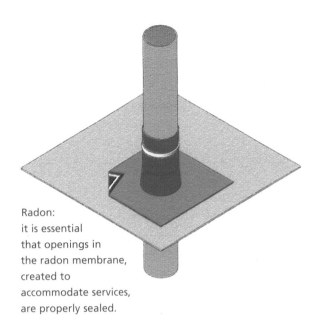

Radon:
it is essential
that openings in
the radon membrane,
created to
accommodate services,
are properly sealed.

A radon sump is a plastic honeycombed chamber that is placed in the hardcore and vented to the outside atmosphere via a 100 mm diameter duct. The sump works by providing an area of low air pressure below the floor that attracts the radon gas. The connecting pipe can be fitted with a fan to improve the effectiveness of the sump.

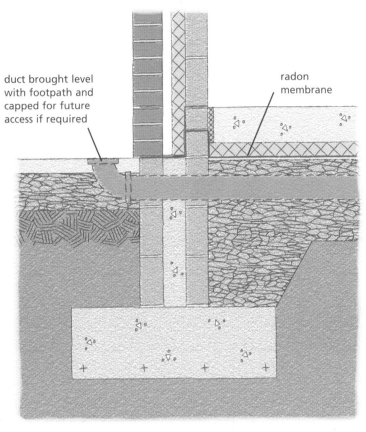

duct brought level
with footpath and
capped for future
access if required

radon
membrane

Radon: solid floating concrete ground floor. Vertical section of floor, strip foundation and concrete cavity external wall showing correct position of radon membrane and radon duct.

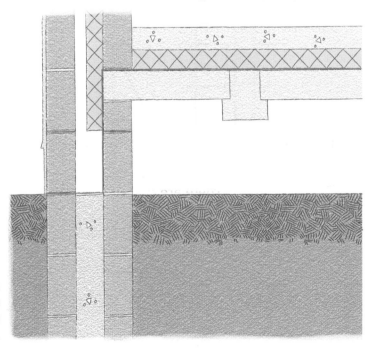

Radon: suspended concrete ground floor. Vertical section showing correct position of radon membrane. Note that because the underfloor void is already ventilated there is no need for a sump/duct.

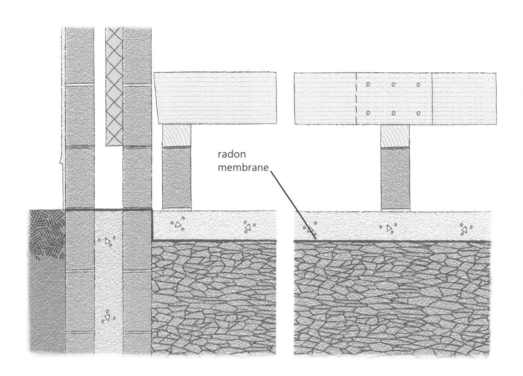

radon
membrane

Radon: suspended timber ground
floor. Vertical section showing
correct position of radon
membrane. Note that because the
underfloor void is already
ventilated there is no need for a
sump or duct.

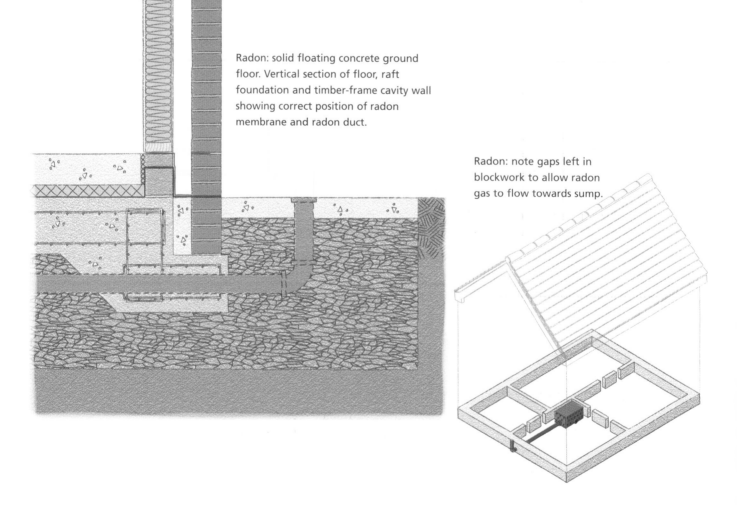

Radon: solid floating concrete ground
floor. Vertical section of floor, raft
foundation and timber-frame cavity wall
showing correct position of radon
membrane and radon duct.

Radon: note gaps left in
blockwork to allow radon
gas to flow towards sump.

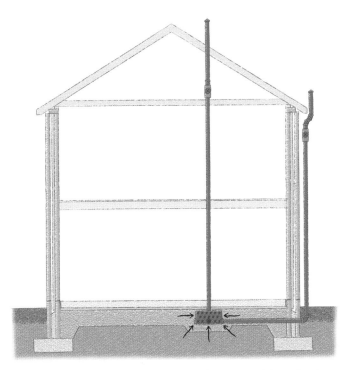

Radon: vertical section of typical house showing position of radon sump and two possible positions (internal and external) for ducting.

Upper Floors

Upper floors serve a number of functions including:

- provide a level usable surface,
- support the imposed loads,
- provide restraint for the external walls,
- limit sound transmission,
- provide resistance to the spread of fire.

An upper floor consists of joists with floor sheeting above and plasterboard below. Joists are usually of solid timber but engineered joists are also used – especially for very wide spans in large houses. These joists, often called I-joists (because of their shape) consist of a plywood or oriented strand board web and solid (or laminated veneer) timber flanges. These joists are light, stable and provide a very firm floor.

The joists are supported by the external walls and in some cases by internal load-bearing walls. The sheeting material was traditionally tongue and grooved flooring boards but is now often plywood or oriented strand board. The plasterboard fixed to the underside of the joists forms the ceiling of the

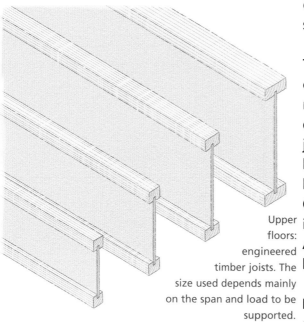

Upper floors: engineered timber joists. The size used depends mainly on the span and load to be supported.

downstairs rooms and also provides resistance to the spread of fire.

The ends of the joists are usually built into the external walls of the dwelling. A gap is left at regular intervals in the blockwork of the inner leaf during construction to accommodate each joist. The joist should have a 90 mm minimum bearing on the blockwork for stability. The end of the joists should be treated with a preservative prior to installation. Once the joist is correctly positioned it can be fixed in place by packing mortar into the space around it. An alternative method is to build galvanised steel hangers into the blockwork to support the joists.

It is essential to provide bridging to prevent twist developing in the joists, especially over long spans. This can be done in two ways. Solid bridging pieces, cut from joist material, can be nailed between the joists or alternatively herringbone bridging can be used. Herringbone bridging can be installed in timber or steel. A packing piece

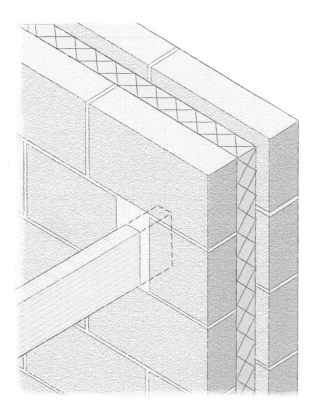

Upper floors: timber joist built into inner leaf of external load-bearing wall.

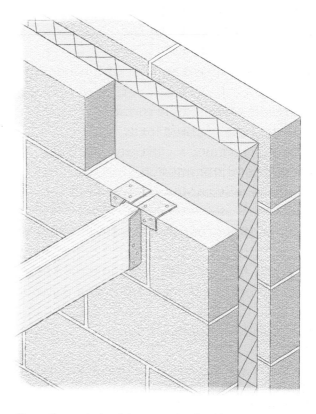

Upper floors: timber joist secured to load-bearing wall using galvanised steel joist hanger.

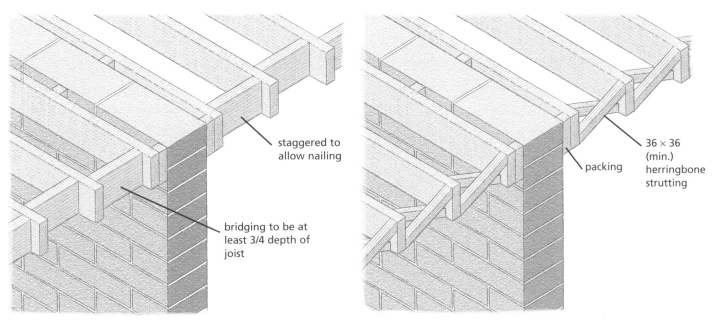

staggered to
allow nailing

bridging to be at
least 3/4 depth of
joist

Upper floors: solid bridging.

36 × 36
(min.)
herringbone
strutting

packing

Upper floors: herringbone bridging.

is secured between the joist and the parallel wall.

The opening needed for the stair is an important part of upper floor construction. The joists used to create the opening for the stair are usually thicker than the regular joists because of the increased loading they have to sustain. Where these joists meet they are butt-jointed and secured using galvanised steel hangers.

The upper floor may also need to be trimmed around the chimney stack. The approach taken here depends on whether the joists are running parallel or perpendicular to the chimney stack. It is essential that the joists are at least 40 mm clear of the chimney stack.

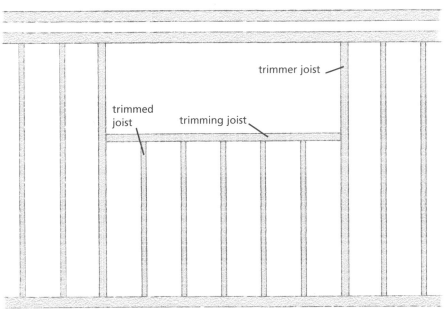

trimmer joist

trimmed
joist

trimming joist

Upper floors: plan view of stair ope. Note the increased thickness of trimming joist and trimmer joists (joist sizes depend on load and span).

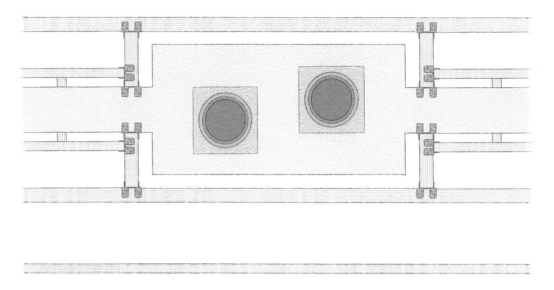

Upper floors: plan view of chimney stack for semi-detached houses showing trimming arrangements when joists run parallel to party wall.

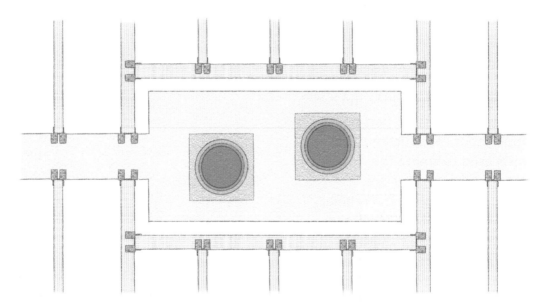

Upper floors: plan view of chimney stack for semi-detached houses showing trimming arrangements when joists run perpendicular to party wall.

Notching and Drilling of Upper Floor Joists

The accommodation of services within upper floors is normal practice in house construction. However, it is essential that the joists are not weakened by this. The bottom portion of the joist is in tension – any notching or drilling in this area could cause the joist to fail. The centre line of the joist is the best place for drilling, while notches can be made in the upper edge only.

Some I-joist manufacturers provide pre-scored knockouts to facilitate the installation of services. Where these are not provided, holes may only be cut along the centre line of the web to a maximum size of 45 mm. The flanges of I-joists must not be notched or cut in any way as this would critically weaken the joist.

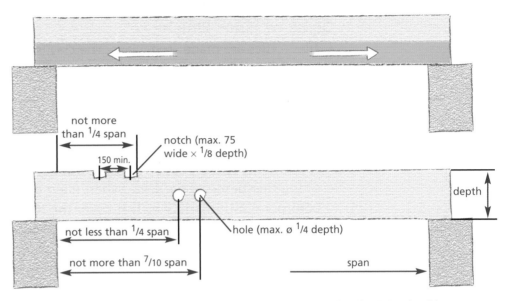

Upper floors (or suspended ground floors): drilling and notching of timber joists should not weaken the joist.

Floor Finishes

The floor finish chosen for a room depends on a number of factors, including:

- Function – Rooms such as kitchens and bathrooms should have a surface finish that is hygienic, reasonably slip resistant and easily cleaned (e.g. tiles and vinyl).
- Aesthetics – The colour and texture of a floor covering can dramatically alter the appearance and feel of a room. Hard surfaces (e.g. timber and tiles) will reflect light and give a room a clean feel, whereas soft finishes (e.g. carpet) can make a room feel cosy and welcoming.
- Durability – Some areas of the home are more exposed to foot traffic than others (e.g. hall and upstairs landing) and for this reason they require a durable floor covering.
- Resilience – Some floor coverings are inherently more flexible than others (e.g. vinyl). This is useful, for example, when covering timber floorboards that are prone to movement, especially in the period shortly after a new house is first occupied.
- Cost – The cost of floor finishes varies widely and is primarily related to the quality of the product. Floor finishes are usually priced per square metre excluding installation.
- Health – Some floor finishes are easier to clean and maintain than others. For example, carpet has become less popular in recent years due to the fact that it harbours dust mites which can aggravate some asthma suffers.

Types of Floor Finish

- Tiles – Tiles are made from clays. The clay is mixed with water, ground, pressed, dried, glazed and fired at high temperature, making them very hard. Tiles are available in a wide range of sizes, colours and surface finishes. Tiles are extremely durable and are especially suitable for kitchens, bathrooms and hallways. Tiles are bonded to the floor using a 3–6 mm bed of synthetic tile adhesive. The spaces between the tiles are then filled with a fine cement called *grout*, which is rubbed smooth and washed off the surface of the tiles before it hardens. When tiling onto a wooden floor (e.g. upstairs bathroom) it is essential to first cover the floor in marine plywood. The plywood should then be coated in PVA (polyvinyl acetate, synthetic resin) to seal it and to improve adhesion. This prevents movement in the floor (caused by changes in temperature and humidity) dislodging or cracking the tiles.
- Timber – This is available in solid and semi-solid form. Solid timber flooring is usually installed on counter battens, whereas semi-solid flooring normally floats on a plastic or rubberised underlay. Semi-solid (laminated) timber flooring typically consists of three laminates or veneers – the top veneer (approx. 6 mm) is a hardwood while the others are usually softwood. Laminated timber flooring is a less expensive and more sustainable solution because it uses less hardwood. Solid timber flooring is sometimes sold unfinished and must be varnished after installation. Laminated flooring is usually factory finished.
- Carpet – This is available in natural and synthetic fibres and in a wide range of colours and patterns. Carpet is particularly good in playrooms because it is soft underfoot and young children find it comfortable for sitting or lying down. Carpets are supplied in roll form and cut to size on the job. Woollen carpets are installed on a carpet cushion (i.e. underlay) using tackstrips which are nailed to the floor around the perimeter of the room, whereas synthetic carpets are usually glued to the floor at their edges. Carpets in a house help to reduce noise levels and minimize heat loss through the floor. Carpets are harder to clean than bare floors, spilled drinks may cause stains and they tend to collect fur from family pets. They should be vacuumed regularly in order to prevent the accumulation of dust. Dust mites can survive very well in carpets, which can be a problem for sufferers of asthma.
- Linoleum – Traditionally linoleum (lino) was made from solidified linseed oil in combination with wood flour or cork dust over a burlap or canvas backing. This was a very popular floor covering because of its hard-wearing properties. A modern version (marketed as Marmoleum) is widely used in commercial buildings. Marmoleum is available in a range of colours and actually gets harder with time – so it is extremely durable.
- Vinyl – Vinyl (polyvinyl chloride, PVC) has largely replaced linoleum for residential applications. It is cheap, flexible and durable. Vinyl is available in

a range of colours and patterns and can also be purchased with a textured surface (e.g. wood-grain effect).

- Laminate flooring – This is a wood-effect floor covering consisting of a melamine, resin-bonded, high-density fibreboard (HDF), with a paper wood-effect image, covered in a hard-wearing plastic finish. Laminate flooring is installed on an underlay provided by the manufacturer. Better quality laminate flooring has a *click* joint that achieves a good quality finish.

Revision Exercises

1. Generate a neat labelled cross-section sketch of a typical concrete ground floor used when a large number of dwellings are being constructed. Indicate on the sketch the key requirements and sizes for each component.

2. Generate a neat labelled cross-section sketch of a typical concrete ground floor used for one-off dwellings. Indicate on the sketch the key requirements and sizes for each component.

3. A self-build home owner is considering installing a concrete ground floor. Discuss the advantages and disadvantages of both approaches to concrete floors.

4. Generate a neat labelled cross-section sketch of a typical suspended timber ground floor.

5. Explain, using neat labelled sketches, the importance of sleeper/tassel walls.

6. Generate a neat labelled cross-section sketch of a typical block and beam suspended concrete ground floor.

7. Describe the possible entry routes of radon gas into a house.

8. Explain, using a neat annotated sketch, the measures taken to prevent radon entering a newly built home in a high radon area.

9. Generate a neat annotated sketch of the stair ope in an upper floor.

10. Describe, using a neat annotated sketch, how floor joists should be notched and drilled to accommodate services safely.

11. Ordinary Level, 2002, Question 1
To a scale of 1:5 draw a vertical section through a suspended timber ground floor, together with an external load-bearing concrete block wall. The wall is of the standard 300 mm insulated cavity type and is plastered on both sides. The section is to be taken from the bottom of the foundation to 300 mm above the level of the floor and should show all relevant constructional details.

12. Ordinary Level, 2003, Question 5, (a) and (b)
The accompanying sketch shows a suspended timber first floor suitable for a dwelling house. The floor consists of wooden joists, tongued and grooved flooring boards with a plasterboard ceiling beneath. Using notes and neat freehand sketches, show a method of:

(a) Supporting the flooring joists at an external block cavity wall.

(b) Bridging and/or strutting the flooring joists.

13. Ordinary Level, 2004, Question 8, (a), (b) and (c)

(a) Show, using notes and neat freehand sketching, the construction details of an insulated solid concrete ground floor of a domestic dwelling.

(b) Include and label in your sketch a design detail which will prevent moisture reaching the inside of the building at floor level.

(c) Recommend a floor covering for the concrete ground floor of a kitchen area and give two reasons for your choice of material.

14. Higher Level, 2000, Question 4, (a), (b) and (c)

A dwelling house has a suspended timber first floor consisting of tongued and grooved flooring boards on wooden joists with a plasterboard ceiling underneath. A room in the house, 4·2 m long by 3·6 m wide, has a stairwell measuring 3·0 m long by 1·0 m wide centrally placed along one of the long walls.

(a) Using sketches, show the layout of the joists at the stairwell. Name the various joists, state their size and show clearly a method of jointing the joists at the stairwell.

15. Higher Level, 2005, Question 1

The sketch shows a combined kitchen and dining space in a single-storey dwelling house. The external wall is a standard 300 mm concrete block wall with an insulated cavity. The kitchen space has a solid concrete floor with a tiled finish and the dining space has a suspended timber floor.

(a) To a scale of 1:10, draw a vertical section through the external wall and ground floor of the house showing both floor constructions. The section should show all the constructional details from the bottom of the foundation to 400 mm above finished floor level and include the abutment of both floors. (For the purposes of this drawing, show a minimum 1·5 m width for each floor type.)

(b) Indicate on the drawing a design detail to show the cross-ventilation of the suspended timber floor through the solid concrete floor.

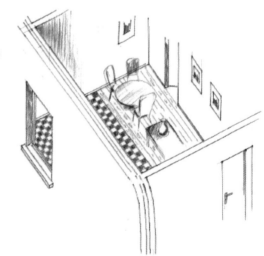

Functions

The primary function of internal walls is to divide the overall space within the house into smaller spaces. Internal walls also have a number of secondary functions, including the accommodation of services (e.g. plumbing), acoustic insulation and fire resistance. Some internal walls are load-bearing. This means they provide structural support to other building elements.

Design Features

Internal walls must be designed to meet a number of performance requirements, including:

- Strength and Stability – Internal walls must be designed to be sufficiently strong and stable to support doors and, in the case of load-bearing walls, the load imposed by the upper floor or roof.
- Fire Resistance – Internal walls should be constructed using materials that will resist the spread of hot gases, smoke and fire within the house.
- Acoustic Insulation – Internal walls should provide a reasonable degree of acoustic insulation so that the activities occurring in one room (e.g. watching television) do not interrupt those occurring in another (e.g. reading a book).
- Accommodation of Services – Internal walls should, for example, provide secure fixing for light switches and sockets and should also accommodate cabling or piping.
- Suitable for Finishing – Internal walls should be suitable for painting or wallpapering.

Load-bearing Internal Walls

A load-bearing internal wall in a house provides support to the upper floor or roof, or both. Internal load-bearing walls must rest on a foundation.

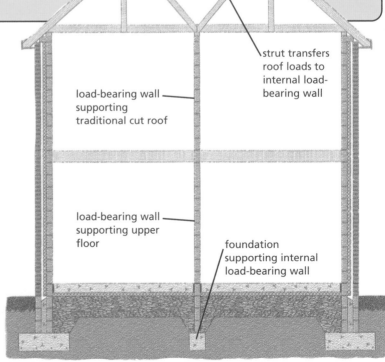

strut transfers roof loads to internal load-bearing wall

load-bearing wall supporting traditional cut roof

load-bearing wall supporting upper floor

foundation supporting internal load-bearing wall

Load-bearing internal walls: vertical section through two-storey house showing concrete block, load-bearing wall supporting a traditional cut roof and upper floor. Note the foundation provided to the internal wall.

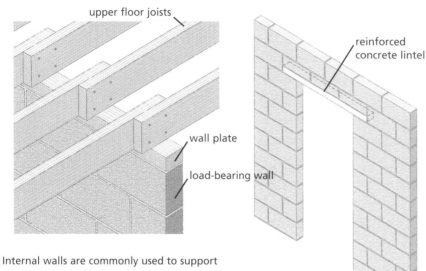

Internal walls are commonly used to support the upper floor of a house. If the wall is constructed from blockwork a timber wall plate is fixed to the top of the wall to provide fixing from the floor joists.

A reinforced concrete lintel is used to span the ope in a internal load-bearing wall.

Concrete Load-bearing Internal Walls

In houses constructed with a masonry external cavity wall, the load-bearing internal walls are usually constructed from blockwork and are raised at the same time as the external walls. In this way they can be built into the inner leaf of the external wall to provide maximum stability.

Where the internal load-bearing wall supports the upper floor only (i.e. not supporting the roof), a wall plate is secured to the top of the internal wall to provide fixing for the joists.

Where an opening is required in a load-bearing internal wall, a reinforced concrete lintel is used to safely transmit the load.

Services, such as electrical cabling, are accommodated by cutting a vertical or horizontal chase (groove) into the surface of the wall. A steel conduit, which will carry and protect the cabling, is then fixed in the chase. The conduit is hidden when the wall is plastered.

Timber Load-bearing Internal Partitions

For timber load-bearing partitions, 44 × 100 mm studs of C14 strength class material are usually used. The soleplate and headplate are doubled to provide increased load-bearing capacity.

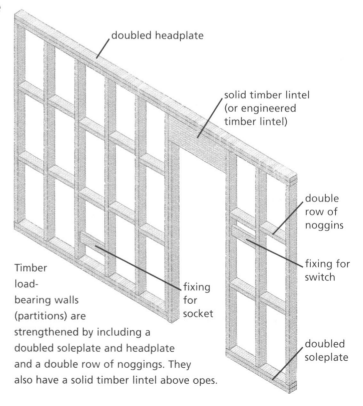

Timber load-bearing walls (partitions) are strengthened by including a doubled soleplate and headplate and a double row of noggings. They also have a solid timber lintel above opes.

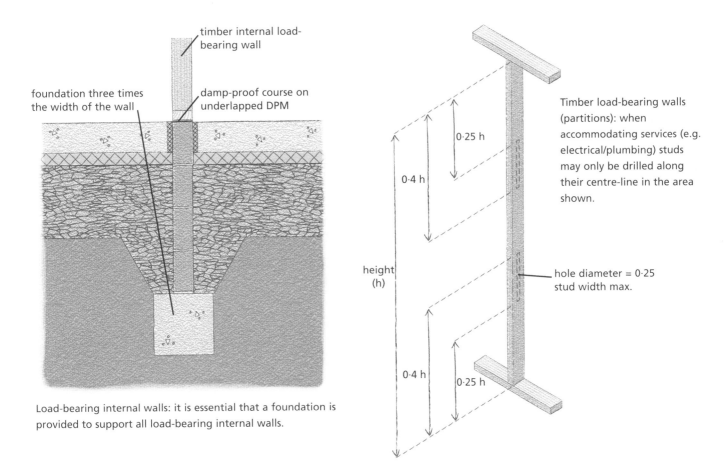

timber internal load-bearing wall

foundation three times the width of the wall

damp-proof course on underlapped DPM

0·25 h

0·4 h

Timber load-bearing walls (partitions): when accommodating services (e.g. electrical/plumbing) studs may only be drilled along their centre-line in the area shown.

height (h)

hole diameter = 0·25 stud width max.

0·4 h

0·25 h

Load-bearing internal walls: it is essential that a foundation is provided to support all load-bearing internal walls.

The load-bearing walls in timber-frame houses, are similar to the load-bearing walls used in the concrete construction. These consist of a stud framework with a doubled headplate. A solid timber lintel is provided over door opes.

In both cases, it is essential that a damp-proof course is provided to the underside of the soleplate for all ground floor partitions resting on a concrete rising wall. This prevents any moisture that might remain in the concrete from soaking into the timber soleplate.

Where services are to be carried in the stud partition, it is important that the strength and stability of the studs is not reduced. To ensure this, holes should only be drilled along the centre line of the stud and the edges of the studs should not be notched. A nogging piece is secured at the required height to provide fixing for sockets and switches.

Plasterboard (12·5 mm) is fixed to both faces of the stud partition. It is held in place using galvanised nails. These nails are driven home firmly leaving a shallow depression to facilitate filling. The edges of the boards should meet on a stud and a slight gap should be left to allow for movement in the wall.

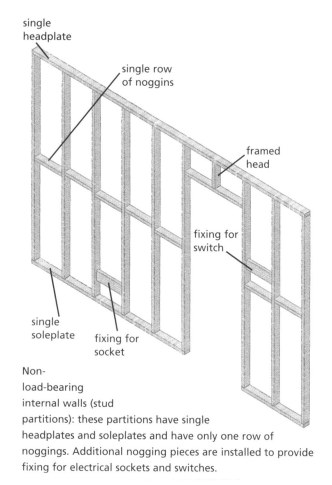

single headplate

single row of noggins

framed head

fixing for switch

single soleplate

fixing for socket

Non-load-bearing internal walls (stud partitions): these partitions have single headplates and soleplates and have only one row of noggings. Additional nogging pieces are installed to provide fixing for electrical sockets and switches.

Non-load-bearing Internal Walls

A non-load-bearing internal wall is often referred to as a stud partition. It consists of a stud framework sheeted on both sides with plasterboard. This internal wall is very similar to the load-bearing type described previously, except in this case lighter timbers can be used (e.g. 36 × 75 mm).

When a stud partition is installed on an upper floor, the joists underneath should be doubled to support the weight of the partition. Where the partition runs perpendicular to the joists the soleplate should be doubled.

The headplate of the stud partition is nailed to the ceiling joists above. Where necessary, additional nogging pieces are installed to provide fixing for the partition.

Non-load-bearing partitions are sometimes constructed slightly shorter in height to allow the ceiling plasterboard to be installed before the partition.

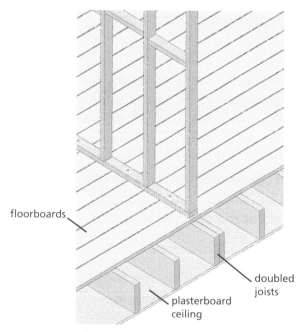

floorboards

doubled joists

plasterboard ceiling

Non-load-bearing internal walls (stud partitions): doubling of joists to support the stud partition (partition running parallel to joists).

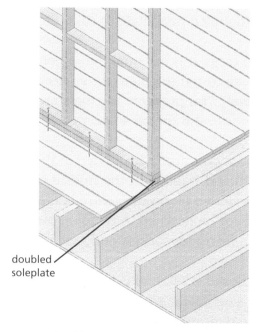

doubled soleplate

Non-load-bearing internal walls (stud partitions): doubling of the soleplate to support the stud partition (partition running perpendicular to joists).

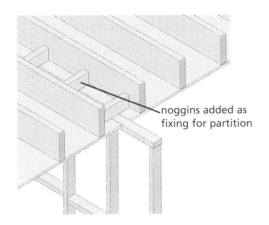

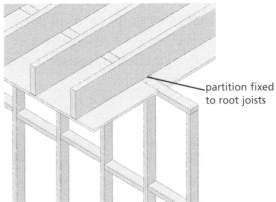

Non-load-bearing internal walls (stud partitions): nogging pieces are installed between the roof joists to provide fixing for the partition.

Non-load-bearing internal walls (stud partitions): the partition is nailed to the roof joists above.

Revision Exercises

1. Describe, using a neat labelled sketch, how an opening is created in a concrete load-bearing internal wall.

2. Generate a neat annotated sketch of a non-load-bearing timber stud partition. Indicate on the sketch the key requirements and sizes for each component.

3. Describe, using neat annotated sketches, the measures to be taken in the following situations relating to the installation of a non-load-bearing stud partition:
   (a) Stud partition running across floor joists.
   (b) Stud partition running parallel to floor joists.
   (c) Stud partition running parallel to ceiling joists.

4. Describe how services may be safely accommodated in a stud partition. Indicate on the sketch the key requirements and sizes for each component.

5. Ordinary Level, 2005, Question 2
   A non-load-bearing timber stud partition with plasterboard finish separates two bedrooms on the first floor of a house.
   (a) Using notes and neat freehand sketches, describe the construction of the stud partition. Indicate on the sketches the names and sizes of all parts.
   (b) Describe two methods of providing a surface finish to the plasterboard prior to painting.
   (c) Recommend one of the finishes described at (b) and give two reasons for recommending this finish.

6. Higher Level, 2002, Question 9
    A load-bearing, timber stud partition with a plaster finish separates a dining room and a living room in the ground floor of a two-storey house.
    (a) Using notes and freehand sketches, describe in detail the construction of the partition.
    (b) Show clearly the design details necessary to accommodate a standard flush panel door.
    (c) Label and give the sizes of each of the components of the partition.
    (d) Discuss in detail the advantages and disadvantages of using either a timber stud partition or a concrete block partition wall.

7. Higher Level, 2005, Question 4
    A new two-storey house has load-bearing and non-load-bearing timber stud partitions. The house has a solid concrete ground floor and a suspended timber first floor.
    (a) Using notes and detailed freehand sketches, compare the design detailing for the construction of each of the following:
        (i) A load-bearing partition to support the first-floor joists.
        (ii) A non-load-bearing partition on the first floor.
    (b) Using notes and freehand sketches, show two design details that ensure that the transmission of sound is reduced through the stud partition constructed on the first floor.

The fireplace (or hearth) has always enjoyed a special place in the Irish home. Our native expression for 'There's no place like home', 'Níl aon tinteán mar do thinteán féin', (literally, there's no hearth like your own hearth) reflects this. In vernacular dwellings the fireplace was the centre of activities. While the sustainability and efficiency of burning solid fuel in the fireplace today is very questionable, the fireplace remains a constant feature of typical houses built in Ireland today.

Functions

The primary function of a fireplace is to provide a safe source of direct heat, including the safe removal of smoke and hot gases. A fireplace also provides a social focal point in the home.

Design Features

Fireplaces must be designed to meet a number of performance requirements including:

Safe Generation of Heat – Fireplace and flue design takes into account the need to safely generate high levels of heat. The heat generated by a fire is safely contained through the use of non-combustible materials such as concrete and ceramics. These materials are capable of safely withstanding high temperatures.

Safe Extraction of Smoke and Hot Gases – A rapid air flow must be created directly above the fire if smoke and other hot gases are to be safely extracted. The up-draught, *draw* or *pull* in a flue results from a combination of the height of the flue and the difference in temperature between the flue gases and the outside air. The column of hot gases in a flue is less dense than an equivalent column of cold air outside, this creates the up-draught required to safely remove the gases. The up-draught created is maximised by having a warm, straight and tall flue.

For this to work efficiently, the shape and volume of the space directly above the fire must be carefully controlled. This is because the volume of gases (smoke etc.) removed by the flue system is controlled by the size and shape of the flue gathering. If, for example, the flue gathering is too big the same amount of gases has to flow through a larger space – this will reduce the velocity at which the gases are travelling and the result may be that smoke will trickle into the room before it can be drawn up the flue.

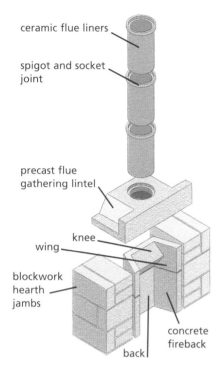

ceramic flue liners

spigot and socket joint

precast flue gathering lintel

knee

wing

blockwork hearth jambs

back

concrete fireback

Fireplaces: pictorial view of main components of a typical fireplace.

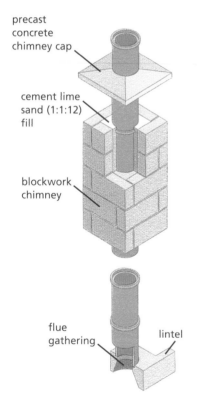

precast concrete chimney cap

cement lime sand (1:1:12) fill

blockwork chimney

flue gathering

lintel

Chimneys: pictorial view of main components of a typical chimney.

Efficient Burning of Fuel – One of the ways in which the design of modern fireplaces differs from traditional fireplaces is that the fuel is raised above the hearth on a *grate* or *basket*. This, coupled with the flow of air created by the flue, leads to the achievement of higher temperatures.

Despite this, fireplaces are inherently inefficient; approximately eighty per cent of the heat generated by burning solid fuel in an open fireplace is lost. For this reason, fireplaces are seen as feature rather than as a heat source in newly built houses. With this in mind, it should be noted that wood is one of the best fuel sources for use in open fires. Burning any solid fuel (e.g. wood, coal, briquettes) releases carbon dioxide and other harmful gases into the atmosphere. However, the carbon released when wood is burned is *cancelled out* by the carbon absorbed when the tree was growing (trees absorb carbon dioxide from the atmosphere as they grow). Thus, wood is a carbon neutral fuel – the net carbon release to the atmosphere is zero. Wood fuel supplied from managed forests is a sustainable resource.

Construction of Fireplaces and Chimneys

Chimney Stack and Flue Construction

A fireplace is connected to the outside atmosphere by a flue, which is supported by the chimney. The first stage in the construction of a fireplace is to create a recess. The position of the recess will depend on whether the chimney is being accommodated internally or externally.

For an internal chimney, piers or jambs are built to create a recess. The width and depth of the recess is

normal insulation

non-combustible insulation

precast flue gathering lintel

jambs/ piers

Fireplace construction: it is essential that non-combustible insulation is used in the cavity behind the fireplace and chimney stack, and that the height of the opening should not exceed 550 mm.

constructed to suit the fireback used. The height of the fireplace opening measured vertically from the top of the grate to the underside of the flue gathering lintel should not exceed 550 mm. This is to avoid smoke trickling out into the room before being drawn up the flue. Non-combustible insulation is used in the cavity behind the fireplace and chimney stack.

A precast concrete flue gathering (throat) is then bonded to the piers using mortar. As mentioned earlier, the shape of the flue gathering has a significant impact on the performance of the fireplace. To avoid problems, precast flue gatherings are nearly always used in modern construction.

The chimney stack is then built-up. The flue liners are installed as the stack is raised. Spigot and socket or rebate and socket flue liners should be installed with the socket upwards.

Liners should be jointed with fire-resistant mortar. The cavity around the flue liners should be packed with a damp mix of cement, lime and sand in the ratio of 1:1:12. It is very important that the flue liners are not in contact with the chimney stack.

The principles are similar for timber-framed houses. The combustible nature of timber means that it's essential that a clear gap is maintained between the chimney and the timber-frame panels.

In semi-detached houses, fireplaces are sometimes constructed *back-to-back* on the party wall. When this is done the thickness of

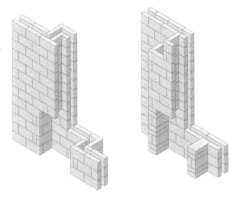

Chimney construction: pictorial views of typical construction details of an external chimney (on the left) and an internal chimney (concrete cavity wall).

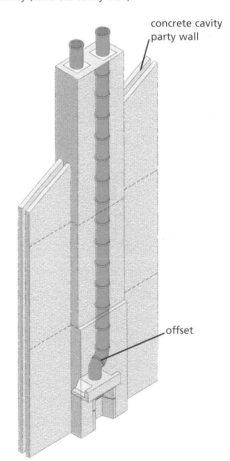

concrete cavity party wall

offset

Chimney construction: pictorial view highlighting the offset flue in a semi-detached house.

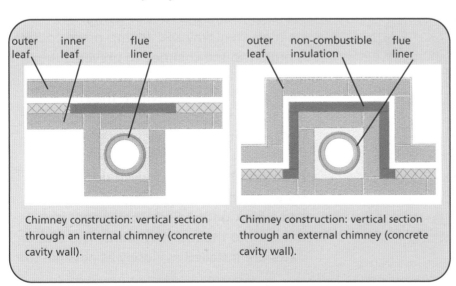

outer leaf | inner leaf | flue liner

outer leaf | non-combustible insulation | flue liner

Chimney construction: vertical section through an internal chimney (concrete cavity wall).

Chimney construction: vertical section through an external chimney (concrete cavity wall).

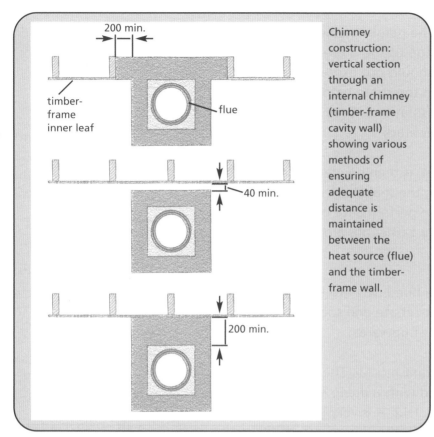

200 min.

timber-frame inner leaf

flue

40 min.

200 min.

Chimney construction: vertical section through an internal chimney (timber-frame cavity wall) showing various methods of ensuring adequate distance is maintained between the heat source (flue) and the timber-frame wall.

the party wall between the fireplaces must be at least 200 mm. A larger chimney stack is built to accommodate both flues. In this case the flue liners are offset. It is important that the offset is 52·5° or greater – otherwise the quality of the up-draught will be impaired.

In timber-frame houses the construction is very similar. The inner-leaf timber panels are cut back to create an ope for the chimney stack, which is constructed in blockwork as described previously.

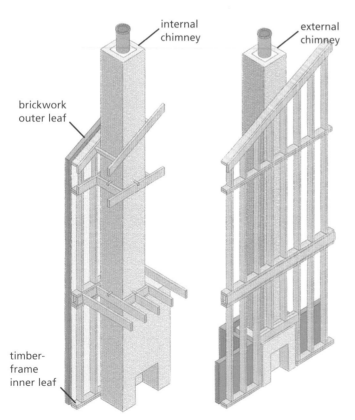

internal chimney

external chimney

brickwork outer leaf

timber-frame inner leaf

Chimney construction: pictorial views of typical construction details of an internal chimney (on the left) and an external chimney in a timber-frame cavity wall.

Weatherproofing

When the roof level is reached it is essential that the junction between the chimney stack and the roof is properly sealed to exclude rainwater. Rainwater can penetrate in two ways: it can soak downward through the masonry of the chimney and it can seep in around the edges where the chimney passes through the roof covering. A stainless steel tray DPC (damp-proof course) is built into the chimney to prevent moisture from soaking down through the masonry. The joint between the chimney and the roof is sealed with flashing to carry the rainwater past the chimney.

A watertight seal is achieved using three components: a front apron, side flashing and a back gutter. Each component is chased into the masonry joints. The tiles or slates are installed as closely as possible to

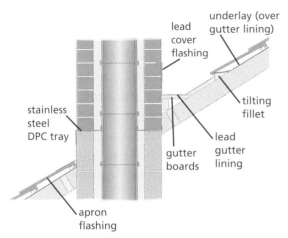

Weatherproofing: vertical section showing typical construction details where chimney passes through the roof. It is essential to prevent the penetration of water between the chimney stack and the adjoining roof surfaces.

the chimney stack. The front apron is dressed down over the tiles or slates. The side flashing can be installed in two ways. For slated roofs, a soaker is fitted under each slate and the flashing is dressed down over this. For tiled roofs, the flashing is dressed down over the tiles in one continuous strip. The back gutter is installed so that it is carried up as far as the first tiling batten – a minimum upstand of 100 mm ensures that rainwater is excluded. While aluminium, copper and other metal flashings are available, lead is still the most commonly used material. This is because it is malleable and ductile, allowing it to be shaped to the profile of the tiles thereby ensuring a good seal.

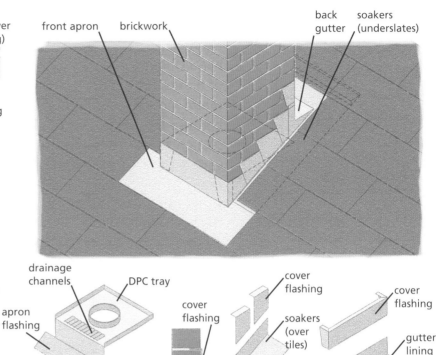

Weatherproofing: pictorial view of weatherproofing installed to a brick chimney in a slated roof. Note how the soakers are fitted under the slates and the flashing is built into the brickwork and dressed down over the soakers.

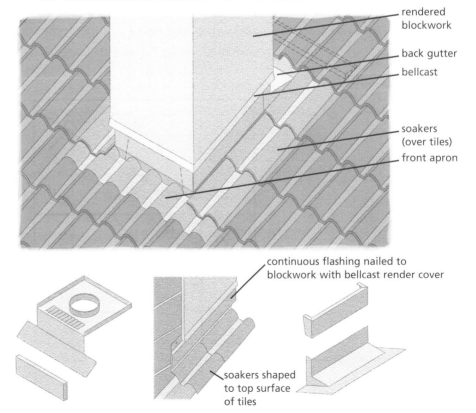

Weatherproofing: pictorial view of weatherproofing installed to a rendered blockwork chimney in a tiled roof.

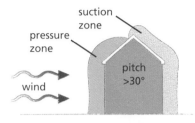

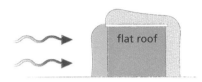

The movement of air (wind) over a roof creates pressure and suction zones.

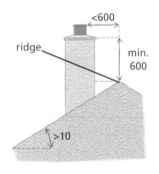

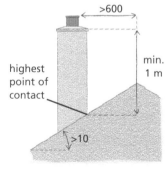

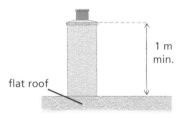

Chimney positioning and height requirements necessary to ensure sufficient draw of air.

Positioning and Height Requirements

For a chimney to function properly there must be a sufficient draw of air up through it. This will only occur if the chimney is at least one metre higher than the highest point of contact with the roof. This is because the movement of air over a roof creates pressure and suction zones. These pressure and suction zones are dependent on a variety of factors including: the weather, the shape of the roof, the shape of the building and the proximity of other buildings. Pressure zones can adversely affect the performance of a chimney by causing a downdraught.

The chimney of a bungalow should extend at least 4·5 m above the top of the fireplace to draw properly. It is also important to ensure that the chimney is properly designed to withstand the force of the wind. The ratio of the allowable height to width varies in different parts of the country.

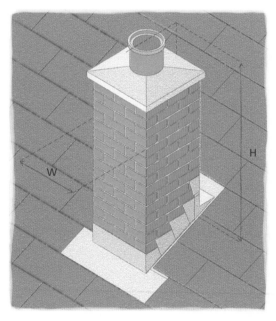

Chimney dimensions: the height to width ratio must be sufficient to withstand dynamic wind loads.

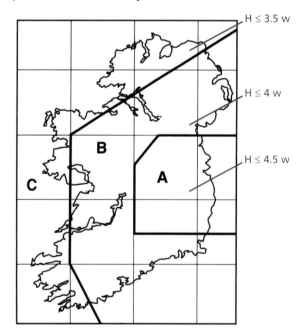

Wind zone map.

Revision Exercises

1. Explain the key design principles governing the performance of fireplaces and chimneys.
2. Explain why wood is the preferred fuel for open fires.
3. Describe, with the aid of neat annotated sketches, the design and construction of a typical internal fireplace and chimney from ground floor to chimney pot level in a concrete cavity-walled house.
4. A red brick chimney passes through a slated pitched roof. Explain, with the aid of neat freehand sketches, how the junction between chimney and roof is made watertight.
5. Explain, with the aid of neat freehand sketches, the importance of chimney design in terms of wind zones, chimney height and chimney position.
6. Explain, using neat annotated sketches the requirements in relation to the height of the chimney stack above roof level. Give three examples.
7. Higher Level, 2002, Question 1
   An open fireplace in a single-storey dwelling is located on a 300 mm external concrete block wall with an insulated cavity. The house has a solid concrete floor.
   (a) To a scale of 1:5 draw a vertical section through the wall and fireplace. The section should show all the constructional details from the bottom of the foundation to the top of the first flue.
   (b) Note on the drawing two design details that ensure the efficient functioning of the fireplace.
8. Higher Level, 2003, Question 6
   A concrete block chimney stack with a sand and cement rendering passes through a pitched slated roof, as shown in the following sketch.
   (a) Using notes and freehand sketches, show the design details necessary to prevent the penetration of water between the chimney stack and the adjoining roof surfaces.
   (b) Poor design detailing may result in the occurrence of a down-draught in a chimney. Outline one situation in which a down-draught might occur and using notes and freehand sketches, show the design detailing that would prevent the occurrence of such a downdraught.

Windows provide a visual link between internal space and the outside world. This is very important to the mental wellbeing of the people in the house – imagine what it would be like to live in a windowless concrete box! Windows also allow natural light and fresh air to enter the house.

Functions

The primary functions of windows are to admit light and fresh air into a building. Light and fresh air are essential to a healthy living space and most people would agree that a bright, airy atmosphere makes a positive impression.

Design Features

Windows must be designed to meet a number of performance requirements including:

Light – The amount of light required in any room depends on the activities occurring there but there is a minimum requirement for the window area of a room to be at least ten per cent of the floor area. Apart from the requirement for natural light, the size and proportion of windows depends on the type of view, the size of the internal space, and the position and mobility of the occupants. In houses, kitchen windows tend to be large, ensuring adequate light is available for the preparation of food and other tasks such as cleaning. Bedroom windows, on the other hand, tend to be smaller. In cases where the home owner is a wheelchair user, it is desirable to lower the bottom of the window to provide a better view and to facilitate opening and closing. Light is examined in detail in chapter 20.

Ventilation – Windows have an important role in the supply of fresh air to the internal spaces of a house. The rooms of a house to which ventilation requirements apply are called habitable rooms. A habitable room is a room with minimum floor area of 6·5 m² that is used for living or sleeping. There are two types of ventilation required for these rooms: background ventilation and rapid ventilation. Adequate ventilation is essential to ensure that water vapour and the products of combustion appliances (e.g. open fires, gas cookers) are dispelled.

Background ventilation is ventilation provided via a permanent vent through an external wall. This vent should have a minimum cross-sectional area of 6,500 mm² (or for larger rooms, 650 mm² for every 10 m² of floor area). Background ventilation is particularly important in living rooms to ensure adequate air supply to fireplaces. Vents should never be blocked or deliberately filled in.

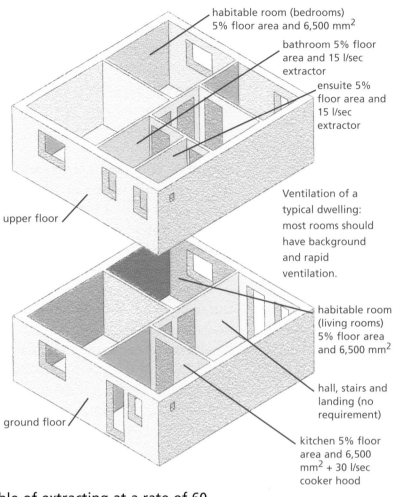

habitable room (bedrooms) 5% floor area and 6,500 mm^2

bathroom 5% floor area and 15 l/sec extractor

ensuite 5% floor area and 15 l/sec extractor

upper floor

Ventilation of a typical dwelling: most rooms should have background and rapid ventilation.

habitable room (living rooms) 5% floor area and 6,500 mm^2

hall, stairs and landing (no requirement)

ground floor

kitchen 5% floor area and 6,500 mm^2 + 30 l/sec cooker hood

Ventilation: background ventilation is provided via a permanent vent through an external wall with a minimum cross-sectional area of 6,500 mm^2

Rapid ventilation is provided by window openings. The size of these openings is also related to the floor area of the room. For habitable rooms the opening must be equal to at least five per cent of the floor area. This provides the opportunity for the home owner to ventilate the room in a short period of time.

For kitchens and utility rooms the use of mechanical extract ventilation is recommended. This system should be capable of extracting at a rate of 60 litres per second (or at a rate of 30 litres per second where the ventilation extract is incorporated in a cooker hood). Mechanical extract ventilation is only used when moisture vapour is being created during cooking or washing.

Thermal Insulation – The thermal insulation of windows is important for human comfort and for the conservation of energy. Poorly insulated, draughty windows are uncomfortable to sit near and waste heat energy. The thermal insulation of windows is achieved primarily through the use of double-glazing and weather stripping. Double-glazing is designed to reduce the amount of heat lost through the glass, while weather stripping is designed to reduce the amount of heat lost due to air infiltration (draughts). Double-glazing can significantly reduce the amount of heat lost through a typical window.

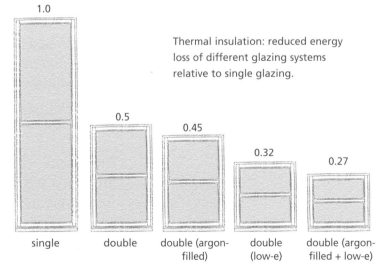

Thermal insulation: reduced energy loss of different glazing systems relative to single glazing.

1.0 — single

0.5 — double

0.45 — double (argon-filled)

0.32 — double (low-e)

0.27 — double (argon-filled + low-e)

Acoustic Insulation – The acoustic insulation requirement of windows is usually solved by the thermal insulation measures. Double-glazed windows are also very effective at reducing the transmission of sound due to the gap between the panes of glass. The use of synthetic rubber seals or gaskets around the openings also reduces the penetration of unwanted noise. (Sound is examined in detail in chapter 22, Sound and Acoustic Insulation.)

Security – Windows rely on locking mechanisms to ensure that they are secure. Modern windows usually have multi-point locking systems that secure the window sash to the frame at several points. These systems also usually offer the option of locking the window in the ventilate position. This secures the window in a slightly open position to permit the circulation of fresh air through the house, when it is unoccupied (see hardware).

Durability (Low Maintenance) – It is important that the windows chosen for a new house perform adequately for the lifetime of the building. Generally, windows should require minimal maintenance. Traditionally, window frames and sashes were made from softwoods (deal) or hardwoods (teak). Teak is a very suitable timber for external components like windows and doors because it is naturally resistant to rot. This durability of the teak family of timbers is related to the warm moist environment (rainforest) where the trees grow. In this environment there is a high risk of insect and fungal attack. To combat this, the teak tree produces high levels of natural oils and chemicals. These naturally occurring oils and chemicals make teak an ideal timber for use in outdoor applications. Softwoods may also be used for window construction, however, the timber should be pressure-treated to ensure that it does not decay. Softwood is a more sustainable and economical material. All types of timber windows require regular protection with a paint or varnish. Unplasticised polyvinyl chloride (uPVC) windows are very common in modern construction. The frames and sashes of these windows are moisture resistant, durable and maintenance free. However, the manufacture and disposal of uPVC windows is harmful to the environment.

Fire Escape – The building regulations require that windows (or doors) suitable for external escape or rescue are provided from every inner room and bedroom in all newly constructed houses. These windows must have a minimum area of 0·33m² and a minimum opening width and height of 450 mm.

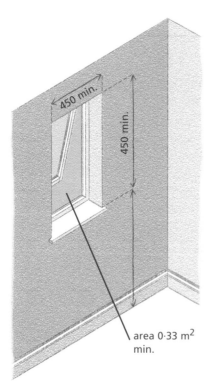

area 0·33 m²
min.

Fire escape from windows: habitable rooms must have a minimum area of 0·33 m² and a minimum opening width and height of 450 mm.

Prevention of Glare and Excessive Solar Heat Gain – One of the disadvantages of having large windows in a home is the problem of solar heat gain.

Solar heat gain is the term used to describe the warming of a living space caused by the sun shining through the windows. If uncontrolled, this can lead to uncomfortably high internal temperatures.

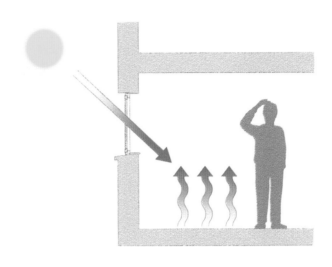

Solar heat gain: temperature rise caused by the sun shining through the windows.

Solar heat gain should be controlled at the design stage of the project. In other words, the orientation of the building to the sun, the size and position of windows and landscape features such as deciduous trees should be used to regulate the amount of solar energy entering the living spaces.

Glare is the term used to describe visual discomfort caused by excessive brightness. Glare can be direct or indirect (reflected). In either case, glare is avoided by providing blinds to south-facing windows and by avoiding the use of glossy surface finishes on which the light will reflect.

Aesthetics – Windows have a significant impact on the aesthetic appeal of buildings. The relationship between the *solid* wall and the *void* window is critical to the overall appearance of a building – because the windows are the main openings of a building, the size, shape and positioning of windows significantly influences the sense of the scale and proportions of a building.

Fenestration is the term used to describe the arrangement of the windows on the façade of a building. Fenestration is a product of a number of features including:
– the pattern of windows on the façade of a building,
– subdivisions of the window,
– proportion of panes,
– recessing and projection of the windows.

The appropriateness of a particular pattern depends on the function of the building.

In general the following guidelines will provide an aesthetically pleasing appearance:

- The variety of opening sizes should be kept to a minimum.
- The shape and arrangement of openings should be simple.
- Openings should be arranged in order to maximise a high solid-to-void appearance where it matters – i.e. concentrate large openings in the main living areas which ideally will have the most sun and external view.
- Highly glazed façades are best located where they are not fully in public view.
- Centre the windows on the natural axis generated by the shape of the wall.
- Vertical emphasis of openings generally looks better than horizontal emphasis.
- First-floor windows are traditionally smaller than, and centred over, ground-floor windows.

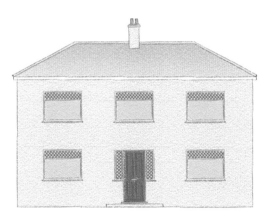

Fenestration: in residential construction windows should generally have a vertical emphasis.

Window Types

The main types of windows used in housing are casement windows, sliding sash windows and pivot windows. The sliding sash window was very common in Irish houses traditionally. However, from around the 1950s onward the casement window became more popular because of its suitability to wider openings and low maintenance requirements. Sliding sash windows are now becoming more popular again, especially in some one-off houses built in rural settings, because they reflect Irish tradition.

In terms of materials, timber was the primary material used for window manufacture for many years. However, the susceptibility of timber to rot in damp conditions, led manufacturers to look for other materials. This briefly led to the use of aluminium windows, however, aluminium was soon surpassed by uPVC products. With the development of penetrating timber-treatment processes, and an increasing awareness of the questionable sustainability of plastics, timber windows are enjoying a comeback.

Casement Windows

A casement window consists of a frame and a number of fixed lights (non-opening) and casements (opening). Casements can be either side hung or top hung. The extent to which the casement opens is usually limited by human reach and the possible effect of the wind. Casement windows were traditionally made in timber but uPVC is much more common nowadays.

The frame of timber casement windows is manufactured from machine-profiled timber and jointed using glued mortice and tenon joints. Many joinery manufacturers today produce machine-fabricated windows that use high-strength combed (bridle) joints. Even though these joints are not in themselves strong enough for the purpose, the use of modern adhesives makes this possible. Combed joints that are fixed using steel pins and weather resistant adhesives are just as strong as traditional mortice and tenon joints.

Synthetic rubber (or PVC) weather stripping is used in timber casement windows to reduce draughts and resist wind-driven rain. Either a flexible strip of synthetic rubber or a strip of nylon filament pile is used. These seals are fitted to the back face of the rebate or the inner face of the frame, so that the rebate acts as the first line of defence against wind and wind-driven rain.

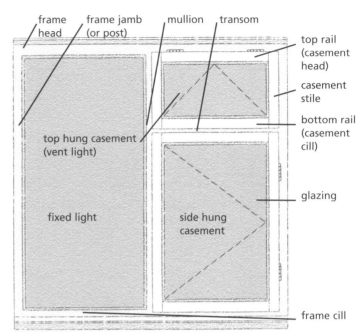

Casement windows: typical casement window.

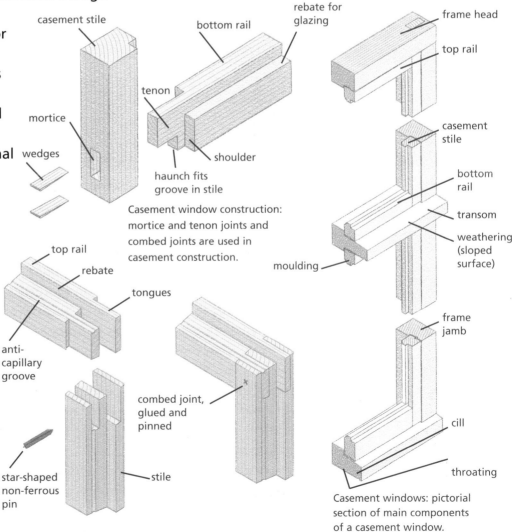

Casement window construction: mortice and tenon joints and combed joints are used in casement construction.

Casement windows: pictorial section of main components of a casement window.

These strips are usually self-adhesive (or tacked) and can be replaced in time when they lose their elasticity, alternatively PVC weather stripping is sometimes installed in purpose-made grooves in the frame.

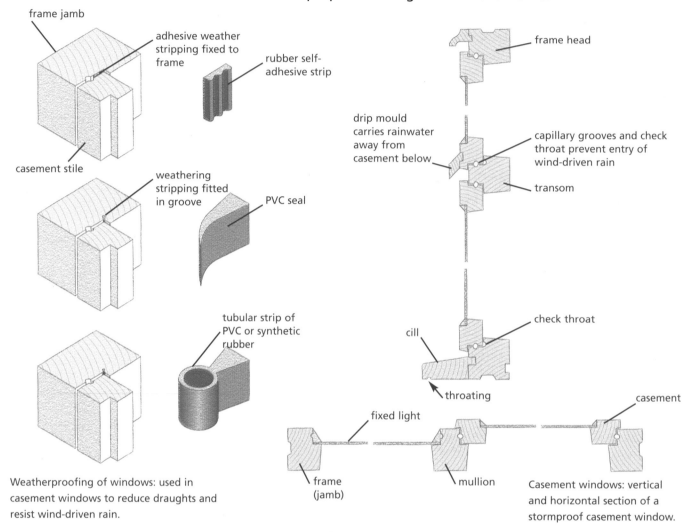

Weatherproofing of windows: used in casement windows to reduce draughts and resist wind-driven rain.

Casement windows: vertical and horizontal section of a stormproof casement window.

uPVC windows are made from hollow section plastic frames and casements (see polymers, chapter 4, Construction Materials). These are manufactured using mitred butt joints that are heat-fusion welded. uPVC window frames and casements are manufactured in bulky sections because they are weaker and less rigid than timber frames. Some manufacturers use aluminium box section to reinforce frames over 1,500 mm in length and casements over 900 mm in length. The bulkiness of uPVC frames and windows is most noticeable in small windows (and heritage buildings) where the bulky frames look out of proportion.

Window frames (both timber and uPVC) are secured in place using L-shaped galvanised steel straps. The straps are nailed to the masonry and later hidden by internal plaster. When completed the outer render holds the window frame securely in place. The window frame is then sealed to the masonry around its outside perimeter.

uPVC casements have synthetic rubber gaskets which form an external weather-tight seal against the face of the frame.

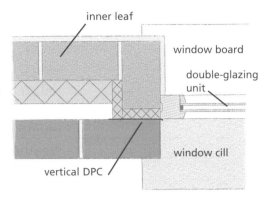

Installation of windows: horizontal section of a typical window jamb showing the position of the window when installed. Note this wall has a brick outer leaf and a block inner leaf (which has been plastered).

Sliding Sash Windows

A vertical, sliding sash window consists of a frame and two sashes, both of which slide up and down to open. The sashes slide up and

uPVC casement window: pictorial section of main components of a uPVC casement window.

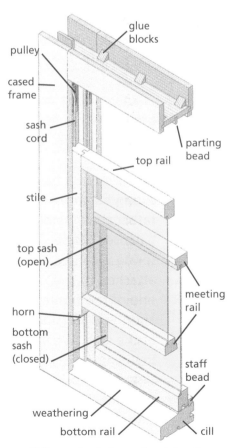

Sliding sash windows: pictorial section of traditional (weight balanced) sliding sash window.

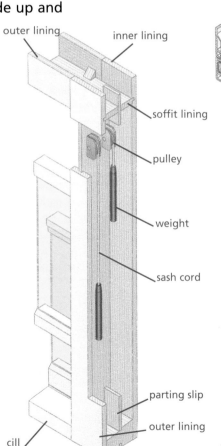

Sliding sash windows: pictorial cut-away view of the weight balances of a traditional sliding sash window.

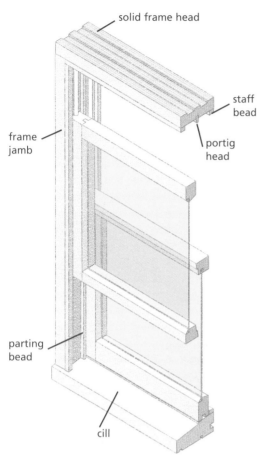

solid frame head

frame jamb

staff bead

portig head

parting bead

cill

Sliding sash windows: pictorial section of modern (spring balanced) sliding sash window.

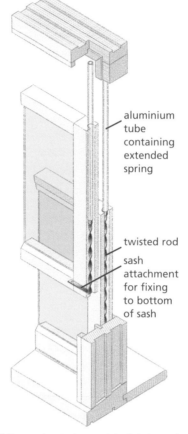

aluminium tube containing extended spring

twisted rod

sash attachment for fixing to bottom of sash

Sliding sash windows: pictorial view of the spring balances of a modern sliding sash window.

down in a recess created in the frame. Traditionally, each sash would have a sloped meeting rail so that they interlock when closed, however, many modern sliding sashes rely instead on weather-stripping seals to create a seal between the sashes when closed.

Traditional timber sliding sash windows had hollow box (cased) frames which incorporated cast-iron weights to act as counterbalance to the sashes – this made them easier to raise and lower. The problem with this design is that the cords connecting the sash to the weight tended to wear and eventually break, requiring time-consuming maintenance.

While sliding sash windows are still made with weight balances, most modern sliding sash windows have spring balances rather than weight balances. These spring balances are attached to the side of the frame and the underside of the sash. When the window is closed the spring is at its maximum extension ready to take the weight of the sash when in motion.

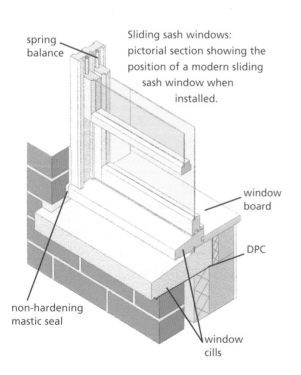

spring balance

Sliding sash windows: pictorial section showing the position of a modern sliding sash window when installed.

window board

DPC

non-hardening mastic seal

window cills

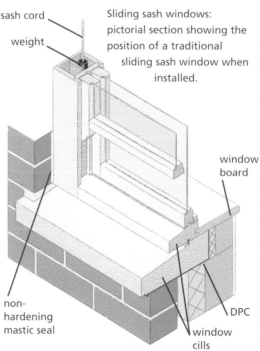

sash cord

weight

Sliding sash windows: pictorial section showing the position of a traditional sliding sash window when installed.

window board

non-hardening mastic seal

DPC

window cills

Sliding sash windows are weather-sealed to exclude wind and rain using vertical wiping seals against the stiles and compression seals at the head and cill. Double-glazing is fitted as standard in new timber sliding sash windows.

Sliding sash windows are also available in uPVC form. These are framed using bulky hollow sections that are reinforced with aluminium box sections and are designed to take double-glazing units.

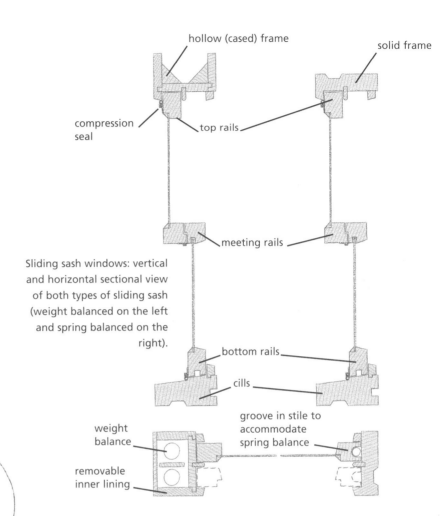

hollow (cased) frame

solid frame

compression seal

top rails

meeting rails

Sliding sash windows: vertical and horizontal sectional view of both types of sliding sash (weight balanced on the left and spring balanced on the right).

bottom rails

cills

groove in stile to accommodate spring balance

weight balance

removable inner lining

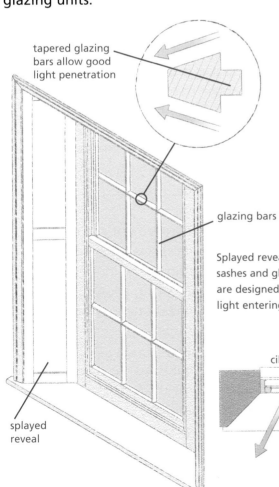

tapered glazing bars allow good light penetration

glazing bars

splayed reveal

The frames, sashes and glazing bars are tapered to maximise the amount of light entering. This is also a feature of the reveals and windows of older Georgian style windows.

Splayed reveal and tapered frames, sashes and glazing bars. These elements are designed to maximise the amount of light entering.

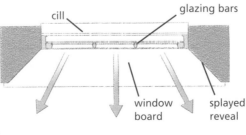

cill

glazing bars

window board

splayed reveal

Pivot (Velux®) window: vertical section through a typical pivot window. These windows are supplied with a flashing kit to ensure a watertight seal with the roof is achieved.

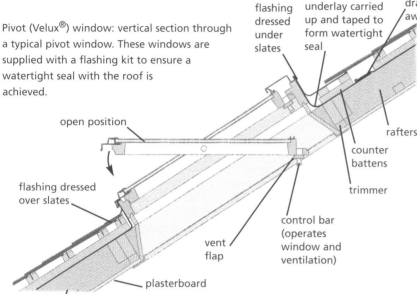

flashing dressed under slates

underlay carried up and taped to form watertight seal

drainage trim to divert any moisture away from head of window

open position

rafters

counter battens

trimmer

flashing dressed over slates

vent flap

control bar (operates window and ventilation)

plasterboard

rafters (insulation omitted for clarity)

Pivot Windows

The most common type of pivot window used in the construction of houses in Ireland is the centre-pivot roof-light. Roof-lights are commonly referred to as Velux® windows. This is because the Velux® brand name has become associated in the public mind with roof-lights in much the same way as Hoover® has for vacuum cleaners. Roof-lights are commonly used to allow light into living spaces in roofs (see chapter 10, Roofs). One of the major advantages of pivot windows is that they can be rotated for easy cleaning from the inside.

Double-glazing

Sealed double-glazing units are used to ensure human comfort and energy conservation in all new houses. These units consist of two panes of glass that are hermetically sealed (i.e. sealed using heat) to a continuous spacer (e.g. self-adhesive, aluminium reinforced, butyl-based bar) around their perimeter.

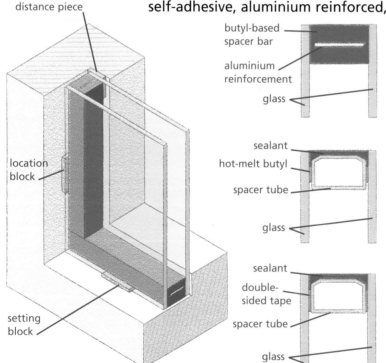

distance piece

butyl-based spacer bar

aluminium reinforcement

glass

sealant

hot-melt butyl

spacer tube

glass

location block

sealant

double-sided tape

spacer tube

glass

setting block

The spacer is typically 12–16 mm wide. This space between the panes reduces the transfer of heat (and sound) energy through the window. The space between the panes is filled with dehydrated air or with a special insulating gas, called argon.

A further improvement can be made by applying a special low-emissivity (low-e) coating to the inner pane of glass. Low-e glazing allows the short-wavelength heat energy from the sun to enter the house but acts as a barrier to the escape of the long-wavelength energy from internal heat sources. About 60% of the heat energy lost between the panes of glass in a double-glazed window is long-wave radiation. A low-e coating will reflect approximately 75% of this long-wavelength heat

Double-glazing: plastic setting blocks (spacers) are used to position the hermetically sealed double-glazed units in the frame.

energy. The low-e coating is usually applied to the outer surface of the inner pane of glass. While these windows are initially more expensive to install, they usually pay for themselves (through the money saved on home heating bills) in a short period of time.

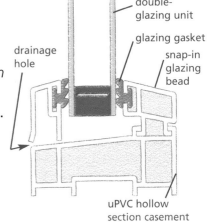

outer pane of glass

inner pane of glass

air or argon in sealed void

low-e coating

short-wave radiation penetrates

short-wave radiation from sun

long-wave radiation reflected back

Thermal insulation: a low-emissivity (low-e) coating applied to the inner pane of glass allows the short-wavelength heat energy from the sun to enter the house but acts as a barrier to the escape of the long-wavelength energy from internal heat sources, e.g. warm floor tiles (summer), radiators (winter).

Double-glazed units are installed using setting and location blocks to ensure a 3 mm clearance between the glass and the casement or sash. These blocks, usually made from plasticised PVC or neoprene, hold the glass in place while allowing room for thermal expansion and contraction of the glass, casement or sash.

In uPVC windows, the glazing unit is held in place using internal *snap-in* beading. Both the frame and the beading have an integral synthetic rubber gasket which forms a watertight seal to the surface of the glass. This beading can be removed if the glass needs to be replaced.

double-glazing unit

glazing gasket

snap-in glazing bead

drainage hole

uPVC hollow section casement

Glazing of uPVC windows: the glazing is held in place using snap-in beading.

In timber windows the glazing unit is usually held in place using a bedding sealant and timber beads fixed with screws. A typical bedding sealant is a non-setting synthetic rubber compound. The sequence of installation typically follows the list below.

- Seal the rebate with primer and allow to dry fully.
- Spread a generous amount of sealant in the rebate.
- Place the setting blocks along the bottom section of rebate.
- Position the glazing unit on the setting blocks.
- Centre the glazing unit using the location blocks.
- Press the glazing unit firmly into place.
- Put the timber beading into position (ensure a thin bead of sealant under the beading and between the beading and the glass).
- Secure the beading using screws at 200 mm centres (max. 75 mm from corners).
- Trim excess sealant (inside and out) and finish with a smooth chamfer.

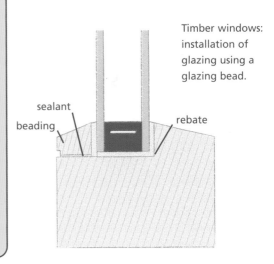

Timber windows: installation of glazing using a glazing bead.

sealant

beading

rebate

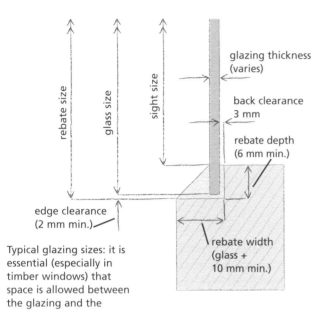

Typical glazing sizes: it is essential (especially in timber windows) that space is allowed between the glazing and the window frame for the seasonal movement of materials.

It is common practice nowadays for factory-glazed windows to be delivered to site. The advantages of factory-glazed windows include time saved on-site and a guaranteed standard of workmanship (clean, dry, work environment). However, they are susceptible to damage during transport and installation. Under normal conditions, the typical service life of sealed double-glazed units is 15–20 years.

Window Hardware

Hardware is the term used to describe the hinges, stays, fasteners and catches used to operate windows.

Timber casements are usually hung using galvanized or lacquer-coated steel butt hinges. The casement is held in the ventilate (open) position using a casement stay. The casement stay is fitted to the bottom rail of the casement and a peg, fitted to the cill of the frame, anchors the stay to provide a variety of opening positions.

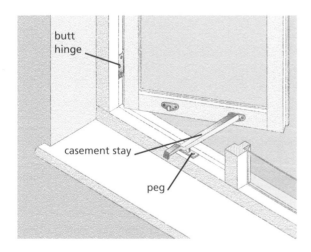

Window hardware: timber casement sash hung using butt hinges and secured using a casement stay.

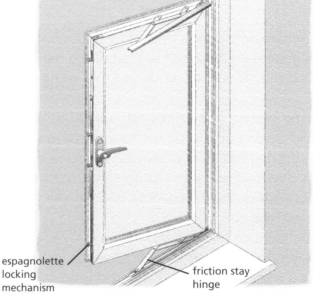

Window hardware: uPVC casement sash hung using friction-stay hinges and secured using an integral espagnolette locking mechanism.

A friction-stay hinge may be used as an alternative to the butt hinge and casement-stay method. This is a two-fold device that acts as both hinge and stay. An adjusting screw is used to ensure adequate friction exists to hold the casement open in the desired position. Friction stay hinges are used by some timber casement manufacturers and nearly all uPVC window manufacturers. Irrespective of which type of hinge is used, it is essential that the casement is held firmly against the weather stripping when closed.

Timber casements are usually secured using a window fastener which is secured at the midpoint of the stile. The fastener locks into a *keep* fixed to the frame. Fasteners are also available with push-button release mechanisms and with key-locking mechanisms, for added security.

uPVC casements are usually secured using an espagnolette-type locking mechanism. This consists of a rail (housed into the stile of the casement) with two (or more) locking pins and two keeps housed in the frame. When the handle is turned the rail slides the locking pins into the keeps, securing the casement. The keeps usually have two locking positions to allow secure ventilation of the house. Espagnolette mechanisms have the benefit of holding the casement more uniformly against the frame thereby improving the weather resistance of the window. Some timber casement manufacturers also fit espagnolette mechanisms.

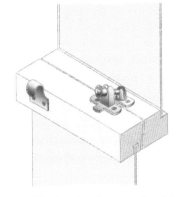
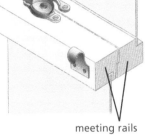

Window hardware: sash catches.

meeting rails

Sliding sash windows are secured by means of a sash catch. This is a simple mechanism (often supplied in stainless steel or brass) that is fixed to the top rail of the lower (inside) sash. It rotates into the locking position to secure the sashes in place.

Revision Exercises

1. Outline briefly the performance requirements of windows in a typical house.
2. Discuss, with the aid of neat freehand sketches, the concept of fenestration.
3. Describe, with the aid of neat freehand sketches, three types of windows.
4. Generate a neat freehand sketch of a sliding sash window. Label each part.
5. Explain the importance of thermal insulation in relation to windows. Compare the effectiveness of various approaches to double-glazing.

6. Describe the requirements, under the building regulations, for the ventilation of habitable rooms.

7. Describe the requirements for windows under the building regulations, in relation to ease of escape from habitable rooms in the event of fire.

8. Explain, with the aid of neat annotated sketches, the procedure for installing a double-glazed unit in a timber sash.

9. A timber window frame with an outward-opening, double-glazed sash, is fixed in a 300 mm concrete cavity wall, which has external rendering and plaster inside. Draw, to a scale of 1:5 a neat freehand vertical cross-section sketch through the head and cill showing all relevant construction details and surrounding structure.

10. Ordinary Level, 2003, Question 4

    (a) Using notes and neat freehand sketches show how a window opening, as shown in the sketch, is formed in a 300 mm concrete block wall with an insulated cavity. Show the details at the top and sides of the window opening.

    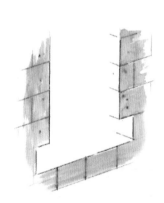

    (b) On a separate sketch, show the correct position of a window frame in the opening, and on the sketch show a method of securing the window frame to the wall.

11. Ordinary Level, 2004, Question 1

    A timber casement window is fixed in a standard 300 mm external block wall with an insulated cavity, as shown in the sketch. The wall is plastered on both sides. To a scale of 1:2 (half full-size) draw a vertical section through the lintel, window head and sash. Show all the construction details from 350 mm above to 200 mm below the top of the window frame.

12. Higher Level, 2003, Question 9

    A stormproof casement window made from softwood is located in an external wall and provides natural lighting to a kitchen area.

    (a) Describe in detail, using notes and freehand sketches, two design details that ensure that the window is weatherproof.

    (b) Discuss two advantages and two disadvantages of using softwood in the manufacture of windows.

    (c) An illuminance of 300 lux is required on a working plane in the kitchen. The daylight factor at a point on the working plane in the kitchen is five per cent. Show by calculation if the illuminance is sufficient, assuming an unobstructed view and the illuminance of a standard overcast sky to be 5,000 lux.

The design of doors has evolved considerably. Early doors consisted of timber planks nailed together to form a crude gate at entrances – modern doors are usually delivered to site in door sets. A door set consists of a door frame, door and hardware. These doors are quickly installed because the door is pre-hung in the frame and most of the hardware is pre-fitted. All the site carpenter has to do is to secure the door in the ope, ensuring it is square, plumb and functioning properly.

Functions

The primary function of any door is to permit access to a building or to a space within a building.

Design Features

Doors must be designed to meet a number of performance requirements including the following:

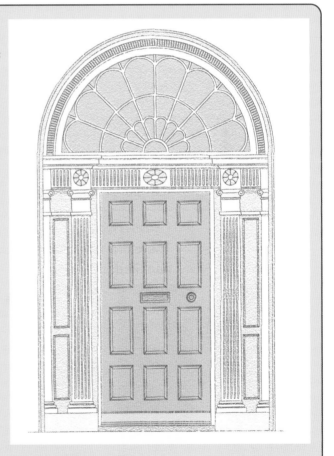

A typical Georgian door.

- Accessibility – A doorway should be designed to allow comfortable access by all users. This relates in particular to size of the opening, positioning of handles, door bells and letter boxes and the level of the ground at the threshold. For example, the main entrance to a home is required to be at least 800 mm wide – however, it is worth remembering that an adult using crutches requires an opening of 950 mm to safely access, while a guide dog user would require an opening of 1,100 mm (see chapter 3, Planning Permission and Sustainable Development).
- Security – This applies mainly to external doors. External doors should ideally have a night latch cylinder lock and a mortice deadlock. Alternatively, a multipoint (espagnolette) lock may be fitted by the door manufacturer (see hardware following).
- Durability – External doors have traditionally been made from hardwoods (e.g. teak) because of their durability. While hardwood doors are very common, uPVC doors are also used. It is essential that hardwood external doors are properly maintained if they are to function properly for their expected service life. This

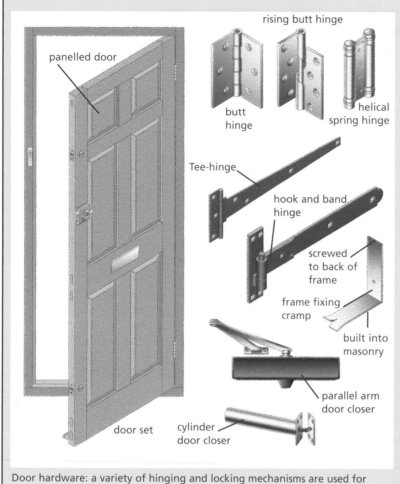

Door hardware: a variety of hinging and locking mechanisms are used for both internal and external doors and gates.

involves varnishing and painting every few years as well as oiling the hardware.

- Thermal Insulation – External doors must be adequately sealed to prevent draughts, heat loss and water penetration. The stiles (sides) and top rail of the door rely on weather seals, while the bottom rail is usually weather-sealed using a proprietary threshold seal. These seals consist of two parts – one part is secured to the ground (threshold) and the other to the bottom rail. When the door is closed the two parts interlock to provide a weather-resistant seal.

- Aesthetics – As the first element of the house which a person has contact with, the front door has an important aesthetic role. An attractive door can greatly enhance the appearance of a house. We know this from experience – the doorways of many heritage buildings, especially Georgian houses, were designed to make the best possible impression on visitors. The proportion, colour and detail of the door contribute to making it an attractive feature of the house.

Construction of External Doors

There are two main types of timber external doors, matchboarded and panelled.

matchboard

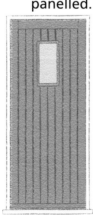

matchboard with small light

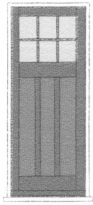

partial glazed, vertical panelled

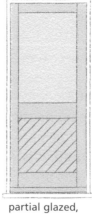

partial glazed, contemporary

partial glazed, panelled

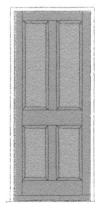

panelled

External doors: a variety of door types exist. Simple, unfussy designs generally suit most houses.

Framed panelled doors are the most commonly used external doors. They are constructed in hardwood using mortice and tenon joints for strength and durability. The panels float (i.e. not glued in place) in grooves in the frame to allow for seasonal movement of the timber.

The bottom of the door should be weathered to resist wind-driven rain and to shed rainwater that runs down the face of the door that could otherwise be blown under the door. This is achieved by using a threshold seal in conjunction with a weatherboard. The threshold seal will resist the wind-driven rain, while the weatherboard ensures that rainwater is carried away from the bottom of the door.

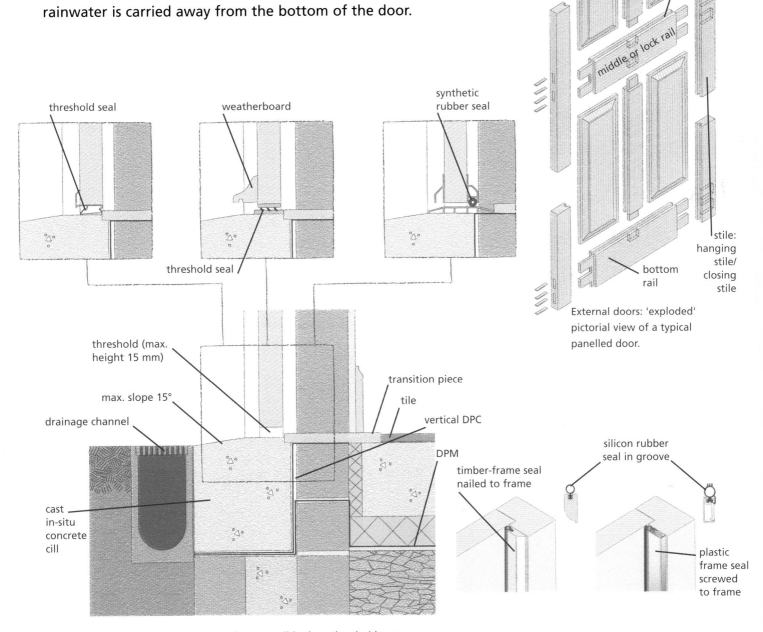

External doors: 'exploded' pictorial view of a typical panelled door.

External door threshold: it is essential that accessible door thresholds are sealed to resist wind-driven rain.

External doors: the frame is weatherproofed to reduce heat loss.

Matchboarded doors are commonly used for outdoor applications such as side gates and shed doors, although they are also used as external doors for houses. Matchboarded doors come in a number of variations, including:

- Ledged – suitable for use in narrow openings only (e.g. side entrances).
- Ledged and braced – commonly used for sheds and other outbuildings.
- Framed, ledged and braced – suitable for use as external doors of houses.

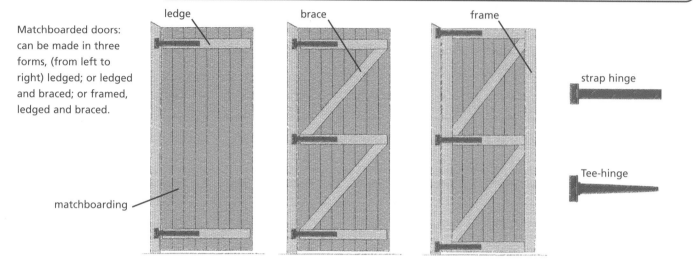

Matchboarded doors: can be made in three forms, (from left to right) ledged; or ledged and braced; or framed, ledged and braced.

Matchboarded doors have the advantage of being relatively quick and simple to make. It is essential that a matchboarded door is hung so that the braces resist the tendency of the door to sag under its own weight – the braces are fixed so that they create a *V* shape with the hanging stile.

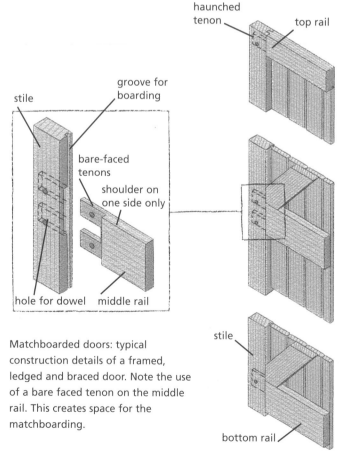

Matchboarded doors: typical construction details of a framed, ledged and braced door. Note the use of a bare faced tenon on the middle rail. This creates space for the matchboarding.

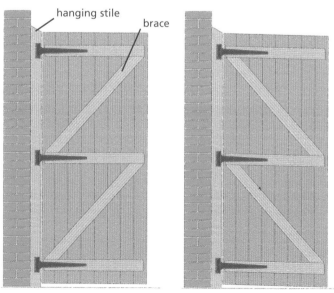

Matchboarded doors: it is essential that a matchboarded door is hung so that the braces resist the tendency of the door to sag under its own weight.

External uPVC doors consist of a frame and a panelled or glazed door. The frame, stiles and rails of the door are manufactured from hollow uPVC sections which have a reinforcing aluminium core. The door panels consist of press-moulded plastic with an insulating core. The panels may be reinforced with sheet aluminium for security. Glazing panels are held in place using internal snap-in beading (see chapter 14, Windows). These doors are commonly supplied in a smooth white finish,

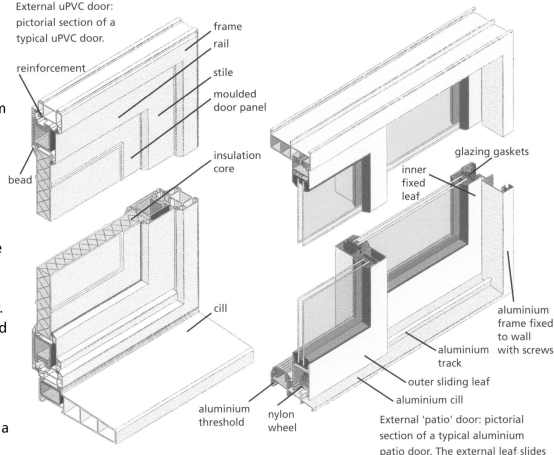

External uPVC door: pictorial section of a typical uPVC door.

frame
rail
reinforcement
stile
moulded door panel
bead
insulation core
cill
aluminium threshold
nylon wheel

glazing gaskets
inner fixed leaf
aluminium frame fixed to wall with screws
aluminium track
outer sliding leaf
aluminium cill

External 'patio' door: pictorial section of a typical aluminium patio door. The external leaf slides while the internal leaf is fixed.

although they are also available in colours and with wood grain effect. uPVC doors require little maintenance, although any scratches or dents that might occur over time cannot be hidden by painting or varnishing as they could be with a timber door.

Patio doors are fully glazed aluminium doors that provide a large area of clear glass for an unobstructed view, usually of the rear garden. A typical patio door consists of one fixed and one sliding leaf. The frame is manufactured from extruded hollow aluminium. The doors are double-glazed to reduce heat loss and have weather stripping to exclude wind and rain. The bottom rail of the opening leaf contains a number of nylon wheels which run on a track in the cill. It is important that the track is cleaned periodically to ensure the door runs smoothly.

Construction of Internal Doors

There are two main types of internal doors: framed and flush. The framed type of internal door is similar to the external version but is commonly made from softwood and is generally of lighter construction. Flush doors come in a variety of forms: cellular core, skeleton core and solid core.

Cellular-core flush doors are made with a cellular fibreboard or paper core in a light softwood frame with lock and hinge blocks. These doors are mass-produced in standard sizes. They are quite flimsy, and provide poor acoustic insulation, security and fire resistance. Skeleton-core flush doors are made with a core of small section timbers, with lock and hinge blocks. The internal timbers usually occupy 30–40% of the core of the door. These doors are sturdier than cellular-core doors and will better withstand normal domestic use. Solid-core flush doors are made with a core of small section timbers (or particleboard strips) glued face to face. The door is edged or lipped to give a neat finish. These doors are much stronger than cellular or skeleton-core doors.

Internal doors: flush doors come in a variety of forms (from left to right): skeleton core, cellular core and solid core.

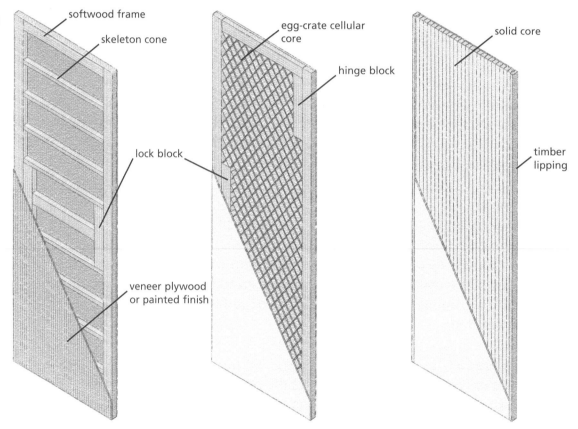

softwood frame

skeleton cone

lock block

veneer plywood or painted finish

egg-crate cellular core

hinge block

solid core

timber lipping

Traditionally, flush doors have a decorative plywood (e.g. mahogany and teak) facing. However, many modern flush doors have a fibreboard surface which has been press-manufactured to look like a panelled door. These doors often have a wood grain effect and are supplied pre-finished.

Frames and Fixings

Doors are traditionally hung from a timber frame, consisting of two side posts (jambs), a head and a cill. The side posts and head are usually machine-rebated to take the closing edge of the door. A door frame is usually less than the full width of the wall. The frame is commonly fixed to masonry walls using steel straps (frame fixing cramps) which are screwed to the frame and secured to the walls using nails. Alternatively, steel screws secured in plastic plugs can be used.

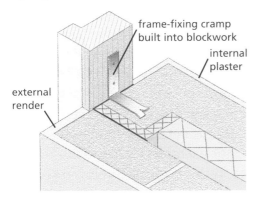
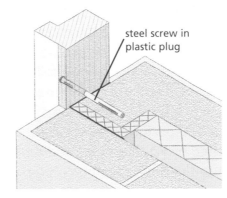

external render

frame-fixing cramp built into blockwork

internal plaster

steel screw in plastic plug

Fixing door frame using steel straps/brackets. Fixing door frame using steel 'hammer-in' screws with plastic plugs.

Internal door frames are lighter than those used for external doors and are sometimes referred to as *linings*. A lining is usually the full width of the partition. As an alternative to a rebated frame, a planted-on stop may be used for internal doors. When securing the frame in a timber stud partition the frame is fixed directly to the studs.

A gap (approx. 3 mm) is left between the door edge and the frame to accommodate dimensional changes that are caused by changes in the frame and door moisture content. This is essential to ensure the door does not stick.

Traditionally the frame is strengthened across the bottom by the saddle. However, in many new houses the saddle is omitted in order to make the house more user-friendly for everyone (see lifetime use, chapter 3, Planning Permission and Sustainable Development). In this case a temporary strut is fixed across the bottom of the frame to keep it square during installation.

The frame provides an edge for the plasterer to work to when finishing the walls. Once the plastering is complete the architrave can be fitted. *Architrave* is a decorative moulding that emphasises or frames the opening (like a picture frame). It is fixed around the sides and tops of the door to provide a neat finish and to hide any gaps that might appear between the plaster or frame due to movement in the timber.

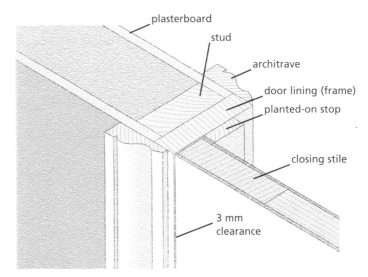

plasterboard

stud

architrave

door lining (frame)

planted-on stop

closing stile

3 mm clearance

Internal door linings are fixed to the internal walls using screws. Architrave is used to frame the door and to hide the joint between the lining and the wall.

stud partition   fire-resisting
                 frame

planted-   typically
on stop   44 mm

architrave (15 mm
min. thickness +
15 mm min. overlap
onto wall and
frame)

intumescent seal
fitted in groove
in door lining

softwood
laminated
core

veneer
facing
(3 mm)

Fire door construction: typical
30-minute fire door (FD30).

intumescent
smoke seal
                 typically
                 54 mm

hardwood
lipping (8 mm)

hardwood
outer stile

softwood
inner stile

core material
(fire-resistant
particleboard)

timber substrate
(5 mm)

veneer facing
(3 mm)

Fire door construction: typical
60-minute fire door (FD60).

Fire Doors

A fire door is a special type of internal door which is designed to resist the spread of flames and hot gases. An example of where a fire door is fitted in a house would be where access to the house is possible from an adjoining garage. Fire doors are classed by the length of time they provide protection. The precise classification of a fire door provides two pieces of information – insulation and integrity. Insulation in this context means the resistance to thermal transmittance between the surfaces of the door. Integrity means resistance to the penetration of flame and hot gases. The performance standard of the door is measured in minutes and denoted by FD (fire door) and a number. For example, FD30 is a 30-minute fire door. Common materials used in the manufacture of fire doors include plasterboard, compressed mineral wool and high-density (fire-resistant) particleboard.

Fire doors can also be designed to withstand the passage of heat and smoke between the door and the frame. This is done by installing an intumescent (swelling) strip in the frame. At temperatures of around 150°C the intumescent strip expands creating a seal around the door

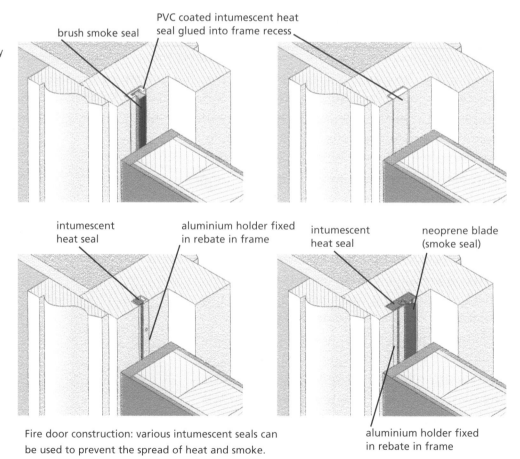

brush smoke seal

PVC coated intumescent heat
seal glued into frame recess

intumescent
heat seal

aluminium holder fixed
in rebate in frame

intumescent
heat seal

neoprene blade
(smoke seal)

aluminium holder fixed
in rebate in frame

Fire door construction: various intumescent seals can
be used to prevent the spread of heat and smoke.

edge. The seal prevents the passage of heat and smoke while still allowing the door to be opened if necessary.

Door Hardware

Hardware is the term used to describe the hinges, locks, bolts, latches and handles used to operate doors. Doors are hung using hinges. The simplest hinge is the steel or brass butt hinge. These consist of two flaps held securely with a pin. The pin is held inside the knuckle and in some types can be tapped out from below to facilitate removal of the door for maintenance. The flaps are housed flush into the edge of the door and the face of the frame and secured with screws.

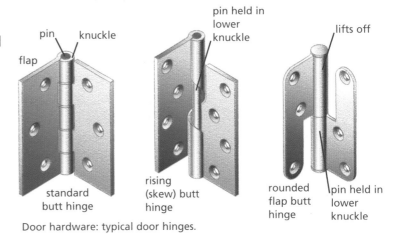

Door hardware: typical door hinges.

One of the disadvantages of butt hinges is the time taken to remove a door when maintenance is necessary. This can be overcome by using lift-off hinges. A lift-off butt hinge allows the door to be lifted off the hinge without having to remove any screws. The pin is fixed to the lower half of the knuckle, so that when the door is lifted the upper half of the hinge slides up off the pin.

A rising butt hinge is similar, except the knuckles are skewed so that the door rises as it is opened. This is useful as it reduces wear on floor coverings (especially deep pile carpets) caused by the door and also has the effect of making the door self-closing.

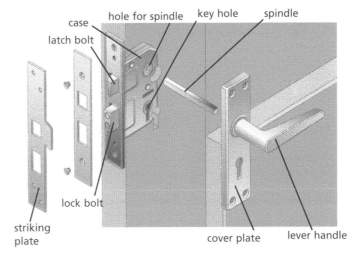

Door hardware: typical two-bolt vertical (upright) mortice lock.

Factory-fitted butt hinges (supplied in door sets) commonly have rounded rather than square flaps. This is because the cutter head used to create the trench for the hinge has a rotational cutting action. Rounding the flaps speeds up the fitting process by eliminating the need to square out the trench.

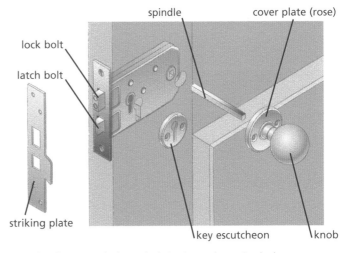

Door hardware: typical two-bolt horizontal mortice lock.

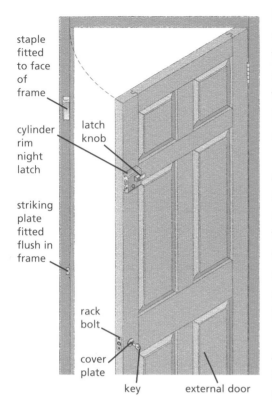

staple fitted to face of frame

cylinder rim night latch

latch knob

striking plate fitted flush in frame

rack bolt

cover plate

key   external door

Door hardware: typical cylinder-rim night latch and rack bolt.

Mortice locks are commonly used in external and internal doors. These locks are set into a mortice (square hole) cut in the lock (middle) rail of the door. Mortice locks typically consist of a latch bolt and a lock bolt. The latch bolt is opened using the door handle, whereas the lock bolt is operated using a key. Both the latch and lock bolts fit into a striking plate which covers the receiving mortices in the door frame.

A cylinder-rim night latch is commonly used on the external doors of houses. This type of lock is designed, for convenient use of front doors, as a latch from the inside and a lock from the outside. Cylinder-rim night latches are not very secure and should be used in conjunction with a mortice deadlock at night and when the house is unoccupied. A mortice deadlock is a mortice lock that has a single key operated bolt. It is usually installed lower down in the door to resist *kicking-in* of the door.

A rack bolt is another type of single-locking device. It can be used on timber casement windows as well as doors. It consists of a cylindrical case and bolt that is fitted into a mortice cut in the stile or casement. A striking plate is fitted over a receiving mortice in the frame. It is operated using a special star-shaped key.

Many timber and uPVC doors are now supplied with espagnolette-type locking mechanisms. These are very similar to those described in window hardware (see chapter 14, Windows). When fitted to a door they usually include one or two bolts to augment the locking pins. The door is normally locked (from either side) by pulling the handle upwards and turning the key. A common disadvantage of these types of mechanism is that once unlocked they can usually be opened from the outside and so do not offer any security unless the key is turned. Furthermore, in the event of a fire, escape is only possible if the key is to hand – this leads many people to leave the key in the door at night, which is not very secure if there are sidelights (side windows).

760 (min.)

1,070 (ideal)

1,450 (max.)

Door hardware: required position of letter plate to avoid back injury to postal workers.

Revision Exercises

1. Outline briefly the performance requirements of doors in a typical house.
2. Describe, with the aid of neat freehand sketches, the construction of a framed, panelled door.
3. Describe, with the aid of neat freehand sketches, the construction of a framed, ledged and braced matchboarded door.
4. Explain how the hanging stile of a braced matchboarded door can be identified.
5. Describe, with the aid of neat freehand sketches, the various types of flush door.
6. Discuss, briefly, the concept of fire doors. Include in your answer an explanation of the classification of fire doors.
7. Generate a neat cross-sectional sketch of the jamb of an external panelled door. Indicate how the frame is held securely in place and the components used to ensure the exclusion of moisture.
8. Explain, using a neat annotated sketch, the hardware used on a modern timber external door.
9. Ordinary Level, 2000, Question 6
   (a) With the aid of notes and sketches show the main constructional details of:
      (i) an internal flush door,
      (ii) an external wooden door.
   (b) Sketch a joint suitable for jointing the top rail to the stile of an external wooden door.
10. Ordinary Level, 2005, Question 1
    A flush panel door is fixed in a standard 100 mm internal solid concrete block wall of a house, as shown in the sketch. The wall has a hardwall plaster finish on both sides. The house has an insulated solid concrete ground floor. To a scale of 1:5, draw a vertical section through the foundation, floor and door. The section should show all the constructional details from the bottom of the foundation of the internal  block wall to 600 mm above finished floor level.

11. Higher Level, 2003, Question 1
An external wooden door with four panels is
shown in the accompanying sketch. The upper
panels are glazed and the lower panels are
solid. The door opening is located in a
standard 300 mm external concrete block wall
with an insulated cavity. The house has a solid
concrete ground floor.

(a) To a scale of 1:5, draw a vertical section
    through the external wall and door,
    showing clearly the threshold, the door
    and the door frame. The section should
    show all the constructional details from
    300 mm below the bottom of the door to
    300 mm above the top of the door frame.
(b) Indicate on the drawing two design details
    that ensure that moisture does not
    penetrate to the inner surfaces surrounding the door.

12. Higher Level, 2006, Question 1
The main entrance to a dwelling house, as shown in the accompanying
sketch, provides access for a person in a wheelchair. The door opening is
located in a 300 mm external block wall with an insulated cavity and the
door is a solid wooden door. The house has
a solid concrete, ground floor with a 20 mm
quarry tile finish.

(a) To a scale of 1:5, draw a vertical section
    through the external wall and door,
    showing clearly the threshold and the
    door.
    The section should show all the
    constructional details from the bottom
    of the foundation to 300 mm above
    finished floor level.
(b) Indicate on the drawing the specific
    design detailing that ensures that rainwater is removed from the
    threshold area and does not penetrate to the inner surfaces
    surrounding the door.

Functions

The primary function of a stair or stairway is to provide safe, comfortable and easy walking access between floor levels in a building.

Design Features

A stair must be designed to meet a number of performance requirements including:

- Safety – Stair safety is primarily dealt with at the design stage. The building regulations provide very specific guidelines in relation to the height (rise) and depth (going) of steps. This ensures that stairs everywhere have a similar *feel*. It also ensures that stairs are designed to be comfortable and safe for all users.
- The factors influencing the safe use of stairs include:
  - Rise and going – steps that are too high can be difficult for some users (children, older people, people on crutches etc.), also the step should provide adequate space for the secure placing of the foot when walking up and down.
  - Pitch (slope) – a steep pitch can be dangerous and a shallow pitch will lead to a waste of space within the house.
  - Provision of handrails – a handrail is essential to provide support in a wide range of situations (e.g. a parent carrying a baby would always hold the handrail when descending the stair) – the handrail should not have any gaps or parts where fingers or clothing might be trapped.
  - Design of railings – children should not be able to climb railings – this is achieved by ensuring that the rail does not have horizontal components that would create a ladder effect.
  - Provision of landings – similar to the rise and going, providing a landing gives less able users a chance to rest and makes the stair safer by reducing the potential fall distance.
- Strength, Stability and Durability – A typical stair accommodates a lot of traffic throughout its service life. For this reason, a stair must reliably support dead loads and live loads: it must be securely jointed to the structure of the house and stair steps and railings must be constructed to resist the wear and tear of everyday use. Each of these factors is influenced by the materials chosen, the jointing methods used and the quality of workmanship.
- Compactness – A quarter or half turn may be used to reduce the floor area taken up by a stair or to locate it in a less obstructive position. When this is done a landing, rather than winders (tapered steps) should be used because winders are not as safe.
- Aesthetics – In many new houses the stairway is prominently located just inside the front door. For this reason, it is important that due consideration is given to its aesthetic quality (e.g. appearance).

Stair Terminology

Stair terminology: typical components of single flight stair in a house.

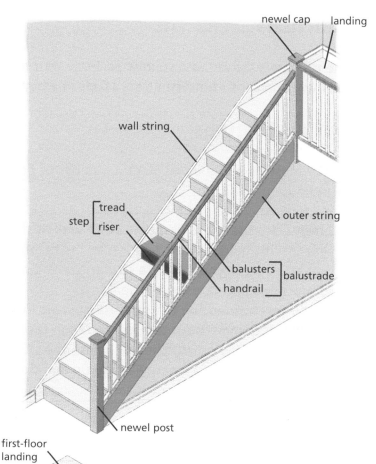

Stair Types

Stair types: depending on the floor plan, various types of stairs can be used to best suit the layout.

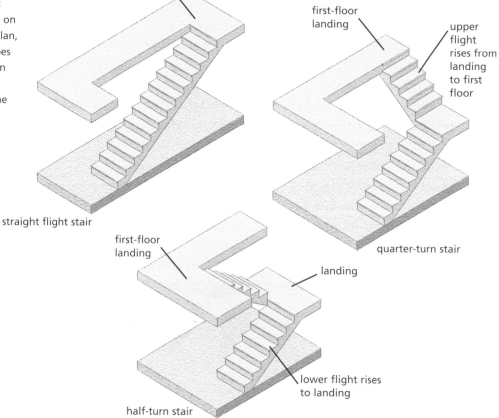

Stair Building Regulations

The following guidelines are derived from the requirements of the building regulations to ensure every stair is properly designed:

- Rise (× 2) + going should be equal to between 550 and 700 mm.
- Maximum allowable rise = 220 mm (175 mm optimum) – any higher may be difficult for young or older people.
- Minimum allowable going = 220 mm (250 mm optimum) – to ensure sufficient space for foot on step.
- Maximum of 16 steps without landing.
- Nosing of 16 to 25 mm – ensures steps overlap.
- Maximum pitch of 42° – not too steep, optimum 35°.
- Handrail 840 to 900 mm above pitch line.
- Headroom 2 m minimum vertically from pitch line.
- Balusters not to allow 100 mm sphere to pass.

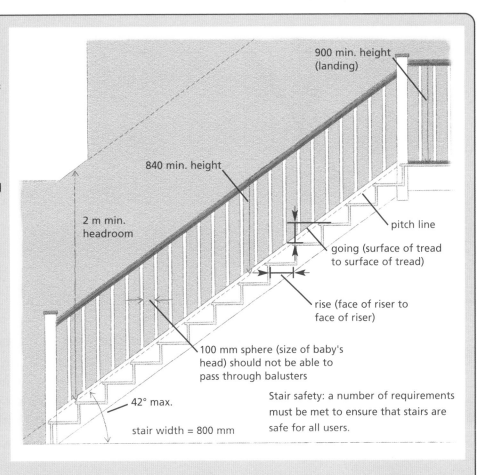

900 min. height (landing)

840 min. height

2 m min. headroom

pitch line

going (surface of tread to surface of tread)

rise (face of riser to face of riser)

100 mm sphere (size of baby's head) should not be able to pass through balusters

Stair safety: a number of requirements must be met to ensure that stairs are safe for all users.

42° max.

stair width = 800 mm

Stair Construction

A staircase consists of two strings and a number of steps. The strings are the wide sloping boards, at either side of the stair, that support the steps. The steps are made up of two parts – the tread (on which we tread or walk) and the risers. The loads imposed on a stair are mainly supported by the strings. There are two types of string:

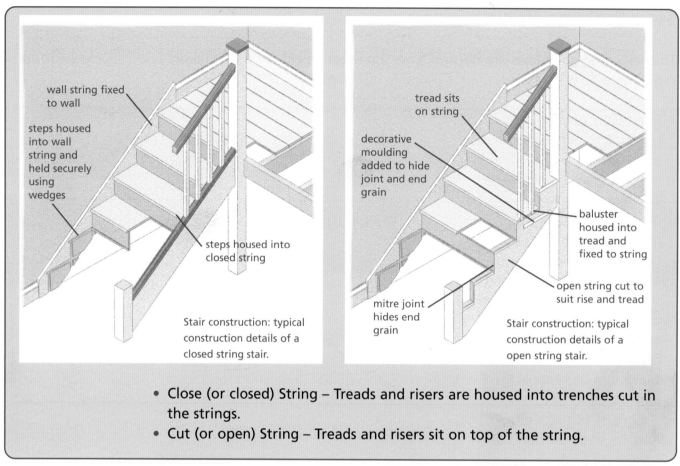

wall string fixed to wall

steps housed into wall string and held securely using wedges

steps housed into closed string

Stair construction: typical construction details of a closed string stair.

tread sits on string

decorative moulding added to hide joint and end grain

baluster housed into tread and fixed to string

open string cut to suit rise and tread

mitre joint hides end grain

Stair construction: typical construction details of a open string stair.

- Close (or closed) String – Treads and risers are housed into trenches cut in the strings.
- Cut (or open) String – Treads and risers sit on top of the string.

The opening in which a stair is constructed is called a *stairwell*. Usually a stairwell is located beside an external wall. When this is the case the string against the wall is called the *wall string* and the string on the outside is called the *outer string*.

When a close string is used, the tread and riser are held firmly in place using wedges. When a cut string is used, the treads and risers are secured using triangular brackets. The tread and riser are jointed to improve the strength, stability and durability of the stair. This is usually done using a tongued joint, although a variety of methods exist. Creaking is one of the most common faults of stairs. This happens when the tread bends under the weight of the user causing the tongue to move in the trench causing a creaking sound. This can be avoided by securing the treads to the risers using glue blocks and brackets. Alternatively, a carriage may be installed to provide additional support to the centre of each step. This consists of a length of timber fixed to the underside of the steps using brackets.

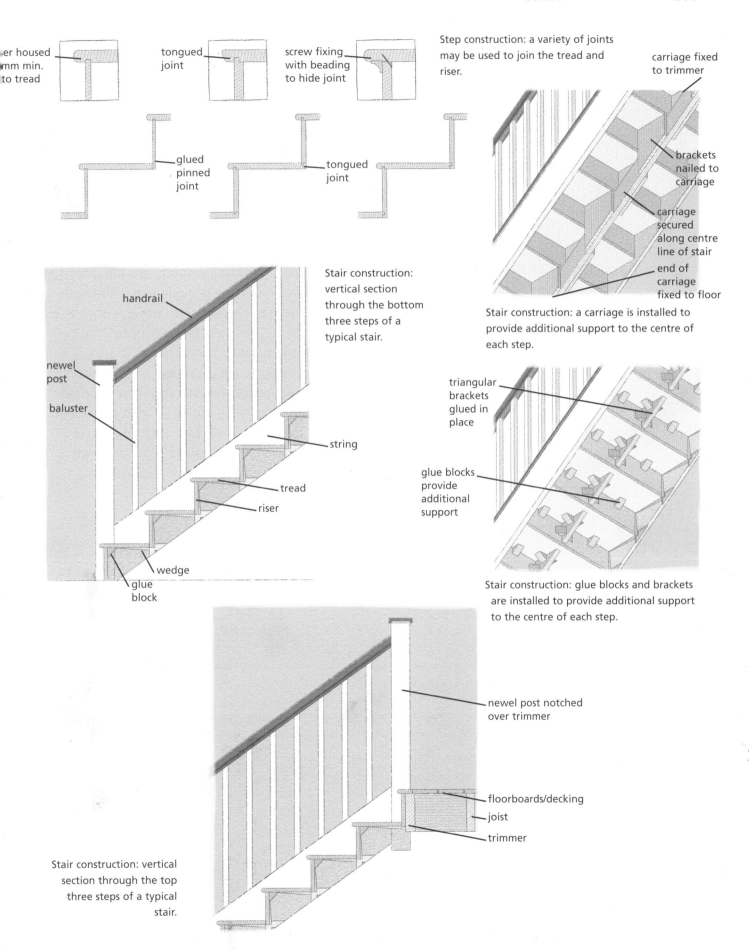

...er housed
...mm min.
...to tread

tongued joint

screw fixing with beading to hide joint

Step construction: a variety of joints may be used to join the tread and riser.

carriage fixed to trimmer

glued pinned joint

tongued joint

brackets nailed to carriage

carriage secured along centre line of stair

end of carriage fixed to floor

Stair construction: a carriage is installed to provide additional support to the centre of each step.

handrail

Stair construction: vertical section through the bottom three steps of a typical stair.

newel post

baluster

string

tread

riser

wedge

glue block

triangular brackets glued in place

glue blocks provide additional support

Stair construction: glue blocks and brackets are installed to provide additional support to the centre of each step.

newel post notched over trimmer

floorboards/decking

joist

trimmer

Stair construction: vertical section through the top three steps of a typical stair.

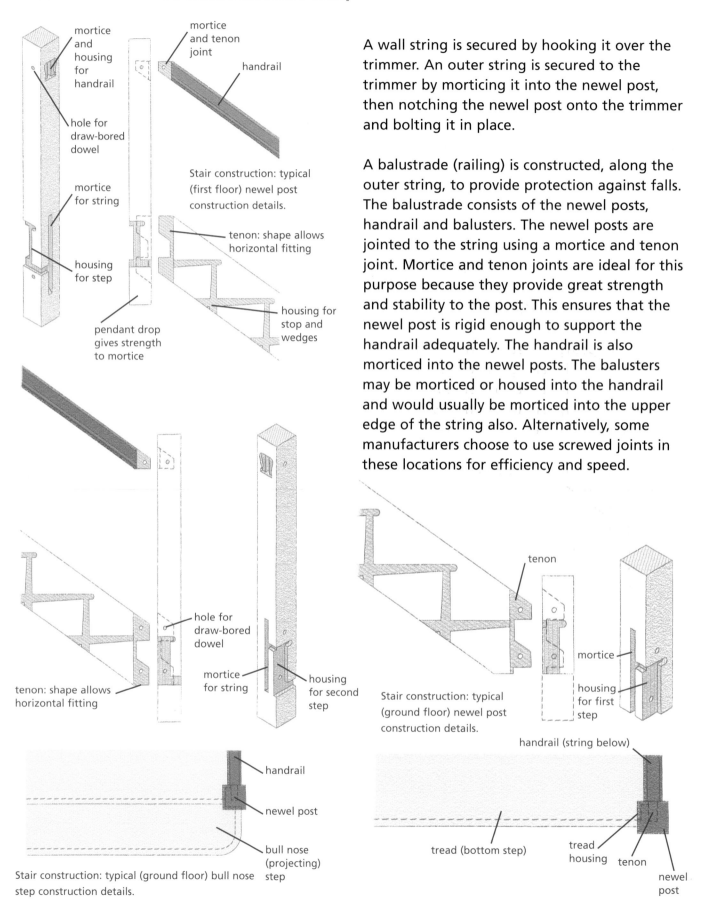

A wall string is secured by hooking it over the trimmer. An outer string is secured to the trimmer by morticing it into the newel post, then notching the newel post onto the trimmer and bolting it in place.

A balustrade (railing) is constructed, along the outer string, to provide protection against falls. The balustrade consists of the newel posts, handrail and balusters. The newel posts are jointed to the string using a mortice and tenon joint. Mortice and tenon joints are ideal for this purpose because they provide great strength and stability to the post. This ensures that the newel post is rigid enough to support the handrail adequately. The handrail is also morticed into the newel posts. The balusters may be morticed or housed into the handrail and would usually be morticed into the upper edge of the string also. Alternatively, some manufacturers choose to use screwed joints in these locations for efficiency and speed.

mortice and housing for handrail

hole for draw-bored dowel

mortice for string

housing for step

mortice and tenon joint

handrail

Stair construction: typical (first floor) newel post construction details.

tenon: shape allows horizontal fitting

housing for stop and wedges

pendant drop gives strength to mortice

tenon: shape allows horizontal fitting

hole for draw-bored dowel

mortice for string

housing for second step

tenon

mortice

housing for first step

Stair construction: typical (ground floor) newel post construction details.

handrail

newel post

bull nose (projecting) step

Stair construction: typical (ground floor) bull nose step construction details.

handrail (string below)

tread (bottom step)

tread housing

tenon

newel post

It is common practice today for a stair to be delivered to site ready for installation, especially when large numbers of similar houses are being built. The construction of the stair in factory conditions allows for the use of automated machinery thereby saving time and ensuring consistent standards and less waste of timber. A stair is usually delivered to the site wrapped in polythene to protect against moisture and with sacrificial timber boards or cardboard secured to the surface of the treads to prevent damage during construction.

It is essential that when a stair is installed, sufficient unobstructed landing space be provided at the top and bottom of the stair. Doorways should be at least 400 mm clear of the landing area.

Stair Calculations

The rise and going of a stair varies from building to building. It is important to calculate the rise and going of a stair to suit the specific building in which the stair is to be installed to ensure that every rise is exactly equal. Remember that if there is any variation in the rise and going of the steps the user is much more likely to fall.

The dimensions of the rise and going of a stair are calculated in three stages. Stage One begins with the fact that the maximum allowable number of rises is 16. This means that the total rise must be divided into 16 or less individual rises. Stage Two requires that the formula for the rise and going be satisfied – twice the rise plus the going must equal between 550 mm and 700 mm. Stage Three involves checking that the rise and going meets the regulatory requirements (i.e. rise less than or equal to 220 mm, going more than or equal to 220 mm). If each of these three requirements are met, the stair design is correct.

Example

Calculate the rise and going of a stair to span between two floors of a house which are 2,660 mm apart.

Solution:

Stage One
Remember, maximum number of rises = 16

2,660 ÷ 16 = 166·25 mm
(This is no good because it would be very difficult to cut a rise to such a precise specification.)

2,660 ÷ 15 = 177·33 mm
(Again, too awkward a dimension.)

2,660 ÷ 14 = 190 mm
(Perfect – a nice round number.)

Note: 2,660 ÷ 10 = 266 mm or 2,660 ÷ 7 = 380 mm (While these are also even dimensions, they are above the maximum allowable rise of 220 mm.)

Stage Two
Formula: Rise (× 2) + Going = 550 to 700 mm

- (190) (2) + Going = 550 to 700 mm
- 380 + 250 (optimum going) = 630 (This is between 550 and 700 mm.)

Stage Three
A rise of 190 is less than 220 mm and a going of 250 is more than 220 mm, therefore this step meets the requirements of the building regulations.

Therefore, a rise of 190 mm and a going of 250 mm is recommended for the stair of this house.

Revision Exercises

1. Explain, with the aid of neat freehand sketches, the following terms: riser, tread, string, newel post, baluster, handrail.
2. Explain, with the aid of neat freehand sketches, the following terms: rise, going, pitch line.
3. Explain, with the aid of neat freehand sketches, the requirements of the building regulations in relation to stair design.
4. Explain, with the aid of neat freehand sketches, three methods of joining treads and risers.
5. Generate, to a scale of 1:5, a neat annotated cross-sectional sketch of the bottom three steps of a typical domestic stair. Include the newel post, handrail and balusters in your sketch.
6. Generate, to a scale of 1:5, a neat annotated cross-sectional sketch of the top three steps of a typical domestic stair. Include the trimmer and one joist in your sketch.
7. Discuss some of the design features that make a stair safe for use by all users.
8. Calculate the rise and going for a flight of stairs to span between two floors which are 2,560 mm apart. The steps must comply with the relevant building regulations.
9. Calculate the rise and going for a flight of stairs to span between two floors which are 2,450 mm apart. The steps must comply with the relevant building regulations.
10. Calculate the rise and going for a flight of stairs to span between two floors which are 2,800 mm apart. The steps must comply with the relevant building regulations.
11. Ordinary Level, 2003, Question 7
    The sketch shows a portion of a closed-string wooden stair.
    (a) To a scale of 1:10 draw a vertical section through the bottom three steps of the stair. (It is not necessary to show the newel post and handrail.)
    (b) On a separate sketch show a method of joining the risers and treads to the string.

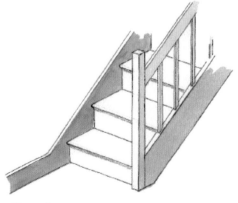

12. Higher Level, 2001, Question 4
    (a) To a scale of 1:5, draw a vertical section through the bottom three steps of a closed-string timber stair suitable for a domestic dwelling. Show the newel, handrail and balusters.
    (b) Indicate, using notes and freehand sketches, three design considerations that would ensure that a stair is safe for all users.

13. Higher Level, 2003, Question 8
A closed-string timber stair leads to a first-floor landing. The landing has a suspended timber floor with tongued and grooved flooring boards on timber joists and a plasterboard ceiling beneath.
(a) To a scale of 1:5 draw a vertical section through the top three steps of the stairs and the landing. Show the newel, balusters and handrail of the stair.
(b) Using notes and freehand sketches show the design details necessary to support the stair at the abutment of the stair and landing.
(c) Using notes and freehand sketches show two design details that ensure that the landing is safe for all users.

It is essential that every home has a clean supply of fresh drinking water. Water is used in a wide variety of everyday activities in the home including food preparation, cooking, washing of all sorts and gardening. The amount of water used in a home depends mainly on the number of people living there. In Ireland, an average of 370 litres of water is used per person per day. This compares with an EU average of 200 litres per person per day!

One of the reasons why we take having an abundance of clean water for granted is that we don't have to pay for it. While domestic water charges were abolished in Ireland in 1997, it is possible that household water meters (similar to electricity meters) will be introduced to control water usage and wastage in the future. Ireland remains the only EU country where households are not directly charged for water. It is worth remembering that 97% of the earth's water is in the salt-water oceans – that leaves only 3% in freshwater form. Of this, 2·15% is frozen in ice sheets and glaciers – that leaves only 0·85% available as a source of drinking water.

Potable water (suitable for drinking) has the following qualities:

- clear appearance,
- crisp taste,
- no odour,
- no sediment or suspended matter,
- no bacteria, viruses or parasites.

Hydrological Cycle

The earth's water is involved in a continuous cycle of precipitation (rainfall) and evaporation. This hydrological cycle is made up of a number of stages:

- Evaporation – When the sun shines on the surface of a body of water, such as the ocean, some of the water evaporates from the liquid state to water vapour. The sun warms the water vapour causing it to rise into the atmosphere where it is circulated by weather patterns.
- Condensation – When the water vapour cools it condenses and the liquid water appears as clouds of droplets.
- Precipitation – When the water droplets in a cloud become large enough they fall as rain, snow or hail.
- Run-off or Percolation – Some of the precipitated water falls on the land where most of it flows toward the sea by two main routes; as surface water (e.g. rivers and streams) or as groundwater.

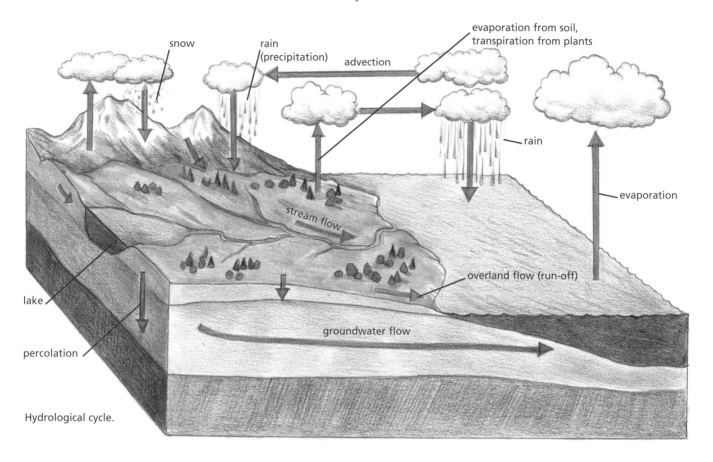

snow

rain
(precipitation)

advection

evaporation from soil,
transpiration from plants

rain

evaporation

stream flow

overland flow (run-off)

lake

groundwater flow

percolation

Hydrological cycle.

Water Sources

Water for supply to most houses comes from two main sources – the local authority (92%) or group water schemes (8%). Local authority water supplies are provided by the local county council or similar group. A group water scheme is a water supply, usually sourced from a well, that is shared between a group of households. Approximately 80% of drinking water supplied to Irish homes is sourced from surface water, 11% from wells, 7% from springs and the remainder from other sources.

Wells and springs are examples of underground water sources. When rain falls on soils or porous rocks, such as limestone or sandstone, some of the water sinks into the ground until it reaches a lower layer of impervious material, such as firm clay or rock. These water-bearing layers are called *aquifers*. The water table or *plane of saturation* is the natural level of the underground water. The water in some aquifers is *confined* and held below the water table by an impermeable layer on top of the water.

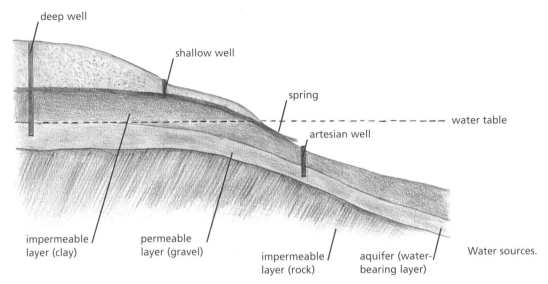

Water sources.

Surface water is water found in rivers, streams and lakes. Water collected in upland areas tends to be soft and of good quality. As a stream or river flows along its course it receives drainage from farms, roads and towns and becomes progressively less pure. Many rivers receive untreated sewage and industrial waste from towns and factories. It is very important that the level of pollution in rivers is controlled. A flowing river tends to purify itself, especially if the flow is brisk and shallow. This self-purification is due to a combination of factors including oxidation (chemical reaction with oxygen) of impurities, sedimentation of suspended material, the action of sunlight and dilution with cleaner water. Even so, water taken from a river for use in a large public supply needs to be treated and carefully analysed for its chemical and bacteriological content.

Water Source Pollution

It is essential that measures are taken to control surface and groundwater pollution. Farming is by far the most important *potential* source of contamination of water in Ireland because:

- It produces large volumes of organic wastes.
  - About 80% of the organic wastes generated in Ireland comes from agriculture.
  - A 40-hectare dairy farm produces wastes equivalent to a small town.
- It uses large amounts of inorganic fertilisers and pesticide.
- It takes place above most of our aquifers.

Septic tank effluent is another major source of pollution, particularly to wells:

- There are over 300,000 septic tank systems serving about 1·2 million people in Ireland.
- These discharge approx. 80 million m³ of effluent into the ground annually.
- Although farmyard wastes have a greater potential to pollute than septic tank systems, septic tank effluent is potentially more harmful because the pollutants are more threatening to human health (faecal coliforms) and because it bypasses the topsoil and some of the subsoil. The pollution is also less obvious because it occurs out of sight (i.e. underground).

Water Treatment

Water quality in Ireland is governed by the European Union Drinking Water Regulations, 2000. These regulations established strict quality standards. They provide maximum and guideline values for various different physical, bacteriological and chemical contaminants, including:

- aluminium,
- ammonium,
- total and faecal coliforms,
- colour,
- fluoride,
- heavy metals (arsenic, chromium, copper, lead, mercury, uranium, zinc),
- iron,
- manganese,
- nitrate,
- nitrite,
- odour,
- taste,
- pH,
- trihalomethanes (e.g. chloroform, created during the chlorination of drinking water),
- turbidity (presence of very finely divided solids).

The preparation of water for human consumption is typically a seven-stage process:

Stage 1

Intake – Water is sourced from surface supplies (e.g. lakes and rivers).

Stage 1

Intake – Water is sourced from underground sources (e.g. wells). As the water travels below ground it is naturally filtered.

Stage 2

Screening – The water is passed through a fine wire mesh to remove debris, such as twigs and plants etc.

Stage 3

Settling – The water is moved into large settling/sedimentation tanks. Whilst in the sedimentation tank the water is treated with a chemical known as 'Alum' (aluminium potassium sulphate) to remove any cloudiness or discolouration.

Stage 4

Filtration – The water is passed through filtration beds made up of sand and gravel to ensure that any dirt particles are removed. This is often referred to as the 'polishing effect'.

Stage 5

Chlorination – Once filtration is complete, the water is disinfected by treatment with chlorine. Enough chlorine is added to kill the bacteria present at that time and to prevent re-infection as it travels along the pipeline. The chlorine should not be present in large enough quantities to smell or taste in the water.

Stage 6

Fluoridation – All public water supplies are treated by adding small quantities of compounds such as sodium fluoride to the water. Fluoride is added as an aid to prevent tooth decay and at the request of the relevant Health Boards. The water is now ready and suitable for human consumption.

Stage 7

Pumping and Distribution – In urban areas, the water is pumped to a reservoir for distribution to each individual house via the local authority mains distribution system. This system consists of a 'ring main' network of pipes which allows small areas to be isolated for repair work without shutting down the supply to the wider area. In rural areas the water may be pumped directly to each house.

Revision Exercises

1. Describe the qualities of water that ensure it is suitable for human consumption.
2. Illustrate, using a neat annotated sketch, the stages of the hydrological cycle.
3. Describe, with the aid of a neat annotated sketch, the various possible water sources for a house built in a rural setting.
4. Explain two ways in which a water supply may become polluted.
5. Illustrate, using a neat annotated diagram, the stages of a water treatment process.

Water Supply to Houses

A number of systems are used to supply and distribute water within a house:

- a cold water supply to appliances (*appliance* is the term used to describe a sink, shower, washing machine, dishwasher or similar equipment),
- a hot water supply to appliances,
- a hot water supply for the heating system.

Design Features of a Water Supply System

A water supply system is designed to meet the following performance requirements:

- Leakproof Pipework – Leaks from pipework can cause great damage and be costly to repair. It is essential that every joint is properly sealed to prevent leaks. Also, the pipework used must be capable of withstanding the stresses caused by changes in water pressure and temperature. For example, the pipework will expand when conveying hot water and contract when the water cools – this movement should not cause the pipework to crack.
- Means of Isolating Appliances – It is important that appliances (e.g. radiators) can be isolated to facilitate maintenance or replacement.
- Means of Draining Pipework and Appliances – Once an appliance or circuit of pipework has been isolated there should be a way of draining the water from that area of the system.
- Overflow Mechanisms – Any appliance that stores water (e.g. cistern) should have an overflow device to prevent flooding.
- Prevention of Back Pollution of Public Supply – This applies in particular to older direct systems. It is possible that older systems could contaminate the main supply, e.g. if there was a drop in main pressure and contaminated water from a house flowed back into the public mains.

Cold Water Supply

Functions
The function of a cold water supply is to deliver a continuous supply of clean water suitable for human consumption at constant pressure.

A typical domestic cold water supply begins with the connection to the mains. This connection is usually made near the roadside. This facilitates maintenance access by the local authority, if necessary. The connection to the main is immediately followed by a stop valve which allows the supply to the house to be shut off easily and quickly in the event of a leak or if maintenance is required.

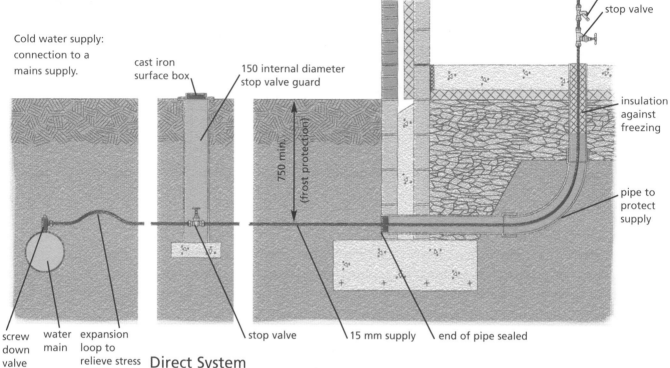

Cold water supply: connection to a mains supply.

cast iron surface box

150 internal diameter stop valve guard

750 min. (frost protection)

drain

stop valve

insulation against freezing

pipe to protect supply

screw down valve

water main

expansion loop to relieve stress on connection due to pipe settlement

stop valve

15 mm supply

end of pipe sealed

Direct System

Traditional cold water supply systems deliver water directly to each appliance. This system has the advantage that drinking water is available from all cold water outlets. However, there are a number of disadvantages including:

- There is no stored water in reserve in case the supply is temporarily cut off.
- There is a risk of back syphonage due to negative mains pressure.
- There tends to be a drop in pressure during peak demand periods (i.e. most households use more water in the evening).
- Higher pressure at appliances tends to cause early wear and tear of fittings (e.g. dripping taps).

The direct system is no longer installed for these reasons. However, it is important to understand how it works because many older homes use this type of system.

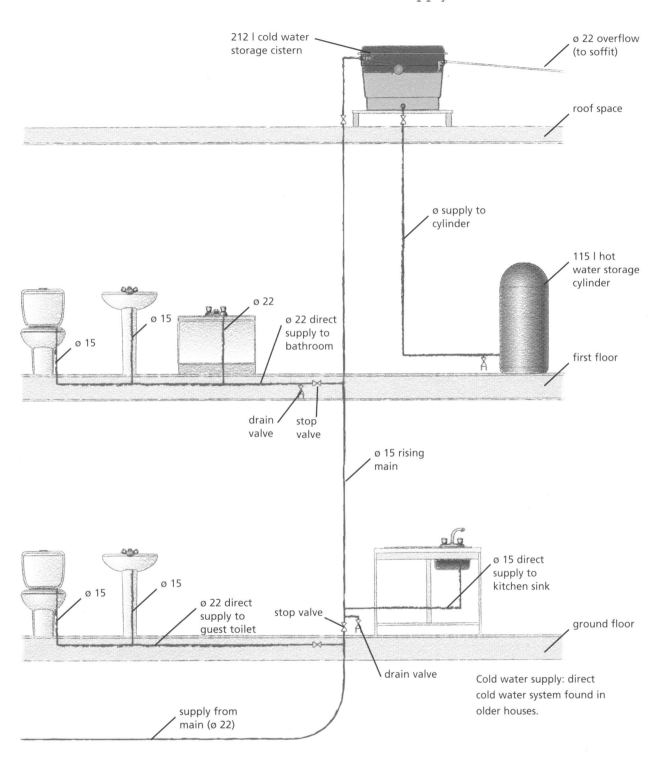

212 l cold water storage cistern

ø 22 overflow (to soffit)

roof space

ø supply to cylinder

115 l hot water storage cylinder

ø 15

ø 15

ø 22

ø 22 direct supply to bathroom

first floor

drain valve

stop valve

ø 15 rising main

ø 15 direct supply to kitchen sink

ø 15

ø 15

ø 22 direct supply to guest toilet

stop valve

ground floor

drain valve

Cold water supply: direct cold water system found in older houses.

supply from main (ø 22)

Indirect System

Modern cold water supply systems deliver freshwater to a storage cistern in the attic. Water is gravity-fed from the cistern to the appliances around the house. This type of system is called an indirect system. The advantages of the indirect system include:

- reserve supply in case of mains failure,
- balanced pressure to all appliances,
- less pressure on the taps and valves resulting in less wear and noise,
- less demand on main supply during peak periods,
- fewer fittings to main – less risk of backflow.

Cistern capacities vary from 212 litres (minimum for a three-bedroom house) to 340 litres (houses with four or more bedrooms). The kitchen sink is the only appliance that is directly fed from the mains. For this reason it is the only source of potable water in a typical house. Cold water should not be drunk from sinks that are indirectly supplied (e.g. bathroom taps) as this water may have been sitting in the cistern for long periods and may have become contaminated.

It is very important that the storage cistern is properly insulated and supported. If the cistern and its associated pipework are not properly insulated there is a risk that a pipe might burst during a freeze in the winter (remember the attic is ventilated to the outside and so can be very cold in winter).

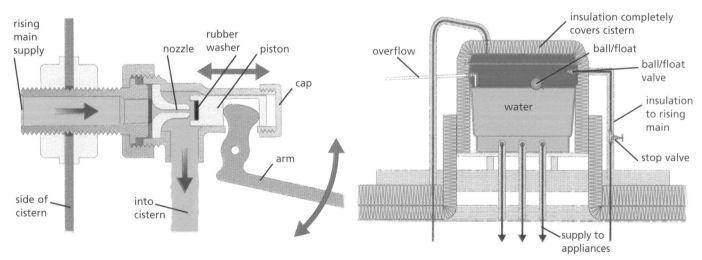

Cold water storage cistern: timber supports must be provided to spread the load over at least four roof joists.

Cold water storage cistern: sectional view of a ball/float valve showing how the up/down movement of the ball/float closes/opens the valve to refill the cistern when required.

Cold water storage cistern: insulation should be provided to the cistern and all exposed pipework. Insulation is not installed under the cistern.

Also, over 200 litres of water is a lot of weight concentrated in a small area of the attic. It is essential that the load is evenly spread over at least four trusses or joists. The flow of water into the cistern is controlled by a float and ball valve to prevent overflow. Should the ball valve malfunction an overflow pipe will carry the water out through the eaves of the house.

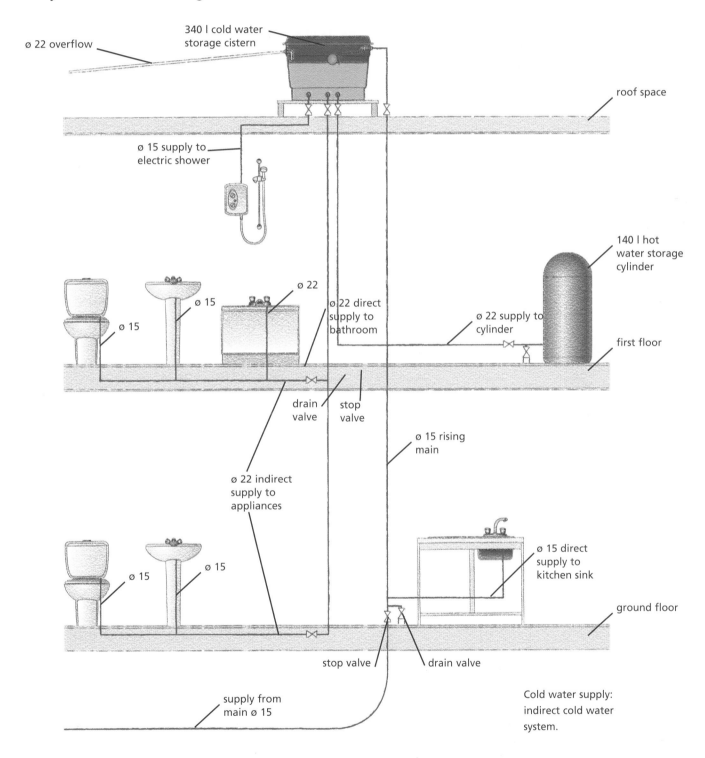

ø 22 overflow

340 l cold water storage cistern

roof space

ø 15 supply to electric shower

140 l hot water storage cylinder

ø 22

ø 15

ø 15

ø 22 direct supply to bathroom

ø 22 supply to cylinder

first floor

drain valve

stop valve

ø 15 rising main

ø 22 indirect supply to appliances

ø 15

ø 15

ø 15 direct supply to kitchen sink

ground floor

stop valve

drain valve

supply from main ø 15

Cold water supply: indirect cold water system.

Many plumbers now prefer to use plastic (polyethylene) components. The advantages of plastic pipework include:

> • flexibility – this means that less joints are required thereby reducing the likelihood of leaks, (e.g. no joints needed at bends),
> • not damaged by freezing temperatures,
> • smooth inner surface resists limescale build-up,
> • lower thermal conductivity – less heat loss and expansion when water is hot,
> • can be joined using push fittings (faster installation).

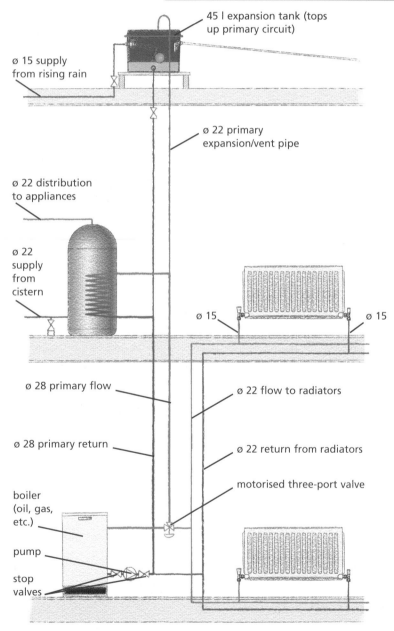

45 l expansion tank (tops up primary circuit)

ø 15 supply from rising rain

ø 22 primary expansion/vent pipe

ø 22 distribution to appliances

ø 22 supply from cistern

ø 15   ø 15

ø 28 primary flow

ø 22 flow to radiators

ø 28 primary return

ø 22 return from radiators

motorised three-port valve

boiler (oil, gas, etc.)

pump

stop valves

Indirect hot water supply system: the primary circuit heats water in a boiler and circulates it to the storage cylinder and radiators.

Hot Water Supply

Functions

Hot water heating systems have two functions: water heating (sinks etc.) and space heating (radiators). In simple terms, a hot water system takes cold water from the cistern in the attic, heats it, stores it in a cylinder and supplies it to the appliances when demanded while also supplying hot water to heat the radiators.

Water Heating

Modern houses use an indirect hot water supply system. This type of system consists of two circuits: a primary circuit and a secondary circuit. The primary circuit heats water in a boiler and circulates this hot water through a coil in the storage cylinder (usually found in the airing cupboard). The coil heats the water in the cylinder.

The primary circuit also sends hot water to the radiators which heat the various rooms around the house. This is a closed circuit – the water never leaves the circuit it just goes round and round. The opposite diagram indicates typical sizes of the components used for the primary hot water circuit of a house.

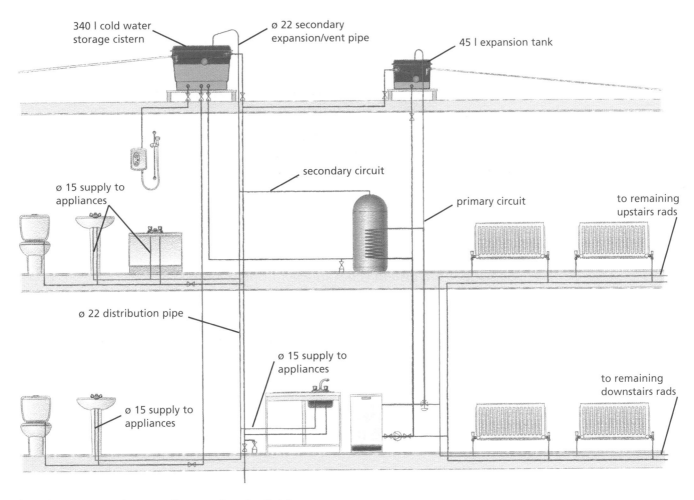

Indirect hot water supply system: the secondary circuit delivers hot water to the various appliances around the house.

The secondary hot water circuit delivers hot water to the various appliances around the house.

The storage cylinder is a vital component of this system. It is here that the heat energy is transferred from the primary to the secondary circuits. It is essential that the cylinder is adequately insulated to prevent heat loss.

Safety Features

Safety is an important consideration when designing hot water supply systems as the build-up of high temperatures and pressures in the storage cylinder, boiler or pipework could be dangerous. To prevent this, a number of safety features are built into the system:

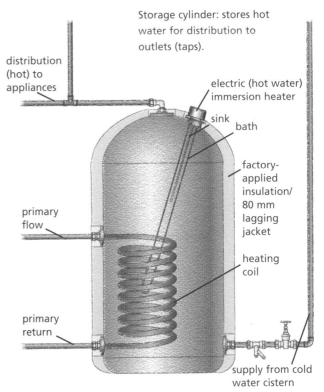

Storage cylinder: stores hot water for distribution to outlets (taps).

- Expansion (Vent) Pipe – prevents pressure build up in storage cylinder.
- Pressure Release Valve – prevents pressure build-up in primary circuit.
- Boiler High Limit Thermostat – prevents overheating of boiler. Factory-set thermostat automatically switches off boiler if excessive temperatures are reached (cannot be adjusted by home owner).
- Output Control Thermostat – controls temperature of water to prevent scalds from hot taps.
- Boiler Control Timer Switch – switches boiler on and off to prevent overuse.
- Vent Valve to Boiler – prevents build-up of pressure in boiler.
- Fire Valve on Fuel Supply – shuts off fuel supply to boiler in the event of fire.
- Frost Thermostat – intermittently switches on boiler to prevent frost damage if the system is not in use during winter.

The indirect system requires the use of a second (45 litre) storage cistern (expansion tank). This tank automatically tops up the primary circuit as necessary. This is necessary due to losses through evaporation in the expansion pipe, bleeding of radiators (i.e. releasing trapped air) and maintenance of the system.

Space Heating

Space heating of houses is traditionally achieved in Ireland using radiators linked to the primary circuit. However, underfloor heating is now commonly used, particularly in larger single houses. The two main methods of supplying hot water to radiators are the one pipe system and the two pipe system.

Space heating: the one pipe system loses heat as the water circulates, whereas the two pipe system ensures that each radiator operates at the same temperature.

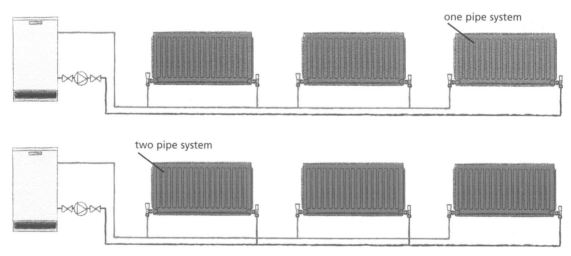

one pipe system

two pipe system

One Pipe and Two Pipe Systems

The pipework supplying water to radiators can be small-bore (12, 15 or 22 mm diameter) or micro-bore (6, 8 or 10 mm diameter). It is important that the correct size of pipe is used to supply the radiators with the correct amount of heat. The size of radiator used depends on the floor area of the room, the number of radiators per room and the temperature at which the radiator will be required to operate.

One pipe system

Advantages	**Disadvantages**
• easier to install,	• water cools as it flows through the system,
• less materials required (cheaper),	• minimal control of individual room temperatures.
• less time to install.	

Two pipe system

Advantages	**Disadvantages**
• equal heat delivered to each radiator,	• slower to install,
• control of individual room temperatures possible,	• more materials required (more expensive).
• warms up whole house faster.	

System Controls

Space heating system controls are fitted to heating systems in buildings to ensure:

- Correct temperatures are maintained in each space.
- Heat is delivered at the required time of the day.
- Energy is conserved by:
  - separate and independent automatic time control of space heating and water heating,
  - space heating where control is based on room temperature,
  - water heating where control is based on stored water temperature.
    - Controls switch off boiler when there is no demand for space or water heating.
  - Separate time and temperature control in two or more zones where floor area of dwelling is greater than 100 m².

Note: These energy conservation measures are requirements under the Building Regulations 2005 (Technical Guidance Document L: Conservation of Fuel and Energy).

Most controls are designed to adjust the rate of flow of water through the radiators – because the temperature of the radiators is directly linked to the rate of flow. Controls usually used include:

- Circulating Pump – positioned on the return pipe beside the boiler, pumps the hot water to all radiators.
- Radiator Valves – each radiator has two valves:
  - A Hand Wheel Valve – adjusts the flow of hot water to the radiator (or thermostatic valve that contains a shut-off valve that responds to the air temperature of the room).
  - Lock Shield Valve – fitted on the return side to balance the amount of heat in each radiator.
- Air Vent – located at the top of the radiator to allow trapped air in the radiator to be released.
- Gate Valves – used to isolate a section of the system without having to drain the entire system.
- Boiler Thermostat – controls the temperature of the water in the boiler.
- Room Thermostat – set to the desired room temperature (electrically controls the circulating pump).

Controls for space and water heating: the room/boiler thermostats 'tell' the system when to send hot water to the rooms/cylinder.

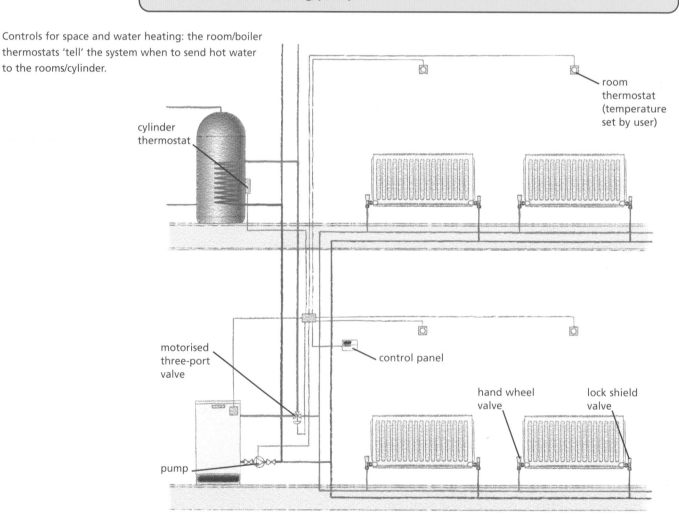

cylinder thermostat

room thermostat (temperature set by user)

motorised three-port valve

control panel

hand wheel valve

lock shield valve

pump

Underfloor Heating

Underfloor space heating systems are usually zoned – each space has its own circuit and is individually controlled by a room thermostat. There are a number of benefits to installing a zoned system, including:

- Sustainability – reduces the amount of fuel burned and hence harmful gases released into atmosphere.
- Comfort – the temperatures in each area can be adjusted to suit the people using each space.
- Timing – different temperatures can be set in specific zones at various times.
- Maintenance – each area can be isolated for maintenance without having to shut down the entire system.
- Control – electronic and remote control technologies can be used to activate the system.

In underfloor space heating systems the hot water is circulated through a circuit of polyethylene pipes laid in the floor at the time of construction. The room circuits are connected to the primary circuit at a manifold. When the room reaches its pre-set temperature the thermostat in the room sends a signal to an electrically operated valve on the manifold closing that particular circuit. The water is distributed to the pipes through the manifold at between 35–50°C and heats the floor surface to a temperature in the range of 18–24°C. The floor effectively becomes one large heat store giving off a gentle, even, radiant heat to the whole room. Unlike a radiator system, which can be programmed to switch on and off throughout the day, underfloor heating operates throughout the heating season maintaining a comfortable even heat all the time. This actually saves energy and costs less to operate because once the floor has become warm it takes a very small amount of energy to keep it warm; whereas with a radiator system, the radiators have to be warmed up from cold each time the system is switched on.

Underfloor heating: typical construction details (solid concrete ground floor).

40 mm concrete cover   heating pipes   steel mesh

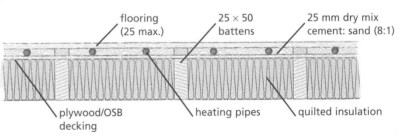

flooring (25 max.)   25 × 50 battens   25 mm dry mix cement: sand (8:1)

plywood/OSB decking   heating pipes   quilted insulation

Underfloor heating: typical construction details (suspended timber ground or upper floor).

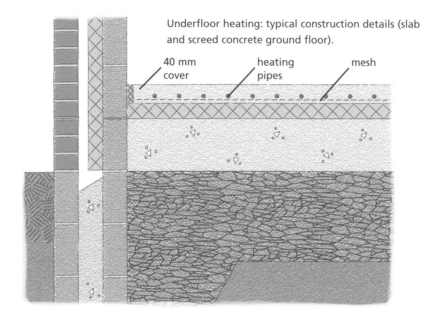

Underfloor heating: typical construction details (slab and screed concrete ground floor).

40 mm cover   heating pipes   mesh

A typical house with a concrete ground floor and a suspended timber upper floor requires different temperatures for each level. The upper floor will typically operate between 50–60°C, while the downstairs floor will be at 35–40°C.

Radiators Advantages

1. The response time (speed of heating up) is suitable for Irish climate.
2. Radiators can be situated to heat cold surfaces, i.e. near single-glazed windows or on poorly insulated walls – reduces down-draughts.
3. Flexibility – radiators can be relocated or replaced and additional radiators can be added to the system.
4. Convenient individual room temperature control is possible.
5. Lower installation costs.
6. Simpler retro-fit in older homes.

Radiators Disadvantages

1. Subject to possible leaks and requires some maintenance.
2. Larger radiators are required to operate effectively with a condensing boiler.
3. Radiators can be ugly and unsightly and will accumulate dirt and dust.
4. Radiators create uneven heating particularly in larger rooms with high ceilings.
5. Furnishing difficulties resulting from location of radiators.

Underfloor Heating Advantages

1. Absence of emitters (radiators) allows freedom for decoration and improves room appearance.
2. Lower temperature, radiant heat provides a stable and comfortable environment. potentially more efficient if properly installed and controlled due to lower temperature of circulating water.
3. Suitable for providing a background level of heating.
4. Additional heat emitters, such as radiators, may be added to ensure comfort in living spaces.
5. Ideal for use with heat pumps or condensing boilers (because lower temperature water circulation is required).
6. More uniform heat distribution throughout the room.

Underfloor Heating Disadvantages

1. High cost of installation (20–25% more expensive).
2. Slow response time is less suited to Irish climate.
3. Controls and design must be of high standard to ensure satisfactory operation.
4. Limited flexibility – considerable building work is required to change the system.
5. Furniture in room may limit heat emitter surfaces available.
6. Low temperature surface of floor may be inadequate to satisfactorily heat poorly insulated spaces.
7. Generally, only appropriate for new homes or new buildings.

Unvented System

The direct unvented system is a modern alternative to the indirect cold and hot water systems. This system is mains-fed avoiding the need for storage cisterns in the attic. This is useful in buildings that have limited attic space (e.g. dormer bungalows, duplex apartments). The principle is simple – the water is delivered at a controlled pressure directly into the hot and cold water systems. Special expansion chambers are used to accommodate the increases in volume and pressure generated when the water is heated. One of the major benefits of this system is the delivery of hot water at consistent temperature and pressure to all appliances.

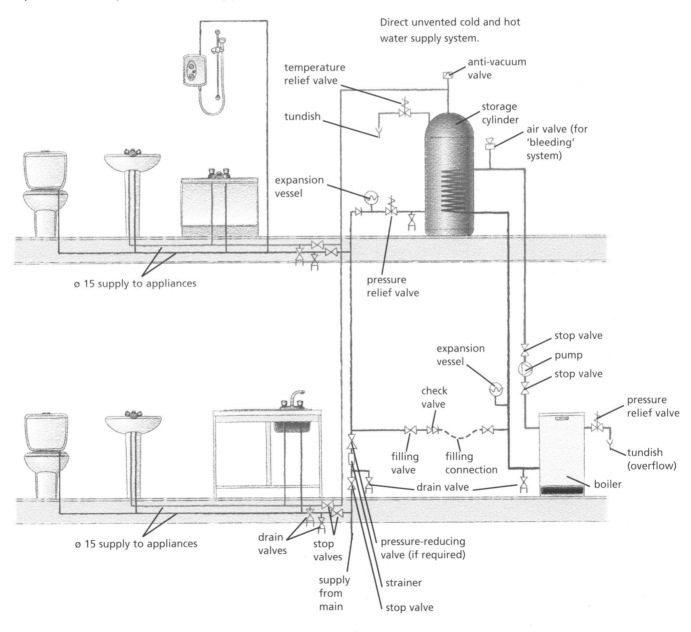

Direct unvented cold and hot water supply system.

Numerous safety valves and devices are used to ensure the system remains safe at all times.

Expansion chamber: water increases in volume when heated. This increase is accommodated by the expansion chamber. The nitrogen gas in the chamber is compressed when the rubber diaphragm is pushed down.

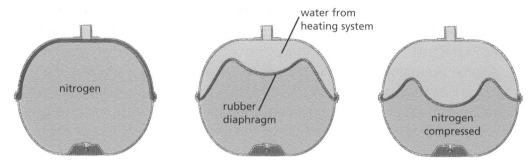

water from heating system

nitrogen

rubber diaphragm

nitrogen compressed

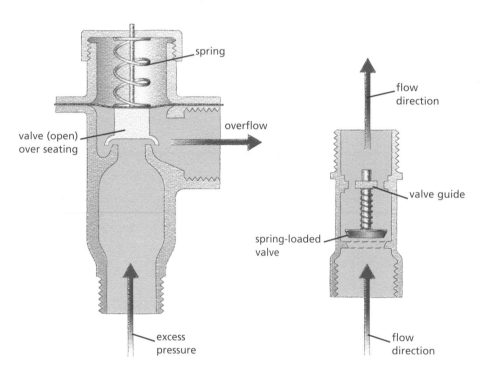

spring

valve (open) over seating

overflow

excess pressure

flow direction

valve guide

spring-loaded valve

flow direction

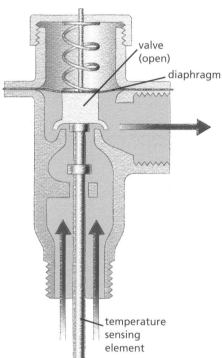

valve (open)

diaphragm

temperature sensing element

Safety valves (unvented system): pressure relief valve. A pressure actuated valve which opens automatically at a specified pressure to discharge water.

Safety valves (unvented system): check (one-way) valve ensures that the water can only flow in the correct direction.

Safety valves (unvented system): temperature relief valve. A temperature actuated valve which opens automatically at a specified temperature to discharge water. Fitted to water heaters to prevent the temperature in the cylinder getting too hot.

Boilers

The term boiler is a general term used to describe oil boilers, gas burners and solid fuel heaters which convert fossil fuels into energy to heat homes. A wide range of fuels are available, including:

- Oil (Kerosene) – Oil is purchased periodically and stored in a storage tank. It is connected to the boiler which is commonly housed in a protective cover or outhouse to the rear of the house.
- Natural Gas (Methane) – Natural gas is popular for home heating because: it is relatively clean burning, it requires no storage, the consumer only pays for it after they use it, it is easier to burn and control than oil. Gas burners are generally simpler than those required for oil and have few moving parts.
- Coal and Peat – These fuels are traditionally used in fireplace back boilers and kitchen ranges. They are no longer used as a primary heat source in newly constructed houses.

The type of boiler installed depends on the fuel type used. The basic principle is the same – the water passes through a series of steel tubes which are heated by the flame of the burning fuel. At best (e.g. condensing balanced flue gas boiler) a boiler is about 90% efficient. That is to say, for every unit of energy burned, 10% of the heat energy created is lost to the atmosphere. Boilers should be serviced annually to ensure they are working as efficiently as possible. This is particularly important for gas boilers because they are usually installed within the house and could release dangerous carbon monoxide gas if they are not functioning properly.

Sustainable Use of Water in the Home

Washing cars, watering plants and flushing the toilet are examples of everyday uses of water that do not require the use of drinking quality water. In fact, over half of the water consumed in a typical Irish home is used on these types of tasks.

The roof of every house in Ireland acts as a rainwater collector – delivering an abundant supply of clean water to ground level. Diverting this water through a filter to a storage tank for reuse can greatly reduce the consumption of treated drinking water. Many products are available on the market that are designed to use this type of water, for

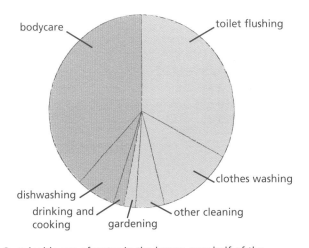

Sustainable use of water in the home: over half of the water used in a typical home does not require water which has been treated for human consumption.

example an attachment hose can be purchased for power washers (used for washing cars etc.) that will draw water from a storage tank. In addition, many water-saving appliances are readily available today. For example, a dual flush toilet, which typically allows the user to flush four or six litres of water, can be used to reduce the amount of water used when flushing the toilet.

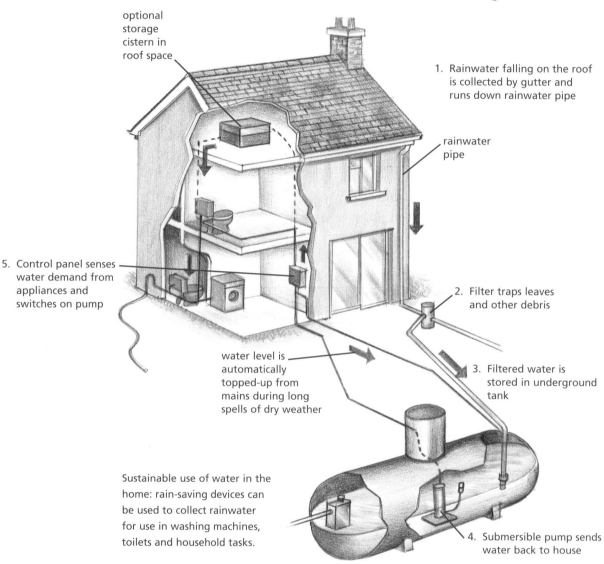

optional storage cistern in roof space

1. Rainwater falling on the roof is collected by gutter and runs down rainwater pipe

rainwater pipe

5. Control panel senses water demand from appliances and switches on pump

2. Filter traps leaves and other debris

water level is automatically topped-up from mains during long spells of dry weather

3. Filtered water is stored in underground tank

Sustainable use of water in the home: rain-saving devices can be used to collect rainwater for use in washing machines, toilets and household tasks.

4. Submersible pump sends water back to house

Revision Exercises

1. Outline the performance requirements of a cold water supply system for a typical house.
2. Generate a neat line diagram of a cold water supply system suitable for a four-bedroom house with a standard bathroom and a downstairs toilet.

3. Explain, using a neat annotated sketch, how the flow of water into the storage cistern is regulated.

4. Generate a neat line diagram of a hot water supply system suitable for a four-bedroom house with a standard bathroom and a downstairs toilet.

5. Generate a neat line diagram of a modern space heating system suitable for a single-storey, four-bedroom house.

6. Explain, using neat annotated sketches, the safety features of a typical indirect hot water system.

7. Explain, using neat annotated sketches, the safety features of an unvented indirect hot water system.

8. Explain, using a neat annotated sketch, a strategy for sustainable water consumption in the home.

9. Ordinary Level, 2002, Question 2
   (a) Using a clear, labelled sketch show the pipework and valves necessary to supply cold water to a bath and wash hand basin.
   (b) Using notes and neat freehand sketches, describe a design detail of a precaution that could be taken to protect copper pipework from the effects of severe frost.

10. Higher Level, 2003, Question 7
    An oil-fired boiler is used as the heat source to provide central heating and hot water for a domestic dwelling.
    (a) Using a single-line diagram, show a design layout for the heating and domestic hot water system. Include three radiators in the proposed layout and indicate suitable dimensions for all pipework.
    (b) Using notes and freehand sketches, show three design details that should be incorporated into the proposed layout to ensure the continuous safe operation of the heating system.

11. Higher Level, 2005, Question 6
    (a) Draw a single-line diagrammatic sketch of the cold water distribution system for a two-storey house. The diagram should show all the design details from the mains supply and include the distribution to the kitchen sink and bathroom. The bathroom includes:
    (i) water closet (WC), (ii) wash hand basin, (iii) bath.
    (b) Include in the proposed layout all the necessary valves and suggest suitable dimensions for all pipework.
    (c) Using notes and detailed freehand sketches, show two design details that regulate the level of water in the storage tank and explain the design principles of each.

Drainage is the term used to describe the system of drains (i.e. pipework) required to convey wastes away from appliances (e.g. sink, washing machine, shower, toilet etc.) to a wastewater treatment facility. The treatment of wastewater may occur on-site or off-site depending on the location of the house. Wastewater is classed as follows:

- wastewater – sinks, showers, dishwashers,
- soil water – toilets,
- foul water – both waste and soil water combined,
- surface water – rainwater from roofs, roads, paved areas.

Functions

The primary function of a wastewater drainage system is to safely and economically convey wastes from the point of collection to the point of disposal.

Design Features

A wastewater drainage system must be designed to meet the following performance requirements:

- Control leakage inwards or outwards.
  - Leaks encourage root penetration.
- Resist deposits of solids and blockages.
- Resist abrasion by grit (scouring) especially road drains.
- Resist corrosion by acid and/or alkali.
  - Sulphate (bog) and chlorine (seawater).
- Accommodate pressure – internal and external.
  - Eliminate undue build-up of gases.
- Allow access for maintenance.

A key concept in the design of wastewater systems is the slope or gradient of the drains. If the gradient is too steep the fluids will flow too quickly and the solids will be left behind causing a blockage – if the slope is not steep enough the fluids won't flow readily.

The wastewater drainage system of a house is usually described in two parts: the above ground system and the below ground system.

Above Ground Drainage

The above ground wastewater drainage system of a house is usually described as a single stack system – this is because there is one main pipe (stack) into which each appliance is connected.

The length, slope and position of each pipe connected to the discharge stack is carefully controlled (by the building regulations) to ensure adequate performance. These requirements greatly influence the design options for a typical bathroom. If the design layout requires the appliances to be positioned outside the recommended distances from the discharge stack, a proprietary macerator unit (e.g. Saniflo™) may be installed. A macerator unit consists of a rotary shredder that reduces solid waste to pulp which is then pumped using the flush water along a small bore waste pipe (18–22 mm) to the discharge stack.

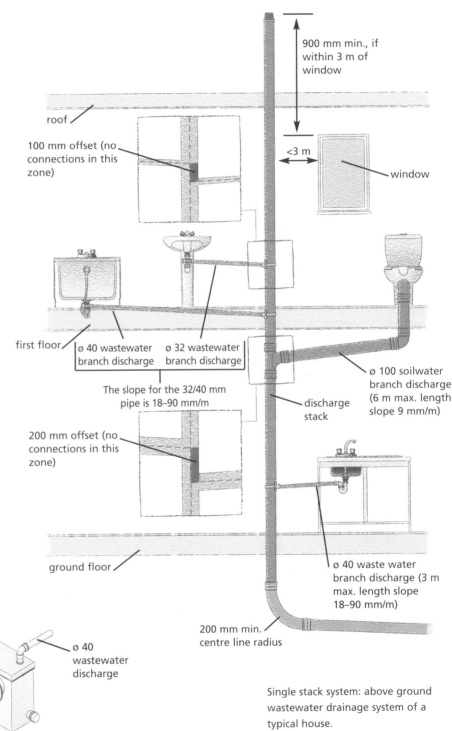

900 mm min., if within 3 m of window

roof

100 mm offset (no connections in this zone)

<3 m

window

first floor

ø 40 wastewater branch discharge

ø 32 wastewater branch discharge

The slope for the 32/40 mm pipe is 18–90 mm/m

discharge stack

ø 100 soilwater branch discharge (6 m max. length slope 9 mm/m)

200 mm offset (no connections in this zone)

ground floor

ø 40 waste water branch discharge (3 m max. length slope 18–90 mm/m)

200 mm min. centre line radius

ø 40 wastewater discharge

Single stack system: above ground wastewater drainage system of a typical house.

Macerator unit: used if the design layout requires the appliances to be positioned outside the recommended distances from the discharge stack.

It is also essential that bathroom designs are suitable for all users throughout their lifetime. This means that the needs of children, elderly people and wheelchair users should be considered at the design stage. The needs of wheelchair users in particular will require the fitting of grab rails near the toilet and bath. (See chapter 3, Planning Permission and Sustainable Development.)

A trap is used at the connection of each appliance to the wastewater drainage system. The purpose of the trap is to prevent foul odours entering the house. A trap is simply a water-sealed bend in the pipework. Traps are available in a variety of shapes for different applications.

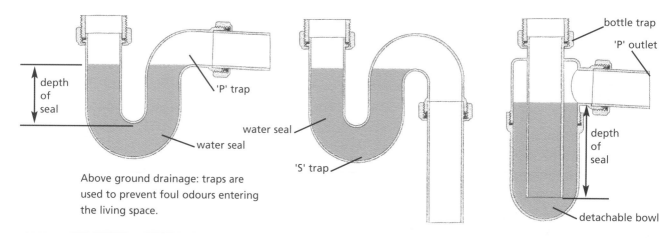

Above ground drainage: traps are used to prevent foul odours entering the living space.

When an appliance is emptied the water flows away through the trap. However, the last portion of the water is left behind in the trap creating a seal until the appliance is next used.

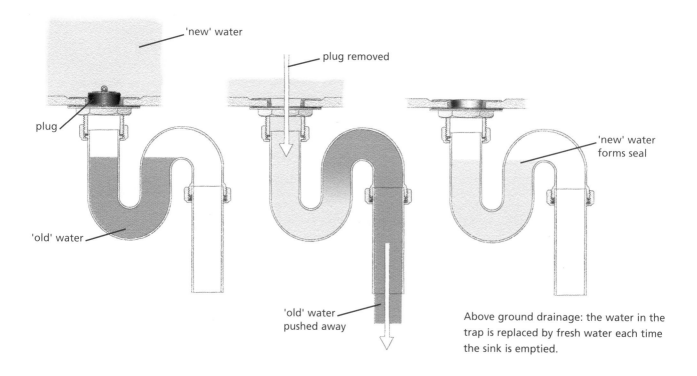

Above ground drainage: the water in the trap is replaced by fresh water each time the sink is emptied.

The depth of the seal is usually 75 mm for sinks, baths and showers. A bottle trap is used in places where the rate of usage is higher and blockages are more likely to occur. Bottle traps allow for quick access for the clearance of blockages.

Poor design can lead to a situation where the water seal in the trap is lost. This can happen if back pressure builds up in the main stack or if syphonage occurs in the pipework.

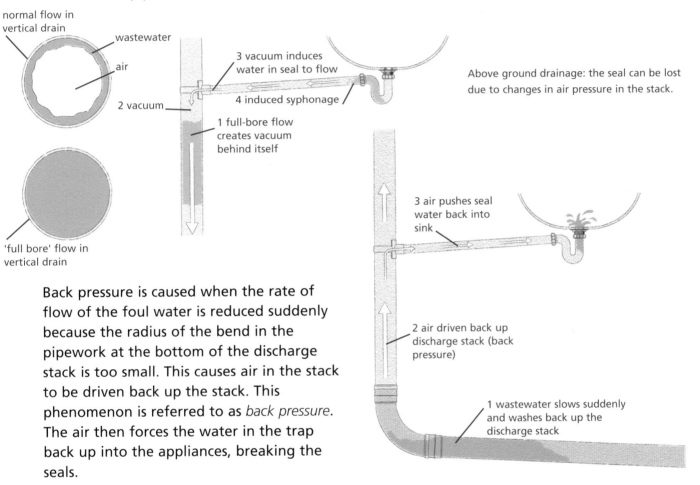

normal flow in vertical drain

wastewater

air

2 vacuum

'full bore' flow in vertical drain

3 vacuum induces water in seal to flow

4 induced syphonage

1 full-bore flow creates vacuum behind itself

Above ground drainage: the seal can be lost due to changes in air pressure in the stack.

3 air pushes seal water back into sink

2 air driven back up discharge stack (back pressure)

1 wastewater slows suddenly and washes back up the discharge stack

Back pressure is caused when the rate of flow of the foul water is reduced suddenly because the radius of the bend in the pipework at the bottom of the discharge stack is too small. This causes air in the stack to be driven back up the stack. This phenomenon is referred to as *back pressure*. The air then forces the water in the trap back up into the appliances, breaking the seals.

Syphonage is a very important concept in drainage. In this case there are two types of syphonage that can occur – induced syphonage and self-syphonage. Induced syphonage occurs when there is full bore flow in the main stack. *Full bore flow* is the term used to describe the flow of wastewater through a pipe that completely fills the pipe. Usually when fluid flows down through a vertical pipe it flows along the inner surface of the pipe – the centre of the pipe remains clear.

In full bore flow the pipe is full and the fluid pushes the air in the pipe ahead of it as it flows downward. This also has the effect of creating a vacuum in the pipe behind the fluid. One way to picture this is to imagine you are standing on the platform in an underground train station as a train passes quickly through – as the train approaches you feel a gust of wind as the air is pushed along by the train. As the train shoots by, the air is pulled out of the station by the vacuum created by the train. When full bore flow occurs in the main stack it creates a vacuum which may pull the water out of the seal in nearby traps.

Self-syphonage occurs when full bore flow occurs in the wastewater branch discharge pipe leading from the appliance. This shouldn't generally happen, but it can occur when the gradient of the pipework is incorrect or in older sinks where the plug hole doesn't have a swirl device to control the rate of flow.

Syphonage should be prevented by correct design and installation. However, when it does occur it can be prevented by providing a source of air when a vacuum is generated. This is done by installing an air admittance valve. Air admittance valves shouldn't be necessary for typical houses where the system is designed to meet the building regulations. However in public toilets, for example, where there are multiple appliances connected to a single pipe the use of an air admittance valve is recommended. An air admittance valve has a rubber diaphragm that gives way in the event of a vacuum forming in the pipework thereby allowing air to enter the pipework before the seal is lost in the trap.

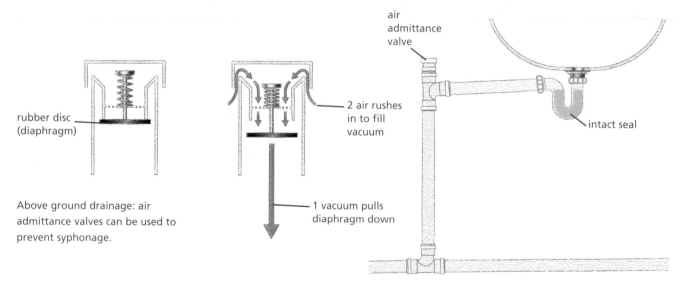

rubber disc (diaphragm)

air admittance valve

2 air rushes in to fill vacuum

intact seal

1 vacuum pulls diaphragm down

Above ground drainage: air admittance valves can be used to prevent syphonage.

Below Ground Drainage

Function
The below ground wastewater drainage system is a system of pipework required to convey wastewater from a building to a disposal site. As

mentioned earlier, the treatment of wastewater may occur on-site or off-site depending on the setting. In most rural areas each house has its own wastewater treatment facility consisting of a septic tank and a percolation area – in urban areas each house is connected to the local authority wastewater system (main sewer).

Design Features

A below ground drainage system should be as maintenance-free as possible. Some important design features of a below ground wastewater system include:

- Drain layout and installation:
  - Layout should be as simple as possible.
  - Changes in direction and gradient (slope) should be minimised.
  - Connections should be made obliquely and in the direction of the flow.
  - Access points should be provided only if blockages could not be cleared without them.
  - System should be ventilated at every house or if the branch is longer than 6 m.
  - Access points should be placed at changes in invert level.
  - Curves (where unavoidable) should have as large a radius as possible.
- Watertightness – This is achieved through the use of synthetic rubber seals on uPVC pipes. The end of the pipe should be lubricated to ensure that the synthetic rubber seal isn't dislodged when joining. It is essential that joints are squarely made to prevent leaks.
- Durability – This depends on a number of factors including: the life span of materials, the effects of building settlement and ground movement on the drain, the effect of chemical attack on the pipes and joints from effluent and soil, possible crushing of pipes, penetration by tree roots.
- Resistance to Blockage – Blockages are avoided by ensuring the following: adequate diameter of drain; smooth internal surfaces; nothing projects into the pipes, joints or fittings that could cause an obstruction; properly aligned joints; no reduction of diameter in direction of flow; pipes laid at correct gradient (slope).
- Ease of Maintenance – Pipework should be laid in straight lines between access points, access points (inspection chambers, manholes, access junctions, rodding eyes) must be located so that all sections can be inspected and cleaned by flexible rods.

Underground pipes are available in a variety of materials including: fibre cement, vitrified clay, concrete, cast iron and unplasticised polyvinyl chloride (uPVC). Concrete drains are commonly used for mains drainage for a number of houses. For example, a concrete drain would be laid along the road outside a row of houses and each house would be connected to it using a uPVC drain. uPVC is the most popular type of drain used for housing applications for a number of reasons, including:

- fast to install – easily cut and assembled,
- strong and flexible,
- resistant to chemical attack,
- stable at high temperature,
- easily stored on-site,
- proven reliability.

Drain Laying

It is essential that drains are laid at the correct depth and gradient if they are to function properly. Pipe laying involves excavating a trench to the required depth and backfilling with aggregate to provide secure bedding. The bottom of each trench is excavated to the required gradient. The depth is established using sight rails and/or a theodolite.

The pipe is laid on a bed of 10 mm aggregate, covered in 40 mm crushed stone and the trench backfilled with the excavated material. A minimum of 300 mm cover should be provided.

The gradients used for drains vary depending on the situation.

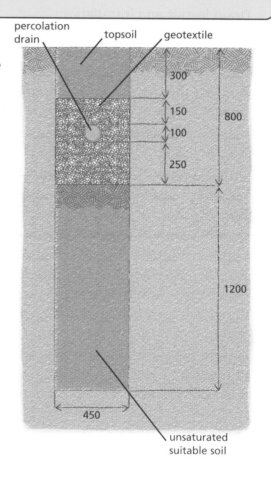

Below ground drainage: drain laying. It is essential that drains are laid at the correct depth and gradient if they are to function properly.

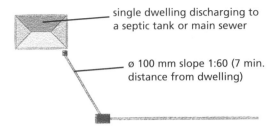
single dwelling discharging to a septic tank or main sewer

ø 100 mm slope 1:60 (7 min. distance from dwelling)

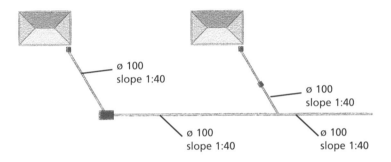
ø 100 slope 1:40

ø 100 slope 1:40

ø 100 slope 1:40

ø 100 slope 1:40

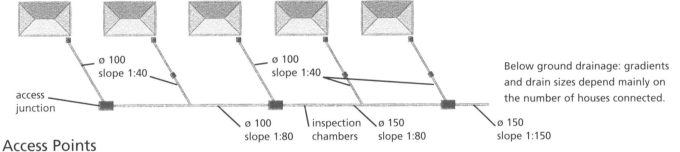
ø 100 slope 1:40

ø 100 slope 1:40

access junction

Below ground drainage: gradients and drain sizes depend mainly on the number of houses connected.

ø 100 slope 1:80

inspection chambers

ø 150 slope 1:80

ø 150 slope 1:150

Access Points

There are five types of access to a below ground drainage system:

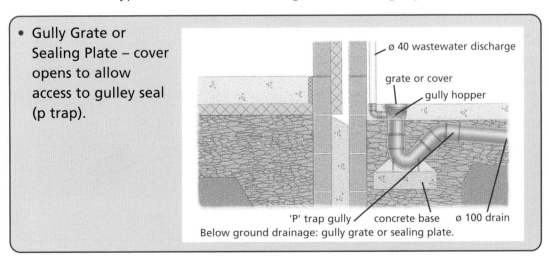

- Gully Grate or Sealing Plate – cover opens to allow access to gulley seal (p trap).

ø 40 wastewater discharge

grate or cover

gully hopper

'P' trap gully   concrete base   ø 100 drain

Below ground drainage: gully grate or sealing plate.

Rodding Point – capped extension to drain used for access to pipework not reachable from other points.

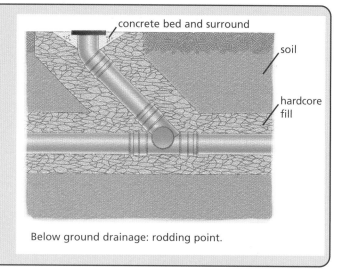

Below ground drainage: rodding point.

Access Junctions – small chamber positioned at joints, changes in direction, changes in invert level etc.

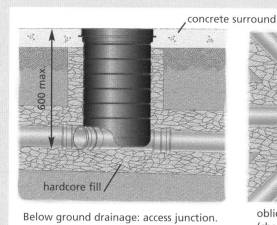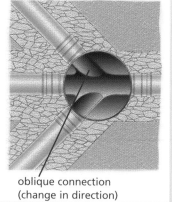

Below ground drainage: access junction.

oblique connection (change in direction)

Inspection Chambers – less than one metre deep chamber for clearing blockages from ground level.

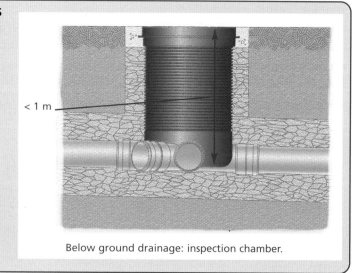

Below ground drainage: inspection chamber.

Manhole – greater than one metre deep for clearing blockages by climbing down into chamber.

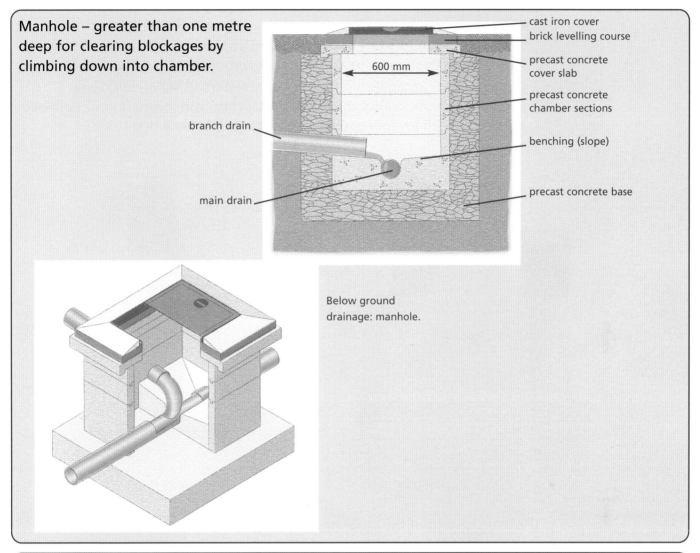

cast iron cover
brick levelling course
precast concrete cover slab
precast concrete chamber sections
benching (slope)
precast concrete base

branch drain
main drain

Below ground drainage: manhole.

Backdrop Manhole – used to ensure correct gradients are maintained when site is very sloped and to reduce excavation costs where a large difference in level between drain and public sewer (or septic tank) exists.

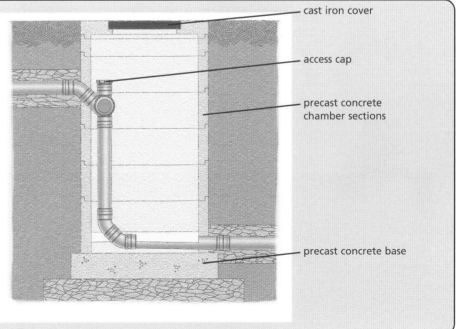

cast iron cover
access cap
precast concrete chamber sections
precast concrete base

Below ground drainage: backdrop manhole.

- Up to now our attention has focussed on disposal of wastewater from within the house. Surface water collected on roofs, footpaths and driveways must also be dealt with. There are two approaches to the disposal of surface water – a separate or a combined system. A separate system provides separate drainage for foul and surface water, whereas a combined system carries both foul and surface water in the same drains.

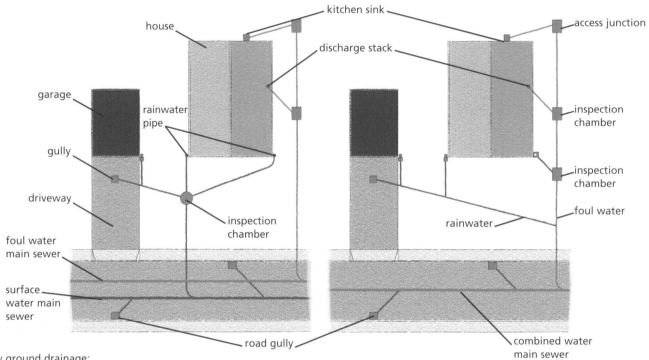

Below ground drainage: combined (on the left) and separate systems.

The main benefit of the separate system is that the relatively clean surface water can be disposed of into a local watercourse (e.g. river) rather than unnecessarily treating it at a wastewater treatment facility. The drawback is the cost of the extra pipework and excavation required.

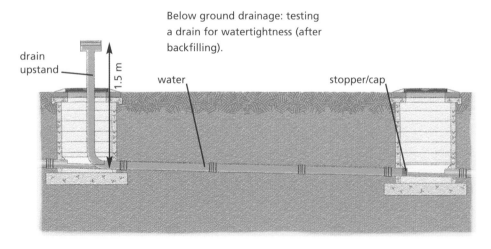

Below ground drainage: testing a drain for watertightness (after backfilling).

Drain Testing

Before trenches are backfilled the drains should be checked for watertightness. Drains are required to withstand a 1·5 m head of pressure above the invert level at the head of the drain.

The procedure for testing is as follows:

1. The end of the drain is temporarily capped.
2. Drain upstand is attached and filled to 1·5 m level.
3. Left to stand for 2 hours and topped up as necessary to allow for porosity and release of air bubbles.
4. Measured for next 30 minutes (i.e. from the 2 hours to 2·5 hours time).
5. Leakage should not exceed 0·05 litres for each metre run ø100 mm pipe (e.g. 5 litres leakage for 100 m run of drain – equates to a drop in water level of 6·4 mm.)
6. If any leakage is noticed it should be found and repaired and the test repeated.

The trench is then backfilled and the test repeated.

Wastewater Treatment Systems for Single Houses

In rural settings, where connection to the local authority main sewer is not possible, each house has its own separate wastewater treatment system for the treatment of foul water. Surface water is not treated in a wastewater treatment system for a single house – it is disposed of in a local watercourse or soakage pit. While a variety of systems are available in Ireland the septic tank system has traditionally been the most widely used. The system chosen for a new house will depend primarily on the site soil conditions but is also influenced by the size and type of house and the installation cost.

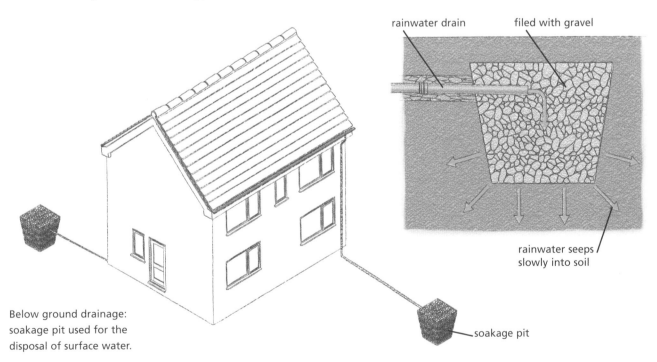

rainwater drain

filed with gravel

rainwater seeps slowly into soil

Below ground drainage: soakage pit used for the disposal of surface water.

soakage pit

Types of Wastewater Treatment Systems

On-site treatment systems for single houses can be divided into two main categories:

> • septic tank system – traditional (conventional) approach used in Ireland,
> • mechanical aeration system – uses a special filter to break down wastewater.

These systems may be supplemented by the use of a filter system. The purpose of a filter system is to further clean the wastewater prior to discharge to the ground. There are two main types of filter systems:

> • wetland system,
> • intermittent filter system.

The final stage in wastewater treatment is disposal to the percolation area. The percolation area is also referred to as a *polishing filter*. The percolation area is where the wastewater seeps into the soil.

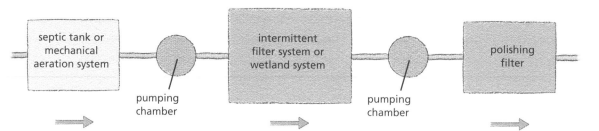

Wastewater treatment systems for single houses: overview of the waste water treatment process.

Function of a Wastewater Treatment System

The function of a wastewater treatment system is to:

> • treat the wastewater to minimise contamination of soils and water bodies (streams, rivers, groundwater),
> • protect humans from contact with wastewater,
> • keep animals, insects and vermin from contact with wastewater,
> • minimise the generation of foul odours,
> • prevent discharge of untreated wastewater to groundwater or surface water (pollution).

The Environmental Protection Agency (EPA) provides guidance on the design, construction and installation of wastewater treatment systems for single houses. Guidance is also available on testing a site for suitability for a treatment system (Percolation Test). The results of this test are usually required to be submitted as part of a planning permission application.

Percolation Test

A Percolation Test is a measure of the time taken for a specific volume of water to be absorbed by the soil at a defined depth. The test involves excavating a hole in the area of the site where the percolation area is to be installed. The hole is filled with water and the percolation time is noted. If the water is absorbed too slowly or too quickly this indicates that the soil is not suitable for a standard septic tank system and alternative or additional measures will be required.

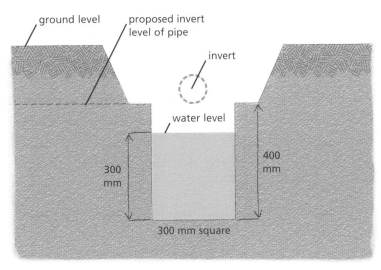

Wastewater treatment systems for single houses: percolation test hole.

Procedure

Day 1

1. Two Percolation Test holes should be dug in the proposed percolation area. Each hole should be 300 mm × 300 mm and 400 mm deep below the proposed invert level of the distribution pipe.
2. The bottom and sides of the hole should be scratched with a knife or wire brush to remove any compacted or smeared soil surfaces and to expose the natural soil surface.
3. Clear water should be carefully poured into the hole at about 10.00 a.m. in order to fill it to the full height of 400 mm.
4. The water should then be allowed to percolate.
5. At about 5.00 p.m. the hole should once again be filled to the full height of 400 mm and allowed to percolate overnight.

Day 2

1. The hole should be filled with clear water at about 10.00 a.m. and the water should be allowed to drop until there is 300 mm of water in the hole.
2. Thereafter, the time in minutes required for the water to drop 100 mm, (i.e. from 300 to 200 mm) should be recorded.
3. The hole should then be refilled to the 300 mm level again and the water allowed to drain to the 200 mm level and the time again recorded.
4. The filling and measurement of the percolation rate through the hole should be carried out once more (i.e. three times in total).
5. The average value in minutes of the three recordings should then be divided by four to give the time required for a fall of 25 mm. This is called the *percolation value* or 't'.
6. The same procedure should be repeated in the second hole in the percolation area.

Test Results

- The time for the 25 mm drop ('t') for each of the two test holes in the percolation area should be averaged to give the value 't'.
- A proposed percolation area whose 't' value is less than 1 second or greater than 50 seconds is deemed to have failed the test.

Septic Tank Wastewater Treatment Systems

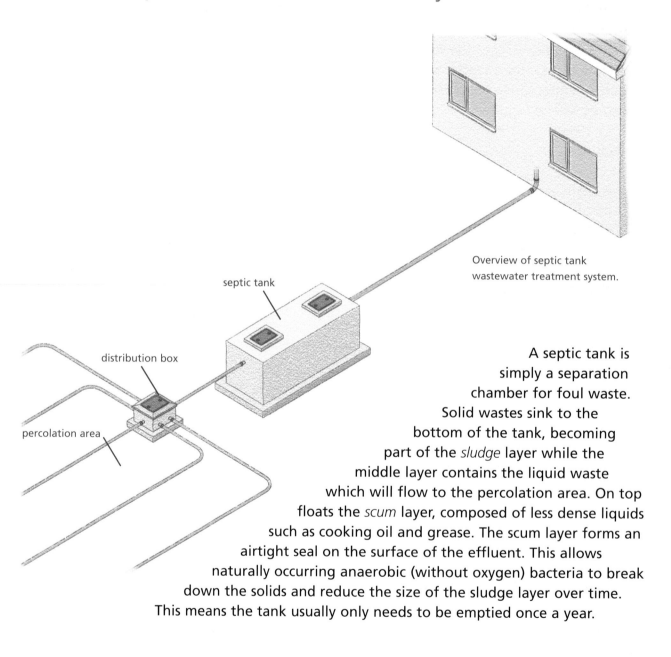

Overview of septic tank wastewater treatment system.

septic tank

distribution box

percolation area

A septic tank is simply a separation chamber for foul waste. Solid wastes sink to the bottom of the tank, becoming part of the *sludge* layer while the middle layer contains the liquid waste which will flow to the percolation area. On top floats the *scum* layer, composed of less dense liquids such as cooking oil and grease. The scum layer forms an airtight seal on the surface of the effluent. This allows naturally occurring anaerobic (without oxygen) bacteria to break down the solids and reduce the size of the sludge layer over time. This means the tank usually only needs to be emptied once a year.

The design and construction of septic tanks must meet the following criteria:

- Adequate Capacity.
  - Calculated by using the formula $C = (180P + 2{,}000)$, where C is the tank capacity (litres) and P is the number of users.
  - Minimum permitted capacity below the level of the outlet is 2,720 litres (2.72 m^3).
- Impermeable to Liquids.
  - Septic tank should not absorb groundwater from surrounding soil.
- Adequately Ventilated.
  - Prevent build-up of gases (methane).
  - Odours not evident from dwelling.
- Not a Danger to the Health of Any Person.
  - Tanks should be covered (buried) or adequately fenced in.
  - Access covers should be of durable quality to resist corrosion and should be secured to prevent removal by children.
  - Not contaminate groundwater or water supply (well).
- Adequate Means of Access for Emptying.
  - Within 10 m of a vehicular access to facilitate desludging.
- Fitted with proper pipework at the inlet and the outlet, and suitable for easy cleaning and desludging (emptying).

The wastewater flows by gravity from the septic tank into the distribution box. The distribution box regulates and disperses the flow of effluent from the septic tank into the percolation area. The distribution box must be watertight, rigidly constructed

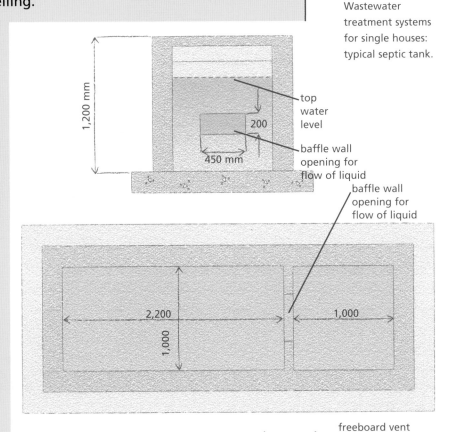

Wastewater treatment systems for single houses: typical septic tank.

and fitted with a durable watertight cover. All disposal pipes must leave the distribution box at the same elevation to ensure an even spread of liquid effluent across the percolation area.

The percolation area is the most important part of a conventional septic tank wastewater treatment system. The septic tank removes most of the suspended solids and grease from the wastewater, but it is in the percolation area that the wastewater is treated. The wastewater flows out through holes in the distribution pipes into a gravel underlay which then distributes it into the soil. While soaking through the soil the wastewater undergoes physical, chemical and biological interactions that eliminate or reduce the contaminants. A standard percolation area consists of four 20 m long, ø100 mm perforated distribution pipes. These pipes have three 8 mm perforations (at 4, 6 and 8 o'clock) every 75 mm along their length which allow the wastewater to seep into the soil.

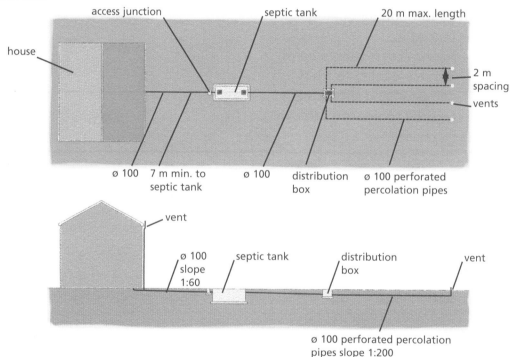

Wastewater treatment systems for single houses: septic tank wastewater treatment system.

Mechanical Aeration Wastewater Treatment Systems

Mechanical aeration systems are used to treat wastewater from a house where a site is unsuitable for a conventional septic tank system (i.e. failed Percolation Test). A wide variety of mechanical aeration systems are available. One such example is the bioCycle™ system manufactured in Dublin.

This system treats the wastewater in a four-stage process: primary treatment, aeration, clarification and disposal. The entire process takes place in a chamber containing a 2·28 m outer diameter outer tank and a 0·95 m diameter inner tank, each subdivided into two sections giving four chambers. The wastewater flows by gravity through each chamber in succession.

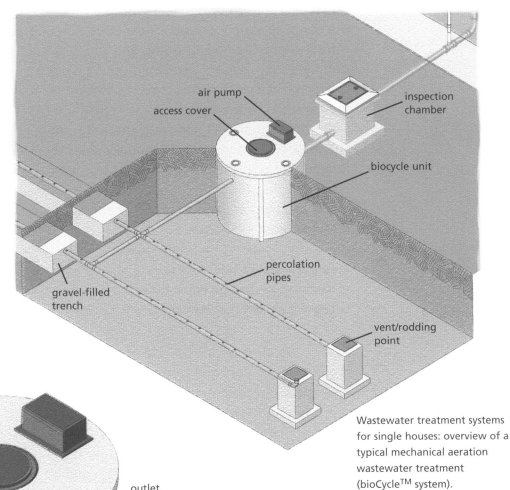

air pump

access cover

inspection chamber

biocycle unit

percolation pipes

gravel-filled trench

vent/rodding point

Wastewater treatment systems for single houses: overview of a typical mechanical aeration wastewater treatment (bioCycle™ system).

outlet

A primary chamber

D pump chamber

inlet

C clarification chamber

C clarification chamber

plastic medium (supports bacteria)

B aeration chamber

In the primary treatment chamber (A) the solids settle and are treated by naturally occurring anaerobic bacteria. Fats and grease rise to the top of this chamber and form a solid crust. The discharge from this chamber flows by gravity to the next chamber (B) where the organic content is almost totally broken

Wastewater treatment systems for single houses: mechanical aeration wastewater treatment (bioCycle™ system).

down by a culture of bacteria within a process known as submerged aerated biological filtration. Oxygen, to support the degradation processes, is introduced by a small air pump and the bacteria are supported on a purpose-designed submerged plastic medium. The discharge from the aeration stage drains to the clarification stage (C), where the solids settle. The settled solids are returned to the primary treatment stage (A). The clarified wastewater flows to the sump (D) and from there it is pumped to the percolation area. Alarms are fitted to alert the home owner if the electricity supply is interrupted or if the air pump or water pump malfunction.

While mechanical aeration systems are very effective they have some disadvantages including:

- require power (electricity),
- low sludge storage capacity,
- require skilled maintenance.

Wetland Wastewater Treatment Systems

A constructed wetland system is another system that will improve the quality of wastewater treatment on a site that is not suitable for a conventional septic tank system. This system essentially consists of an artificially created wetland planted with specially selected species of reed (e.g. Phragmites australis) that have the ability to absorb oxygen from the air and release it through their roots. This creates ideal conditions for the development of aerobic bacteria which are able to break down the contaminants present in the liquid effluent.

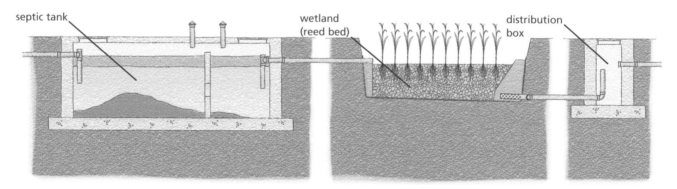

Wastewater treatment systems for single houses: overview wetland treatment system.

Constructed wetlands should be inspected weekly. The flow distribution should be carefully examined for channelling and blockages. Rabbits, weeds and plant diseases can cause damage to the reeds. Solids from the wastewater will reduce the pore space in the gravel especially at the inlet end and it may

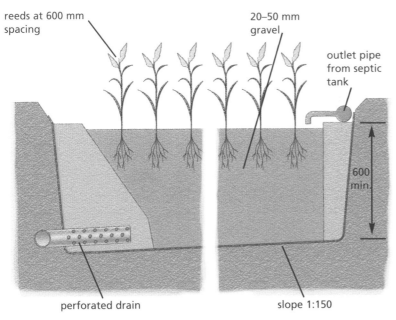

reeds at 600 mm spacing

20–50 mm gravel

outlet pipe from septic tank

600 min.

perforated drain

slope 1:150

be necessary to replace some of the gravel after a period of time. Vegetation growth, flows and wastewater quality should be monitored. The wetland should be securely fenced off.

Wastewater treatment systems for single houses: section through a typical reedbed.

Intermittent Filter System

An intermittent filter system can be soil, sand or peat based. These systems are termed intermittent because stored wastewater is pumped into the filter 3–4 times per day. The filter may be underground or mounded. The wastewater is pumped onto the surface of the filter through lateral pipes. A biofilm of bacteria builds up on the media (soil, sand or peat) and this treats the contaminants in the wastewater as it percolates downward.

The Puraflo™ system developed by Bord na Mona is an example of a peat-based intermittent filter.

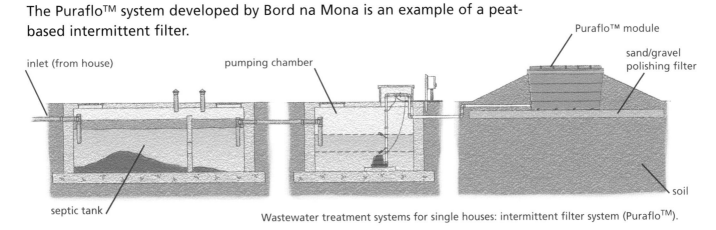

inlet (from house)

pumping chamber

Puraflo™ module

sand/gravel polishing filter

septic tank

soil

Wastewater treatment systems for single houses: intermittent filter system (Puraflo™).

In this system the wastewater flows from the home into a septic tank. The liquid effluent which leaves the septic tank is stored and pumped intermittently into the Puraflo™ modules. Floating sensors located in the storage tank monitor effluent levels to ensure optimal use of the electrically powered pump.

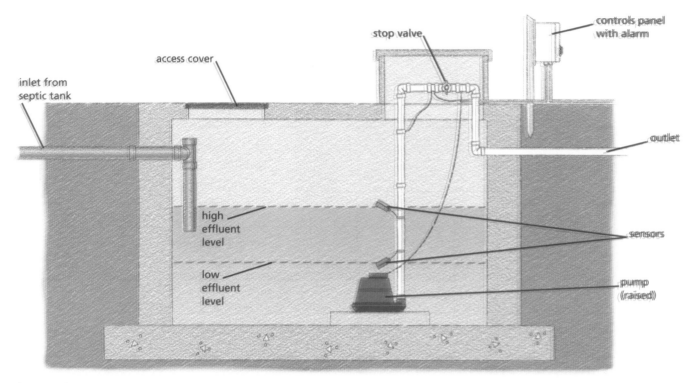

Wastewater treatment systems for single houses: vertical section through typical pumping chamber (Puraflo™).

The compressed peat is contained in plastic modules around 700 mm deep. The wastewater enters the modules through a distribution pipe just below the top surface of the peat. A combination of biological, chemical and physical processes treat the wastewater as it filters down through the peat. The treated liquid effluent emerges from the base of the unit and is dispersed into the ground through a polishing filter.

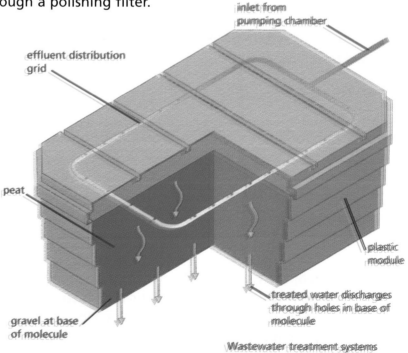

Wastewater treatment systems for single houses: cutaway view of Puraflo™ module.

Revision Exercises

1. Generate a neat cross-sectional sketch of a wastewater trap used in your home. Explain how this trap prevents foul odours from entering your home.
2. Explain, in your own words, how syphonage can occur in a single stack, above ground drainage system.
3. Explain the design features of a below ground system that ensures the system is maintenance free.
4. Generate a neat sketch of any two access points used for unblocking drains.
5. Generate a neat cross-sectional sketch of a typical septic tank. Indicate on the sketch the design features that ensure safe treatment of effluent.
6. Generate a neat cross-sectional sketch of a typical reed bed used in a wetland treatment system. Indicate on the sketch the design features that ensure safe treatment of effluent.
7. Discuss the advantages and disadvantages of septic tank and mechanical wastewater treatment systems.
8. Generate a neat line diagram of a separate wastewater drainage system for a typical house in the countryside.
9. Ordinary Level, 2001, Question 4
   (a) Using notes and neat freehand sketches, show a suitable system of sewage disposal for a dwelling house situated in a rural environment.
   (b) Describe two features of this system of sewage disposal that make it suitable for use in a rural environment.
10. Ordinary Level, 2002, Question 3
    Describe using notes and neat freehand sketches how to set out and lay a drain from a domestic dwelling to a septic tank, under each of the following headings:
    (a) Excavation of trench.
    (b) Depth and width of trench.
    (c) Safety.
    (d) Slope or gradient.
    (e) Pipework.
    (f) Backfilling.
11. Higher Level, 2000, Question 2
    (a) Draw a plan of a layout of a bathroom showing the position of the following appliances and the associated pipework for the disposal of waste:
    (i) Water closet.
    (ii) Wash hand basin.
    (iii) Bath.
    (b) Indicate on your drawing the location of a door and a window and give reasons for your choice of location.

(c) Using notes and freehand sketches, describe the design details that ensure the safe disposal of waste from each of the appliances listed at (a).

12. Higher Level, 2004, Question 6
Inadequate treatment and disposal of sewage creates environmental and health hazards.
   (a) Describe three hazards that could occur in a sewage treatment and disposal system of an individual house, situated in a rural area, if the system is not properly designed.
   (b) Using notes and sketches show how proper design detailing would prevent each of the hazards described at (a) above.
   (c) Outline three considerations to be taken into account when selecting a site for a house in a rural area to ensure that the site is suitable for the proper treatment and disposal of sewage.

13. Higher Level, 2005, Question 8
A properly designed and constructed sewerage system is essential for the safe removal of waste from a domestic dwelling.
   (a) Describe in detail, using notes and freehand sketches, three necessary considerations in the design and installation of a sewerage system from a domestic dwelling to either the main sewer or septic tank.
   (b) The accompanying sketch shows a house situated on a sloping site. When designing the sewerage system a backdrop manhole is necessary to achieve the correct gradient. To a scale of 1:10, draw a sectional elevation through the backdrop manhole. The depth from the top of the manhole to the invert level is 1,800 mm. Show and label all necessary design details.

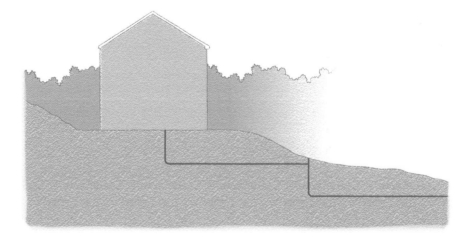

Mankind has been aware of the importance of the sun since prehistoric times. Many of the oldest man-made structures are aligned to the sun at important times of the year. For example, the Neolithic Passage Grave at Newgrange in Co. Meath was so precisely constructed that for just a few minutes after sunrise on the winter solstice (21 December) the sun shines directly down the passageway illuminating the main chamber inside.

Benefits of Natural Light in the Home

Light is essential to the enjoyment of a home. Natural light is good for the mind and the body. There are many benefits to having natural light in a home, including:

- a healthier indoor climate,
- creates good conditions for seeing,
- makes for more enjoyable interiors,
- saves energy and money,
- reduces pollutant emissions,
- conserves the earth's resources,
- contributes to the aesthetic appreciation of the space.

Sun Position and Available Light

The amount and quality of the light we receive in Ireland is influenced by our position on the planet. Ireland's position in the northern hemisphere means that the sun is to the south – therefore, the south facing façade of a building receives the most daylight. The design of buildings to optimise natural light is based on three simple facts:

- The sun is almost due south at noon.
- The sun rises in the east and sets in the west.
- The sun is higher in the sky in summer than in winter.

The position of the sun can be plotted using a sunpath diagram. This diagram provides an accurate picture of the sun's position when viewed from a particular point on earth at various times of the year.

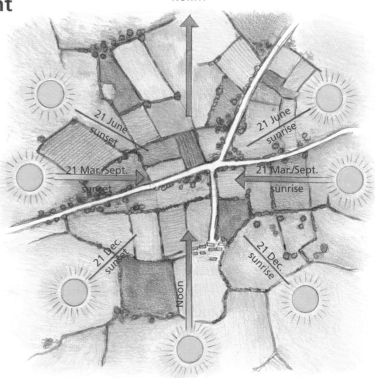

Sunpath diagram: shows the path travelled by the sun from the east in the morning to the west in the evening. The longest day is 21 June; the shortest 21 December.

Design Features

The position of Ireland in the northern hemisphere means that we get less sunlight than countries nearer the equator – especially in winter. For this reason, most people like to maximise the amount of sunlight in their homes. With this in mind the following suggestions in relation to house design and site selection should be considered:

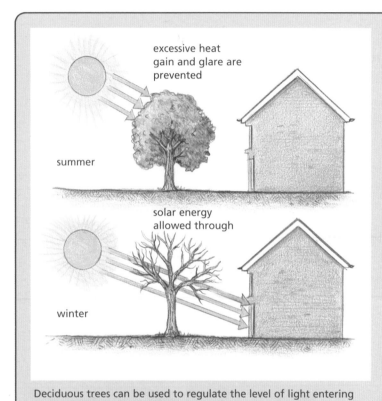

summer

excessive heat gain and glare are prevented

winter

solar energy allowed through

Deciduous trees can be used to regulate the level of light entering the home. This is a useful way of controlling solar heat gain.

- It is pleasant to wake in the morning with the sun streaming in the window – some bedrooms should have an easterly aspect (face east).
- Kitchens are used throughout the day and should have high levels of natural light – a south or south-eastern aspect is ideal.
- Living rooms are mostly used in the afternoon and evening – placing the living room on a south-western or western aspect maximises the afternoon and evening sun.
- The low altitude of the sun in winter makes sites on northern slopes less favourable than sites on southern slopes and therefore southern slopes should be developed before northern slopes.
- Obstructions to the south have a greater impact – where trees are planted on the boundary it is better that they are deciduous so that they lose their foliage in winter to admit sunlight.

Of course, each situation is unique and other factors will also come into play. For example, the home owner may want a view of a nearby mountain range from the kitchen window – if this view is to the east, then that's where the kitchen should be. Also, the possibility of glare on a television screen may mean that the living room is located to the north or east.

Sunspaces

Where the house cannot be oriented to the sun (or for existing houses) the addition of a sunspace can greatly improve the quality of the living space. A sunspace is essentially a room that is either incorporated into or built onto the house to maximise solar heat gain. Sunspaces can be either entirely glass or glass-walled with a tiled roof. A well-designed sunspace will act as a heat collector for the home during the summer. Glass-roofed sunspaces tend to overheat during the summer and are very difficult to heat during the winter – they should be separated from the main living space of the house with well-insulated doors to prevent excessive heat loss during winter.

A well-designed sunspace has a number of features including:

- located on the south side of the house,
- not excessively overshadowed by trees or buildings,
- walls have maximum amount of glazing possible,
- roof is tiled, slated and very well-insulated (prevents overheating in summer and heat loss in winter),
- floor and adjoining wall have high thermal mass and are very well insulated,
- have a well-insulated connecting door to prevent heat loss at night.

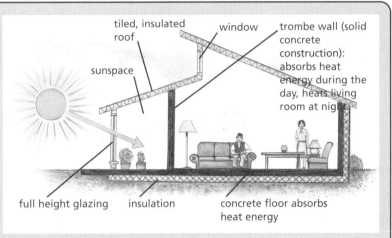

Solar heat gain can be used to warm the living space of a home through the use of a trombe wall and natural ventilation.

Solar Heat Gain

While there are many benefits to having sunlight in a home, excessive exposure can result in problems:

- Sunlight can cause excessive direct glare and even when reflected from shining surfaces it can be uncomfortably bright.
- Direct sunlight generally makes written material too bright to read.
- Shadows cast by direct sunlight can make it difficult to see objects.
- Excessive sunlight will result in overheating.
- It is uncomfortably hot to work in sunlight if there is little or no breeze to cool you down.
- Heat and light can be harmful to materials and cause a deterioration of their physical properties (e.g. paintings and fabric will fade in direct sunlight).

Sunlight Control

A number of elements contribute to the control of sunlight within buildings:

- Choice of Site (see chapter 3, Planning Permission and Sustainable Development, Planning to Build a Single House in the Countryside).
- Grouping of Buildings – Various arrangements of buildings can be used to provide shade at different times of the day. This is of most relevance when planning the layout of housing schemes.
- Height of Buildings – The height of adjacent buildings will determine the amount of light in a building – remember the high position of the sun in summer means that most buildings will not cast a shadow on their neighbours. However, in winter when the sun is low in the sky, the shadow cast might be significant.
- Orientation – Orienting the glazed façade of a building to within 15° of south will reduce energy consumption by as much as 30% due to the higher levels of solar energy received.
- Location of Rooms on Plan – In simple terms, rooms with an easterly aspect will have sun in the morning, rooms with a westerly aspect will have sun in the evening and rooms with a southerly aspect will have sun for most of the day. Unobstructed north-facing rooms may have some sun in the early morning and late evening in the summer.
- Fenestration – The size, shape, position, type of reveal and type of glazing used for windows will have a significant impact on the amount of light entering a building. Windows are generally a trade-off between the amount of light gained and the amount of heat lost. For this reason southern façades are usually 25–30% glazed while northern façades are 20–25% glazed. Even the cleanliness of a window has a significant impact on the amount of light entering – a dirty window will reduce the light entering by up to 10%.
- Devices – Blinds, curtains, brise soleil and other devices are also used to control the amount of light entering a building.

Terminology and Calculations

Up to this point, the term *sunlight* has been used to describe the natural light entering a building from outside. However, sunlight is actually the term used to describe solar radiation that reaches the surface of the earth directly from the sun, while *skylight* is the term used to describe solar radiation that reaches the surface of the earth after being scattered in the atmosphere. Together these types of light are referred to as *daylight*.

Of course, the sky over Ireland changes continuously – it is rarely completely clear of clouds. The sky becomes brighter toward noon and its light distribution changes as clouds form, thicken and disperse. These changes mainly occur in response to the movement of the sun across the sky. To simplify matters the concept of a standard sky, called the CIE Overcast Sky, has been developed. (CIE stands for the Commission Internationale de L'Éclairage – an international body that oversees standards relating to light sources, vision and lighting design.) In simple terms, this standard defines a standard overcast sky as *the degree of horizontal illuminance at a point on the ground from an unobstructed overcast sky as equal to 5,000 lux.* This figure is used by designers for light calculations in relation to buildings.

The light that enters a house is made up of a number of components:

- **Sky Component** – light that enters directly through a window.
- **Internally Reflected Component** – light that reflects off surfaces inside.
- **Externally Reflected Component** – light that reflects off surfaces outside and then enters the house.

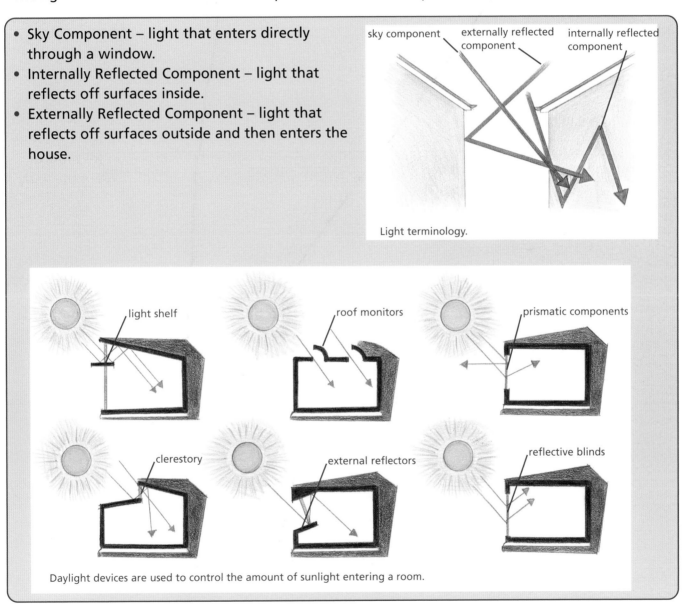

Light terminology.

Daylight devices are used to control the amount of sunlight entering a room.

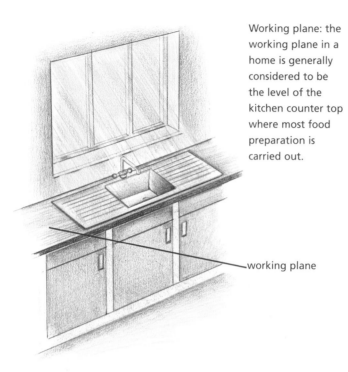

Working plane: the working plane in a home is generally considered to be the level of the kitchen counter top where most food preparation is carried out.

working plane

Daylight factor is another important concept in relation to light. The daylight factor describes the amount of natural light available in a building for everyday activities. In simple terms, the daylight factor is the ratio of the light inside to the light outside. It is defined as *the ratio, expressed as a percentage, of illuminance at a point on a given plane due to light received directly or indirectly from a standard overcast sky, to illuminance on a horizontal plane due to an unobstructed hemisphere of the sky.*

Recommended	Daylight Factors	Lux inside
School classrooms	5.0%	(5,000 x 5.0%) = 250
Assembly halls	1.0%	50
Living rooms	1·5%	75
Bedrooms	1.0%	50
Kitchens	2.0%	100

Light is measured in units called *lux*. The size of a window required to provide the appropriate daylight factor for a room can be calculated in several ways. One example is the *degree of efficiency method* which uses the following formula:

$$\text{lux inside} = \text{lux outside} \times \text{window factor} \times \text{efficiency coefficient} \times \frac{\text{window area}}{\text{floor area}}$$

Where,

> - lux inside (Li) = required level of light inside
> - lux outside (Lo) = standard overcast sky (i.e. 5,000 lux)
> - window factor (WF) = the reduction in incident light due to the fact that the window is in the vertical plane. This reduction is typically assumed to be in the region of 50% and so, is given a constant value of 0·5 for calculations.
> - efficiency coefficient (E) = the reduction in incident light due to factors relating to the glazing (e.g. transmittance, cleanliness, reflection). For calculations, this is given a constant value of 40% or 0·4.
> - window area (WA) = length × height (of window)
> - floor area (FA) = length × width (of room)
>
> *Worked example*
> A room 3 m by 4 m, with an unobstructed view, requires an illumination of 150 lux. Determine, using the degree of efficiency method, the approximate area of the glazing required.
> Step 1. Li = Lo × WF × E × WA/FA
> Step 2. Li = 5,000 × 0·5 × 0·4 × WA/FA
> Step 3. Li = 1,000 × WA/FA
> Step 4. 150 = 1,000 × WA/(3 × 4)
> Step 5. 150 = 1,000 × WA/12
> Step 6. 150 × 12 = 1,000 × WA
> Step 7. 150 × 12/1,000 = WA
> Step 8. 1.8m² = WA

Revision Exercises

1. Discuss the benefits of having natural light in a home.
2. Generate a sunpath diagram for your home. Show the orientation of your house (angle to south) and the path of the sun in summer and winter.
3. Using the sunpath diagram drawn for question 2, generate a neat floor plan of the ground floor of your house and indicate on it which areas of the house have daylight in the morning and in the evening.
4. Outline, using a neat sketch, the features of a well-designed sunspace.
5. Explain, using neat annotated sketches, any three features that are used to regulate the amount of daylight in houses.
6. A classroom 8 m by 20 m, with an unobstructed view, requires an illumination of 250 lux. Determine, using the degree of efficiency method, the approximate size of the glazing required.
7. Higher Level, 2000, Question 7

    (a) Determine, by the degree of efficiency method (or by any other suitable method), the approximate size of a vertical window suitable for a kitchen 4·8 m long by 3·6 m wide requiring an average illumination of 150 lux on the working plane. Assume an unobstructed view and the illumination of a standard overcast sky to be 5,000 lux.

    (b) Select two materials commonly used in the manufacture of window frames and discuss in detail the advantages and disadvantages of each material for window frame manufacture.

8. Higher Level, 2002, Question 9, (c)

An illuminance of 300 lux is required on a working plane in the kitchen. The daylight factor at a point on the working plane in the kitchen is 5%. Show by calculation if the illuminance is sufficient, assuming an unobstructed view and the illuminance of a standard overcast sky to be 5,000 lux.

The design of vernacular buildings all over the world has been based on the need to provide shelter and thermal comfort in situations where there is either too little fuel for heating or no natural means of cooling. In these situations, the buildings themselves have to be the means by which the interiors are made comfortable. Such buildings, which primarily use form and materials to provide comfort, are described as *passive* buildings.

Since the industrial age, societies have made a greater use of heating and cooling systems to keep their buildings at comfortable temperatures. Buildings that rely upon such systems to maintain comfort levels are described as *active* buildings.

A general characteristic of all active buildings is that energy is needed to run their heating and cooling systems. The advantage of using such systems is that the designer is released from constraints on design. The disadvantage is the expense and environmental impact incurred in making and running the active systems. Additionally, there is now a realisation that an overdependence upon active systems can lead to an excessive exploitation of fossil fuels and that this is detrimental to the Earth's environment.

The residential sector consumes over a quarter (26%) of the energy used in Ireland every year. The breakdown of energy use in a home is: cooking 7%, lighting and appliances 11%, water heating 25%, space heating 57%.

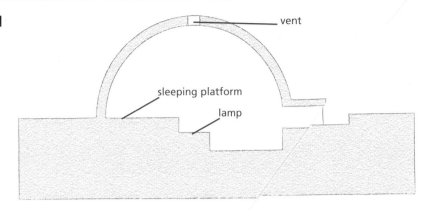

Passive house: the Inuit igloo is a wonderful example of a passive building which relies primarily on form (shape) and material to provide comfort.

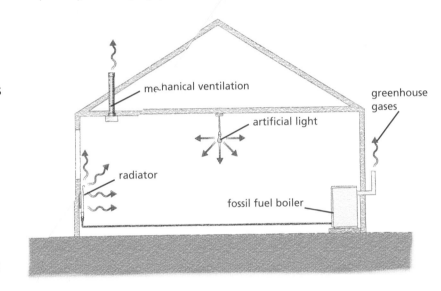

Active house: this type of dwelling relies on artifical systems to provide comfort.

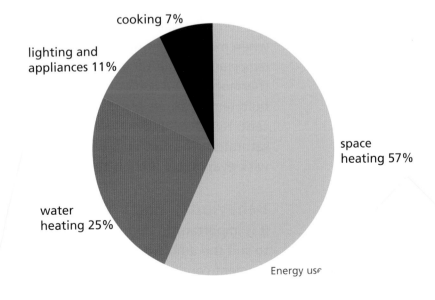

cooking 7%

lighting and appliances 11%

space heating 57%

water heating 25%

Energy use

In doing so, Irish homes release nearly eleven million tonnes of harmful carbon dioxide into the atmosphere. If the demand for energy is to be reduced, buildings must be designed to use less energy. Reducing the energy consumed by buildings may be achieved by fully exploiting the natural advantages of a site, adopting appropriate building forms and using construction methods that are energy efficient.

Thermal Comfort

Thermal comfort is achieved when there is a balance between the generation of metabolic heat within the body and the loss of heat from the body. Where this is in balance, a person will feel neither too hot nor too cold. Of course, people differ in what they find comfortable – what some find warm, others find cool. However, in general people are remarkably similar in the way that they react to different temperatures.

Environmental Factors

There is a range or zone of thermal comfort known as the *comfort zone*. This zone is affected by a number of factors including:

- dry bulb temperature of the air,
- mean radiant temperature – average temperature of various surfaces in a room (e.g. windows, walls),
- relative humidity of the air – amount of moisture held in the air,
- air movement,
- temperature differentials between the feet and the head,
- draughts of cold air,
- excessive solar radiation.

The term *dry bulb temperature* is used to describe the temperature of the air when measured using a thermometer whose bulb is dry. The *wet bulb temperature* is the temperature of the air when measured using a thermometer whose bulb is wet (wrapped in wet wick). If the dry and wet bulb temperatures are equal then the relative humidity is one hundred per cent. However, a lower wet bulb temperature indicates that the air is not fully saturated. The wet bulb temperature is lower because as the water in the wick evaporates it cools the bulb.

Behavioural Factors

It is important to appreciate that people can wear different types of clothing to suit the ambient conditions. Also, if people are able to respond to the ambient conditions by opening a window, adjusting a Venetian blind or moving to a shadier seat, then they will be able to tolerate a wider range of ambient conditions. Furthermore, the level of thermal comfort a person is experiencing will depend on what they are doing. For example, when skiing, the vigorous exercise combined with the solar gain means that a person can be comfortable even when the air temperature is quite low.

Heat Transfer

Heat is transferred via the mechanisms of conduction, convection, radiation and evaporation.

Conduction is the transfer of heat energy through a substance **without** any movement of the substance. For example, when a concrete cavity-walled house is heated, the inner leaf of the wall is heated. The heat energy is conducted through the blockwork until it reaches the insulation.

Convection is the transfer of heat energy through a substance by movement of the substance. Water is moved upwards by convection when heated (the heated water becomes less dense and rises, the cool water falls with gravity). For example, when an immersion heater is used, the heated water in the

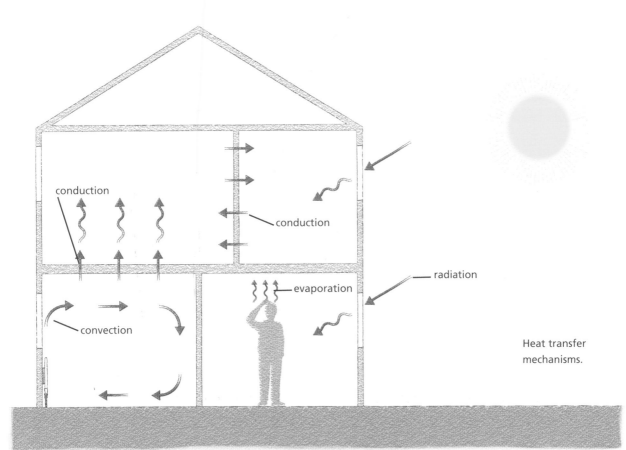

Heat transfer mechanisms.

storage tank rises. This is why the hot water distribution pipe is connected to the top of the cylinder.

Radiation is the transfer of heat energy from one point to another in the form of electromagnetic waves. An example of this is when sunlight shines through a window and heats a room (solar heat gain).

Evaporation is the process by which liquid water is converted into a gaseous state. The human body cools by evaporation – evaporation requires a lot of heat energy (the evaporation of one gram of water at a temperature of 100° Celsius requires 540 calories of heat energy) and so has a cooling effect on the body. Evaporation occurs whenever a damp object is heated. For example, rainwater can sometimes be seen evaporating from the tiles on a roof when the sun shines after a rain shower, or every time a kettle is boiled water evaporates and can be seen as steam.

Building for Energy Conservation

Houses will need to become more energy efficient if the environmental impact of space and water heating in houses is to be reduced. Heat is lost from houses through various routes, as shown below.

The building regulations provide strict guidance in relation to the construction of houses to reduce energy consumption (Building Regulations

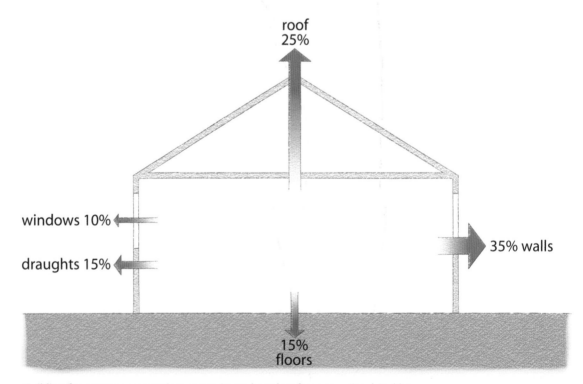

Building for energy conservation: approximate heat loss from an uninsulated home.

2005, Technical Guidance Document L: Conservation of Fuel and Energy – Dwellings). The regulations aim to ensure the conservation of energy by:

- limiting the heat loss and maximising the heat gains through the fabric of the building,
- controlling the output of the space heating and water heating systems,
- limiting the heat loss from pipes, ducts and vessels used for the transport or storage of heated water or air.

Before design for thermal performance is examined it is essential to understand the terminology of heat transfer in buildings including:

- Thermal Conductivity (lambda, λ)– A measure of a material's ability to conduct heat energy. It is defined as the quantity of heat that will pass through a square metre of a material that is one metre thick, for each degree difference in temperature between one side and the other. The benefit of this measure is that it allows various insulating materials to be easily compared. It is measured in W/mK (Watt/metre Kelvin).
- Thermal Resistance (R-value) – A measure of a material's ability to resist the flow of heat energy. Measured in m^2K/W. The relationship between resistance and conductivity can be expressed as $R = d/\lambda$ where R is the thermal resistance of the insulation, d is the thickness of the insulation and λ is the thermal conductivity of the insulation.
- Thermal Transmittance (U-value)– A measure of the rate at which heat passes through a particular element of a building (e.g. a cavity wall) when unit temperature difference is maintained between the ambient air temperatures on each side. It is measured in W/m^2K. The U-value takes into account the resistances of the various materials, the surface resistances and the cavity (if present).

Recent changes to the building regulations have reduced the acceptable U-values for the walls, floors and roofs of houses. This means that houses now require higher levels of insulation than in the past.

Element	U-value 2005 Regulations (W/m²K)	U-value 1997 Regulations (W/m²K)
Floor	0·25	0·45
Wall	0·27	0·45
Pitched roof (insulation at ceiling)	0·16	0·25
Pitched roof (insulation on slope)	0·20	0·25

U-values: maximum acceptable U-values (Elemental Heat Loss Method).

There are several steps that can be taken to improve the thermal performance of a house, including:

- preventing thermal bridging,
- using a better quality of insulation product (i.e. one with higher R-value) in the walls, floors and roof,
- using greater thicknesses of insulation in the walls, floors and roof,
- using low-emissivity double-glazing in all windows,
- preventing air infiltration (draughts) and air leakage.

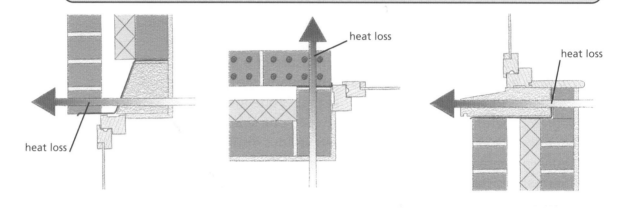

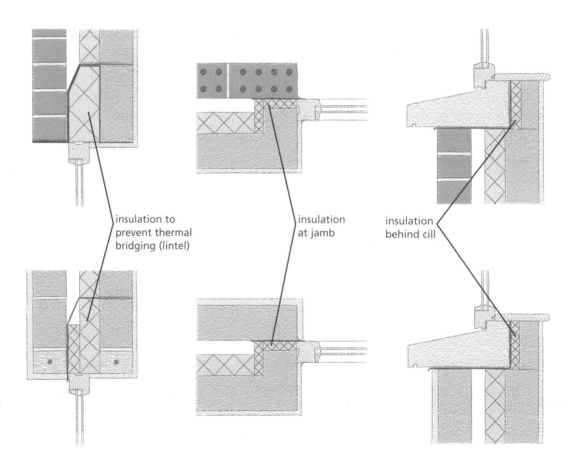

insulation to prevent thermal bridging (lintel)

insulation at jamb

insulation behind cill

Thermal bridging: openings at windows and doors (heads, cills and jambs) must be properly protected using insulation and damp-proof courses to prevent heat loss and dampness.

Thermal bridging (or *cold bridging*) is one of the weaknesses of traditional housing construction techniques. A thermal bridge occurs where an insulation layer is broken by non-insulant material allowing for higher heat loss. Thermal bridging is common around openings in cavity walls, such as lintels, jambs and cills.

The infiltration of cold air from outside should be prevented by ensuring doors, windows and service ducts are properly sealed. Draught stripping should be fitted to the frames of openable elements of doors and windows and to the loft hatch (attic door).

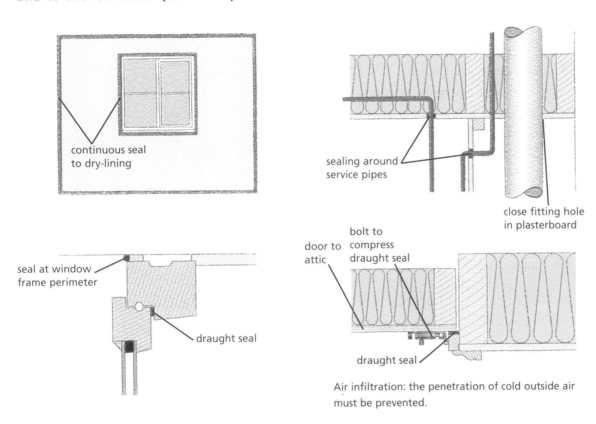

continuous seal to dry-lining

sealing around service pipes

close fitting hole in plasterboard

seal at window frame perimeter

draught seal

bolt to compress draught seal

door to attic

draught seal

Air infiltration: the penetration of cold outside air must be prevented.

Also, the void between dry lining and masonry walls should be sealed at all edges using timber battens. Vapour barriers in timber-frame construction should also be sealed.

Insulation Materials

There are four basic types of insulation: loose-fill, batts and blankets, rigid board, and injected.

Loose-fill Insulation

Loose-fill insulation includes loose fibres or fibre pellets that are blown into building cavities or attics using special equipment. It generally costs more than batt insulation. However, it usually fills nooks and crannies easier and reduces air leakage better than batt-type insulation. Various materials are used including, cellulose fibre which is made from recycled paper. It is chemically treated for fire and moisture resistance. It can be installed in walls, floors or attics using a dry-pack process or a moist-spray technique. Fibreglass and rock wool loose-fill insulation involve blowing insulation into masonry cavities of timber-frame stud panels behind netting.

plasterboard

netting stapled to studs

stud partition

Loose-fill insulation.

loose-fill insulation

Batt and Blanket Insulation

Batt and blanket (also referred to as *quilted insulation*) is made of mineral fibre (either processed fibreglass or rock wool) and is used to insulate below floors, above ceilings, and within walls. Generally, batt insulation is the least expensive wall insulation material but requires careful installation for effective performance. Soft insulation is only suited to applications where it will not be compressed. A common application is in the insulation of attic spaces, where it is placed between and across the ceiling joists. It is best suited to a standard joist, rafter, or stud spacing of 400 or 600 mm. Batts and blankets come in widths to fit tightly between the wood-framing members. Some are available with a foil backing. Batts generally come in lengths of 1,200 or 2,400 mm. Blankets come in long rolls that are cut to the desired length for installation.

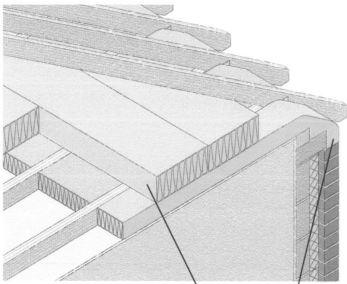

Blanket or quilted insulation.

quilted insulation laid in two layers

quilted insulation dressed down to meet wall insulation

Rigid Board Insulation

Rigid board insulation is commonly made from expanded polystyrene board (EPS). Polyurethane (PUC) and polyisocyanurate (PIR) are also used. These boards are available in a variety of thicknesses and insulating values. It is typically used in the cavity of concrete walls and in concrete floors. It is particularly suited to these applications because of its rigidity and ability to withstand compressive loads. It is easy to install and can be easily cut to size by hand.

Injected Insulation

Injected or blown-in insulation is mainly used to retro-insulate the cavity walls of older houses. While there is a wide variety of foam insulants available including, phenolic foam, polyisocyanurate foam and polyurethane foam, the most commonly injected insulants used in Ireland are glass fibre, rockwool fibre and polystyrene bead. In each case the insulant is sprayed through a series of holes drilled in the external leaf of the cavity wall. Polystyrene bead-type insulant is injected into the cavity with a binding agent which improves the long-term stability of the insulation.

The thermal properties of each of these insulants do not differ much (their thermal conductivities tend to fall within the range 0·025–0·04 W/m K). Insulating the cavity with these materials will typically reduce heat loss through the wall to one-third of its original value. These cavity insulation materials are generally water-repellent, rot-proof and some (e.g. mineral wool) are non-combustible.

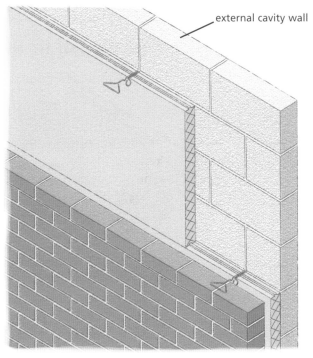

external cavity wall

Rigid insulation.

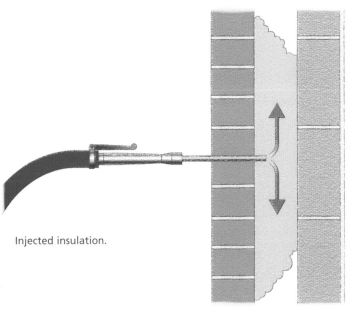

Injected insulation.

Injected insulation should be installed by a trained contractor or licensed installer using special equipment. The cost of installing blown-in insulation is usually recouped within five years (due to reduced heating bills), and greatly reduces the amount of greenhouse gases generated.

Insulation of Houses: Construction Details

Insulation of Newly Built Houses

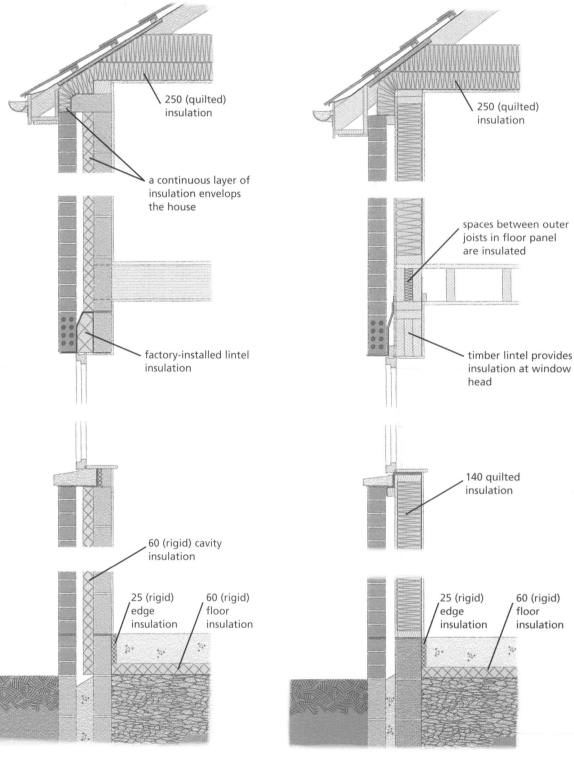

250 (quilted) insulation

a continuous layer of insulation envelops the house

factory-installed lintel insulation

60 (rigid) cavity insulation

25 (rigid) edge insulation

60 (rigid) floor insulation

250 (quilted) insulation

spaces between outer joists in floor panel are insulated

timber lintel provides insulation at window head

140 quilted insulation

25 (rigid) edge insulation

60 (rigid) floor insulation

Insulation of newly built houses: typical construction details for a concrete cavity wall with a brick outer leaf.

Insulation of newly built houses: typical construction details for a timber-frame cavity wall with a brick outer leaf.

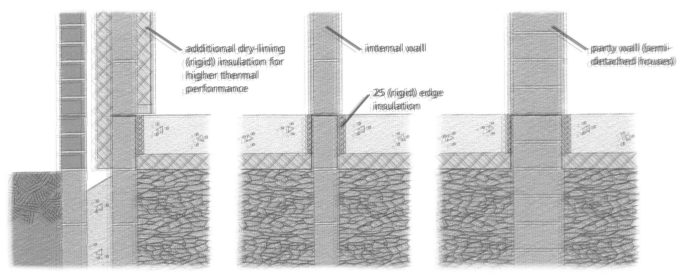

Insulation of newly built houses: typical construction details for solid concrete ground floors.

Improving the Thermal Performance of Older Houses

As people become more aware of the need to conserve energy the need to improve the thermal performance of older buildings is becoming more widespread. Some of the key areas where houses can be improved include:

- fitting draught stripping to external doors and windows,
- re-glazing old windows with argon-filled, low-e double-glazing units,
- insulating attic spaces with blanket insulation,
- insulating cavity walls.

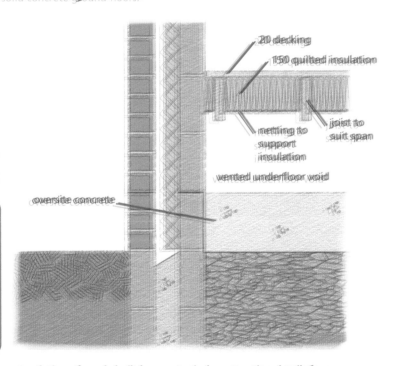

Insulation of newly built houses: typical construction details for a suspended timber ground floor.

Most of these procedures are fairly straightforward, however, external cavity walls present a significant challenge. There are two approaches that can be taken:

- injecting loose-fill or foam insulation into the cavity,
- dry lining the internal surfaces.

Loose-fill or foam injection requires a large number of holes (typically Ø22 mm) to be drilled in the external leaf, through which the insulant can be injected. In brick walls the holes are drilled in the mortar joints and later repointed. In rendered walls the location of the holes is aesthetically less important as they can be easily filled and the whole wall repainted. The main problem with this approach is that it fills the cavity completely thereby increasing the risk of moisture transfer across the cavity. For this reason moisture-resistant insulation is usually used.

Improving the thermal performance of older houses: typical drilling pattern for injecting insulation into a cavity wall.

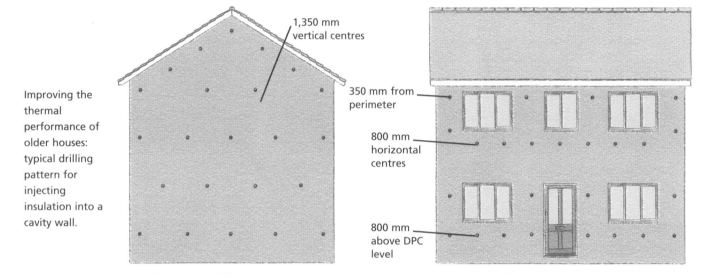

1,350 mm vertical centres

350 mm from perimeter

800 mm horizontal centres

800 mm above DPC level

Dry lining is a more traditional approach that involves fixing insulation and plasterboard to the inner surface of the wall to improve its thermal properties (see chapter 24, Rendering and Plastering). While this method is relatively cheap and requires no specialist equipment it does involve the complete redecoration of the interior. Wall-mounted appliances such as sockets, light switches and radiators will all have to be moved. Also, the home owner will be disrupted for the duration of the work. For some home owners, the injection technique might be more convenient.

Heat Loss Calculations

There are three methods of calculating the overall thermal performance of a house:

- Elemental Heat Loss – This method is carried out in two stages. Stage one establishes the performance of individual elements of the house (e.g. U-value of the external walls, U-value of the roof) and checks that they do not exceed the maximum permitted values. The U-value of each element must be calculated and checked against the maximum permitted values stated in the building regulations.

Element	Maximum permitted value (2005 Regulations)
Floor	0·25 W/m²K
External wall	0·27 W/m²K
Pitched roof (insulation at ceiling level)	0·16 W/m²K

If the requirements for each element have been met, the second stage is carried out. Stage two takes into account the combined area of external door, window and roof-light openings and requires that these should not exceed a certain percentage of floor area, for given U-values, as follows:

U-value (average) (W/m²K)	Maximum combined area of doors, windows and roof lights as a percentage of floor area (%)
2·0	27·9
2·2	25·0
2·4	22·7
2·6	20·7
2·8	19·1
3·0	17·7

Example

1. Total floor area = 120 m²
2. Total window area = 24 m² (average U-value 2·3 W/m²K – data supplied by manufacturer)
3. Total door area = 5 m² (average U-value 2·0 W/m²K – data supplied by manufacturer)
4. Average U-value of opes = (24)(2·3) + (5)(2·0)/2 = 2·24 W/m²K
5. From the table the maximum combined area of doors, windows and roof-lights as a percentage of floor area for an average U-value of 2·24 W/m²K = 25 W/m²K
6. = 29/120 = 24·2 m²
7. => 24·2 m² < 25 m²
8. Therefore, this house meets the building regulation requirements.

- Overall Heat Loss – This method allows the designer greater flexibility because it is related to the shape of the house. Houses that have a simple shape (e.g. square plan) have a low surface area to volume ratio. Whereas houses that have unusual shapes (e.g. L-shaped plan) have a higher surface area to volume ratio. This is important because the overall heat loss method allows a lower average U-value for houses that have a high surface area to volume ratio.

The surface area is the total area of all the heat loss elements (i.e. external walls, ground floors, roofs). To calculate the overall heat loss of a house, the area of each heat loss element is multiplied by its U-value. These are then added up giving the total area U-value. This is divided by the total area. This gives an average U-value for the entire house. The ratio of the total area of the heat loss elements to the total volume is then calculated. These figures are then compared to check if the house design complies with the building regulation requirements.

Maximum average U-value (U_m) as a function of building volume (V) and fabric heat loss (A_t)	
Ratio of area of heat loss elements to building volume (A_t/V) (m⁴)	Maximum average U-value (U_m) (W/m²K)
1·3	0·39
1·2	0·40
1·1	0·41
1·0	0·43
0·9	0·45
0·8	0·48
0·7	0·51
0·6	0·56
0·5	0·62
0·4	0·72
0·3	0·87

- Heat Energy Rating Method – This very detailed method takes into account the size of the house, the heat loss through the building fabric, the heat loss due to ventilation, water heating, space heating and solar energy gain. It calculates the amount of heat energy used per square metre of floor area every year. This value (*heat energy rating*) is then compared to the maximum permitted values specified in the building regulations.

Both the overall heat loss method and the heat energy rating method require that the following maximum elemental U-values are not exceeded:

Ground floors	0·37 W/m²K
External walls	0·37 W/m²K
Roofs	0·25 W/m²K

These U-values are higher (i.e. less insulation required) than the U-values required for the elemental heat loss method. This is because the elemental U-value method is the most stringent method as it does not take the shape of the house into consideration.

In simple terms, if two houses have the same volume they require roughly the same amount of heat. If one of the houses is made up of a number of smaller components, it has a high surface area to volume ratio. This type of house can perform adequately with a lower level of thermal insulation. This is because there is a greater amount of surface area available to *contain* the same amount of heat energy.

Calculating U-values

All three heat loss calculation methods require the U-value of each structural element (e.g. ground floors, external walls, roofs) of the house to be calculated. In order to calculate the U-value of any structural element, the conductivity of each component (e.g. blockwork, insulation, plasterboard) must be known. The conductivity of a building component is normally provided by the product manufacturer. The following measured values of generic building components (i.e. no particular brand) can be used to calculate the typical thermal performance of a structural element.

Material	Density (kg/m³)	Thermal Conductivity, λ (W/mK)
Clay brickwork (outer leaf)	1,700	0·77
Clay brickwork (inner leaf)	1,700	0·56
Concrete block (heavyweight)	1,900	1·33
Concrete block (medium weight)	1,400	0·57
Concrete block (autoclaved aerated)	600	0·18
Aerated concrete slab	500	0·16
Concrete screed	1,200	0·41
Reinforced concrete (1% steel)	2,300	2·30
Reinforced concrete (2% steel)	2,400	2·50
Wall ties, stainless steel	7,900	17·00
Wall ties, galvanised steel	7,800	50·00
Mortar (protected)	1,750	0·88
Mortar (exposed)	1,750	0·94
External rendering (cement sand mix)	1,300	0·57
Plaster (gypsum lightweight)	600	0·18
Plaster (gypsum)	1,200	0·43
Plasterboard	900	0·25
Natural slate	2,500	2·20

Material	Density (kg/m³)	Thermal Conductivity, λ (W/mK)
Concrete tiles	2,100	1·50
Fibrous cement slates	1,800	0·45
Ceramic tiles	2,300	1·30
Asphalt	2,100	0·70
Felt bitumen layers	1,100	0·23
Timber, softwood	500	0·13
Timber, hardwood	700	0·18
Wood wool slab	500	0·10
Wood panels (plywood, chipboard etc.)	500	0·13
Expanded polystyrene beads (injected)	12	0·040
Expanded polystyrene board (EPS) (standard density)	15	0·037
Expanded polystyrene board (EPS) (high density)	25	0·031
Extruded polystyrene	30	0·025
Phenolic foam	30	0·025
Polyurethane board (PUR)	30	0·025
Polyurethane board (PUR) (low-e foil faces)	32	0·023
Polyisocyanurate board (PIR) (low-e foil faces)	29 (avg.)	0·022
Glass fibre quilt	12	0·040
Hemp wool batt	30	0·040
Rock wool loose (injected)	12	0·040
Rock wool batt	24	0·037
Sheep wool quilt	25	0·037
Glass fibre batt	25	0·035

U-value calculations are carried out in a number of steps:

1. Make a quick cross-sectional sketch of the element (e.g. wall).
2. Number and label each layer or surface beginning with the internal surface resistance and work outward.
3. Convert the thickness of each layer into metres (divide by 1,000, e.g. 19 mm = 0·019 m).
4. Calculate the resistance of each layer (beginning with the internal surface resistance) by dividing the thickness of the component by its conductivity (i.e. thickness divided by conductivity) – use a table to keep the calculations tidy.

5. Calculate the total resistance (i.e. add up the layer and surface resistances).
6. Calculate the reciprocal of the total resistance (i.e. one divided by the total resistance) to get U-value.

Note: For accuracy and consistency, all figures should be rounded to three places of decimals (i.e. 0·000) until the U-value is determined. The final answer is then rounded to two places of decimals (i.e. 0·00). U-values are universally stated to two places of decimals.

Remember, in order to simplify the calculations it is best to use a table.

	Table of Calculations			
	Layer/Surface (name)	Thickness (m)	Conductivity, λ (W/mK)	Resistance, R (m²K/W)
1.				
2.				
3.				

Note: It is standard practice today for manufacturers to provide a conductivity λ value for their products. However, if the resistivity of a component is provided instead, the conductivity can be calculated as the reciprocal of the resistivity (i.e. 1/resistivity = conductivity). Alternatively the resistance can be calculated by multiplying the resistivity by the thickness. Also, it is common for the units of temperature to be stated as Kelvin instead of as degrees Celsius (i.e. m² °C/W = m² °C/K), this makes no difference to the calculations.

Example 1: Masonry Cavity Wall
This is a straightforward example of a U-value calculation where the element (i.e. external wall) is made up of a number of uniform layers: the outer leaf and inner leaves are both concrete block, the cavity contains rigid insulation (high-density EPS).

In this case the cavity is increased to 140 mm. Traditionally, the cavity of an external wall is 100 mm wide. However, in this example it has been widened to accommodate a 100 mm layer of high-density expanded polystyrene. The construction of walls with cavities in excess of 110 mm wide requires adjustments to lintels, wall ties, cavity barriers, etc. This is necessary to meet the 2005 building regulations requirement of a maximum U-value of 0·27 W/m²K. It is important to remember that if the width of the cavity is

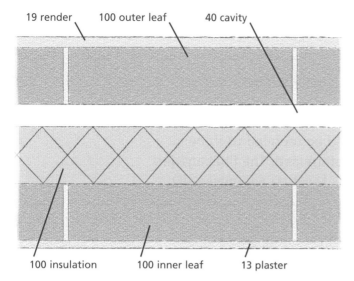

19 render   100 outer leaf   40 cavity

100 insulation   100 inner leaf   13 plaster

Calculating U-values:
masonry cavity wall.

increased the width of a strip foundation (if used) must also increase proportionally – the width of a strip foundation should always be three times the overall wall thickness. Also, the maximum cavity width allowed by Irish standards is 150 mm. In practice there is a variety of measures (other than increasing the thickness of insulation in the cavity) that can be taken to ensure that the external walls meet thermal insulation requirements, such as using a combination of cavity insulation and dry lining or the use of full-fill injected insulation products.

Following the steps outlined above the following data table is created:

Layer/Surface (name)	Thickness (m)	Conductivity, λ (W/mK)	Resistance, R (m²K/W)
1. Internal surface resistance	-	-	0·130
2. Lightweight plaster	0·013	0·18	0·072
3. Concrete block	0·100	1·33	0·075
4. Expanded polystyrene (HD)	0·100	0·031	3·226
5. Cavity	-	-	0·180
6. Concrete block	0·100	1·33	0·075
7. External render	0·019	0·57	0·033
External surface resistance	-	-	+ 0·040
Total resistance (R_{total})			**3·831**

U-value

= $1/R_{total}$

= 1/3·831

= 0·26 W/m²K (meets regulatory requirements: maximum 0·27 W/m²K)

Note: The units for the U-value are the reciprocal of the units of resistance.

Example 2: Timber-frame Cavity Wall
This is a more complex example of a U-value calculation where the element is made up of a number of layers, some of which are bridged. The outer leaf is of uniform construction; (e.g. red brick) and it does not change along its length. However, the inner leaf is bridged by timber studs. Therefore, there are two possible heat-flow paths: across the studs or across the insulation between the studs. The heat-flow path across the studs cannot be dismissed as being negligible, because the resistance at the studs is considerably lower.

The combined method of U-value calculation is based on an average of the upper and lower resistances of the wall. The upper limit of resistance (R_{upper}) is calculated based on the two separate heat-flow paths: across the studs and across the insulation between the studs. The lower limit of resistance (R_{lower}) is calculated based on one heat-flow path – this method views the stud and insulation as one unit, but makes allowance for this in the calculations. These different methods are easier to understand when represented graphically.

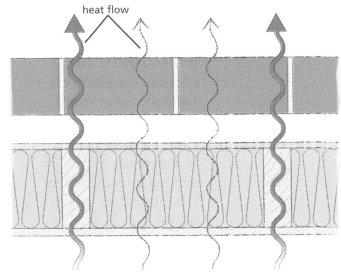

Calculating U-values: a timber-frame cavity wall has two heat-flow paths; through the studs and between the studs (through the insulation).

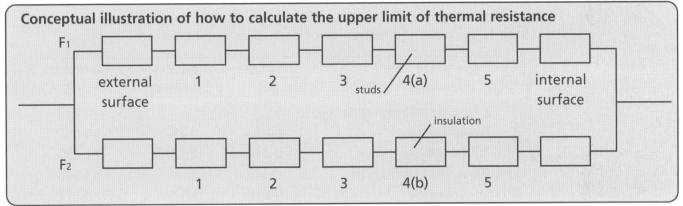

Calculating U-values: combined method of U-value calculation – upper limit of resistance.

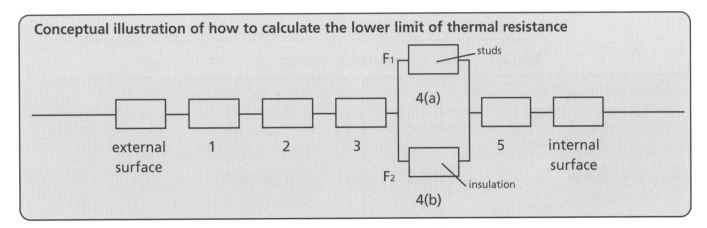

Calculating U-values: combined method of U-value calculation – lower limit of resistance.

While complex, this method provides the most accurate results for structural elements with bridged layers and is recommended in the building regulations.

In this example the external wall consists of a timber panel (10 mm plywood sheathing on 140 mm timber studs) containing 140 mm glass fibre quilted insulation, a 50 mm cavity and a 100 mm red brick outer leaf. The internal surface of the timber panel is sheathed in plasterboard.

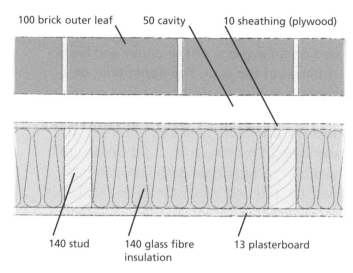

Calculating U-values: timber frame cavity wall.

The calculations are carried out in four stages.

The first stage is to calculate the upper resistance. This is done in four steps:

Step 1: Calculate the resistance of each layer and/or surface.

Layer/Surface (name)	Thickness (m)	Conductivity, λ (W/mK)	Resistance, R (m²K/W)
1. Internal surface resistance	-	-	0·130
2. Plasterboard	0·013	0·25	0·052
3a. Timber studs	0·140	0·13	1·077
3b. Glass fibre insulation	0·140	0·04	3·500
4. Sheathing	0·010	0·13	0·077
5. Cavity	-	-	0·18
6. Brick	0·100	0·77	0·13
7. External surface resistance	-	-	0·40

Step 2: Calculate the resistance through each heat-flow path.

Resistance through timber studs, R_1	
Layer/Surface (name)	Resistance, (m²K/W)
1. Internal surface resistance	0·130
2. Plasterboard	0·052
3a. Timber studs	1·077
4. Sheathing	0·077
5. Cavity	0·180
6. Brick	0·130
7. External surface resistance	+0·400
$R_1 =$	2·046

Resistance through insulation, R_2	
Layer/Surface (name)	Resistance, (m²K/W)
1. Internal surface resistance	0·130
2. Plasterboard	0·052
3b. Glass fibre insulation	3·500
4. Sheathing	0·077
5. Cavity	0·180
6. Brick	0·130
7. External surface resistance	+0·400
$R_2 =$	4·469

Step 3: Calculate the fractional area of each heat-flow path.

In this example, the inner leaf is constructed using 44 mm joists at 600 mm centres. Therefore, the fractional area of the timber studs, F_1 is:

F_1 = 38/400 = 0·095 (i.e. timber studs make up 9·5% of the total surface area of the wall)

Therefore, the fractional area of the insulation F_2 is:

1 – 0·095 = 0·905 (i.e. 90·5% of the surface area)

Step 4: Calculate the total upper resistance.

The total upper resistance is calculated using the formula:

$$R_{\substack{combined \\ fractional \\ areas}} = \frac{1}{\dfrac{F_1}{R_1} + \dfrac{F_2}{R_2}}$$

Where,
F_1 is the fractional area of the first material (studs)
R_1 is the resistance of the first material
F_2 is the fractional area of the second material (insulation)
R_2 is the resistance of the second material

R_{upper}
= 1/((0·095/2·046) + (0·905/4·469))
= 1/(0·0464 + 0·2025)
= 1/0·249
= 4·014 m²K/W

The **second stage** is to calculate the lower resistance. This is done in two steps:

Step 1: Calculate the resistance of the bridged layer (i.e. timber studs and insulation only). This is done using the same fractional resistance formula used earlier:

$R_{bridged\ layer}$
= 1/((0·095/1·077) + (0·905/3·5))
= 1/(0·088 + 0·259)
= 1/0·347
= 2·88 m²K/W

Step 2: Use this resistance to calculate the total lower resistance (R_{lower}) of the wall.

	Layer/Surface (name)	Resistance, (m^2K/W)
1.	Internal surface resistance	0·130
2.	Plasterboard	0·052
3.	**Bridged layer**	**2·88**
4.	Sheathing	0·077
5.	Cavity	0·180
6.	Brick	0·130
7.	External surface resistance	+0·400
	R_{lower} =	3·849

The **third stage** is to calculate the combined total resistance of the wall. As mentioned above, this is the average of the upper and lower resistances:

R_{total}
$= (R_{upper} + R_{lower})/2$
$= (4·014 + 3.849)/2$
$= 7·863/2$
$= 3·93 \ m^2K/W$

The final fourth stage is to calculate the U-value. As always, the U-value is the reciprocal of the total resistance.

U-value
$= 1/R$
$= 1/3·93$
$= 0·25 \ W/m^2K$

Example 3: Domestic Pitched Roof

The combined U-value calculation method is also used in this example. A pitched roof consisting of 100 mm of glass fibre quilted insulation is fitted tightly between 44 × 100 mm timber joists spaced

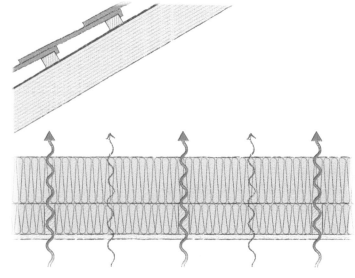

Calculating U-values: domestic pitched roof.

600 mm apart and 150 mm of glass fibre quilted insulation is laid over the joists. The ceiling consists of 12·5 mm foil-backed plasterboard. The roof has a typical underlay and tiled covering. When calculating the U-value of a pitched roof, all layers are taken into account including the ceiling structure and roof space. (A constant value for the roof space and/or covering is used.) In this example, the ceiling is bridged by timber joists. Therefore, there are two possible heat-flow paths – across the joists or across the insulation between the joists.

As before, the calculations are carried out in four stages. The **first stage** is to calculate the upper resistance. This is done in four steps:

Step 1: Calculate the resistance of each layer/ surface.

	Layer/Surface (name)	Thickness (m)	Conductivity, λ (W/mK)	Resistance, R (m²K/W)
1.	Internal surface resistance	-	-	0·100
2.	Plasterboard	0·013	0·25	0·052
3a.	**Timber joists**	**0·100**	**0·13**	**0·769**
3b.	**Glass fibre insulation (between joists)**	**0·100**	**0·04**	**2·500**
4.	Glass fibre insulation (above joists)	0·150	0·04	3·750
5.	Roof space	-	-	0·200
6.	External surface resistance	-	-	0·400

Step 2: Calculate the resistance through each heat-flow path.

Resistance through timber joists, R_1:

	Layer/Surface (name)	Resistance, (m²K/W)
1.	Internal surface resistance	0·100
2.	Plasterboard	0·052
3a.	**Timber joists**	**0·769**
4.	Glass fibre insulation (above joists)	3·750
5.	Roof space	0·200
6.	External surface resistance	+0·040
	$R_1 =$	4·911

Resistance through insulation, R_2:

	Layer/Surface (name)	Resistance, (m^2K/W)
1.	Internal surface resistance	0·100
2.	Plasterboard	0·052
3b.	**Glass fibre insulation (between joists)**	**2·500**
4.	Glass fibre insulation (above joists)	3·750
5.	Roof space	0·200
6.	External surface resistance	0·040
	$R_2 =$	6·642

Step 3: Calculate the fractional area of each heat-flow path.

In this example, the inner leaf is constructed using 38 mm timber studs at 400 mm centres. Therefore, the fractional area of the timber studs, F_1 is:

$F_1 = 44/600 = 0·073$ (i.e. timber studs make up 7·3% of the total surface area of the wall)

Therefore the fractional area of the insulation F_2 is: $1 - 0·073 = 0·927$ (i.e. 92·7% of the surface area).

Step 4: Calculate the total upper resistance.

The total upper resistance is calculated using the formula:

$$R_{\substack{combined \\ fractional \\ areas}} = \cfrac{1}{\cfrac{F_1}{R_1} + \cfrac{F_2}{R_2}}$$

Where,
F_1 is the fractional area of the first material (studs)
R_1 is the resistance of the first material
F_2 is the fractional area of the second material (insulation)
R_2 is the resistance of the second material

R_{upper}
= 1/((0·073/4·911) + (0·927/6·642))
= 1/(0·0148 + 0·140)
= 1/0·1548
= 6·460 m²K/W

The **second stage** is to calculate the lower resistance. This is done in two steps.

Step 1: Calculate the resistance of the bridged layer (i.e. timber joists and insulation only). This is done using the same fractional resistance formula used earlier:

$R_{bridged\ layer}$
= 1/((0·073/0·769) + (0·927/2·500))
= 1/(0·095 + 0·371)
= 1/0·466
= 2·145 m²K/W

Step 2: Use this resistance to calculate the total lower resistance (R_{lower}) of the wall.

	Layer/Surface (name)	Resistance, (m²K/W)
1.	Internal surface resistance	0·100
2.	Plasterboard	0·052
3.	**Bridged layer**	**2·145**
4.	Glass fibre insulation (above joists)	3·750
5.	Roof space	0·200
6.	External surface resistance	0·040
	R_{lower} =	6·287

The **third stage** is to calculate the combined total resistance of the wall. As mentioned above, this is the average of the upper and lower resistances:

R_{total}
= (R_{upper} + R_{lower})/2
= (6·460 + 6·287)/2
= 12·747/2
= 6·374 m²K/W

The **fourth stage** is to calculate the U-value. Again, the U-value is the reciprocal of the total resistance.

U-value
= $1/R_{total}$
= $1/6.374$
= 0.16 W/m²K (meets regulatory requirements: maximum 0.16 W/m²K)

Example 4: Improving Uninsulated Cavity Walls
Another U-value calculation often encountered relates to the improvement of existing uninsulated cavity walls. It is common for the external cavity walls of houses which were built over thirty years ago, to be uninsulated. These calculations involve working out the U-value of each remedial approach in order to decide which option is best for a particular house.

In this example the options include, injecting phenolic foam insulation into the cavity or dry lining the wall with 15 mm plasterboard bonded to 75 mm polyurethane board. The dry lining is secured to the wall using dabs.

Step 1: Calculate the U-value of the existing uninsulated wall. (Although this is not strictly required it does give an indication of how the walls are currently performing.)

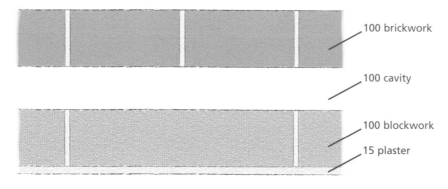

100 brickwork

100 cavity

100 blockwork

15 plaster

Calculating U-values: uninsulated masonry cavity wall.

	Layer/Surface (name)	Thickness (m)	Conductivity, λ (W/mK)	Resistance, R (m²K/W)
1.	Internal surface	-	-	0·130
2.	Lightweight plaster	0·015	0·180	0·083
3.	Blockwork	0·100	1·330	0·075
4.	Cavity	-	-	0·180
5.	Brickwork	0·100	0·770	0·130
	External surface	-	-	+0·040
	Total resistance (R_total)			**0·638**

U-value

= $1/R_{total}$

= 1/0·638

= 1·57 W/m²K (almost six times above the current standard: 0·27 W/m²K)

Step 2: Calculate the U-value of the injected foam option:

Note: The thermal insulation value of the injected insulation is substituted for the value of the cavity.

100 injected phenolic foam insulation

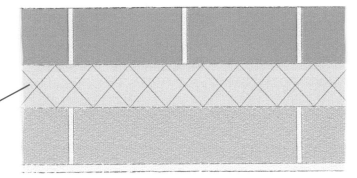

Calculating U-values: masonry cavity wall with injected insulation.

	Layer/Surface (name)	Thickness (m)	Conductivity, λ (W/mK)	Resistance, R (m²K/W)
1.	Internal surface	-	-	0·130
2.	Lightweight plaster	0·015	0·180	0·083
3.	Blockwork	0·100	1·330	0·075
4.	**Phenolic foam insulation**	**0·100**	**0·025**	**4·000**
5.	Brickwork	·100	0·770	0·130
6.	External surface	-	-	0·040
	Total resistance (R_{total})			**4·458**

U-value

= $1/R_{total}$

= 1/4·458

= 0·22 W/m²K (meets regulatory requirements: maximum 0·27 W/m²K)

Step 3: Calculate the U-value of the dry lining option:

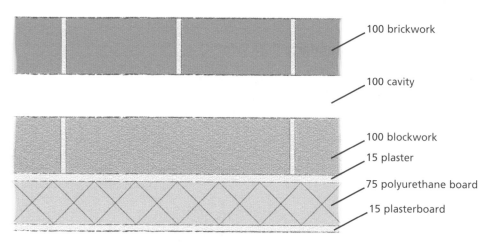

100 brickwork

100 cavity

100 blockwork

15 plaster

75 polyurethane board

15 plasterboard

Calculating U-values: masonry cavity wall with dry lining insulation.

	Layer/Surface (name)	Thickness (m)	Conductivity, λ (W/mK)	Resistance, R (m²K/W)
1.	Internal surface	-	-	0·130
2.	**Plasterboard**	0·015	0·180	0·083
3.	**Polyurethane board**	0·075	0·025	3·000
4.	Lightweight plaster	0·015	0·180	0·083
5.	Blockwork	0·100	1·330	0·075
6.	**Cavity**	-	-	0·180
7.	Brickwork	0·100	0·770	0·130
8.	External surface	-	-	+0·040
	Total resistance (R_{total})			**3·721**

U-value

= $1/R_{total}$

= 1/3·721

= 0·27 W/m²K (meets regulatory requirements: maximum 0·27 W/m²K)

From these calculations it is clear that both methods meet the current building regulation requirement of a maximum elemental U-value of 0·27 W/m²K. However, the injected phenolic foam insulation achieves a lower U-value and will therefore improve the thermal performance of the house to a greater level than dry lining will. Injected insulation is also easier and less disruptive to install. However, although modern injection insulants are water resistant, filling the cavity can potentially increase the risk of moisture penetration.

Energy Performance of Buildings

In 2002, the European Union issued the Energy Performance of Buildings Directive (EPBD). It contains a range of provisions aimed at improving energy performance in newly built and existing buildings. The Directive requires anyone who is offering a property for sale or rent to provide prospective clients with detailed information and advice about the energy performance of the property. The intention is that this information and advice will help consumers to make informed decisions leading to practical actions to improve energy performance.

The most interesting part of the directive is the requirement for Building Energy Rating certificates (effectively an energy label) to be available at the point of sale or rental of a building, or on completion of a new building. The energy rating certificate must be accompanied by a report setting out recommendations for cost-effective improvements to the energy performance of the building. In Ireland, this is expected to impact on over 150,000 sale or rental transactions per year in the residential market. The directive also specifies targets for measures to improve the efficiency of boilers and heating installations.

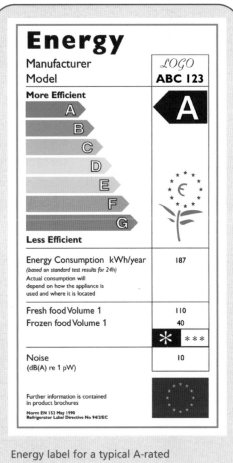

Energy label for a typical A-rated fridge/freezer.

The directive must be brought into operation by EU Member States by 4 January 2006. However, because of a possible shortage of independent experts to carry out certifications, full implementation is not required until 2009.

Energy rating is a tried and tested concept in Europe and Ireland. The energy rating label system introduced for laundry, dishwashing and refrigeration appliances in 1996 led to a thirty per cent increase in the efficiency of fridges and freezers by 2000. This meant that over half of all cold appliances for sale in Ireland were in the class A by 2000.

Energy performance certificates will have to be displayed in large buildings (over 1,000 m^2) regularly visited by the public, to raise awareness among the general public about the issue of energy efficiency in their local community.

Revision Exercises

1. Explain, in your own words, three modes of transfer of heat energy.
2. Explain why it is desirable for an insulation product to have a high R-value.
3. Explain, in your own words, what the U-value tells us about the thermal insulation properties of a structural element.
4. Explain, in your own words, any five measures that could be taken to improve the thermal performance of your home.
5. Describe, using neat annotated sketches, the use of any two insulation products in your home.
6. Describe, using a neat annotated sketch, and also discuss the merits of a method of improving the thermal performance of uninsulated cavity walls.
7. Explain, in your own words, how the energy rating of buildings and heating installations can lead to energy conservation.
8. State whether a masonry cavity wall whose construction consists of 15 mm lightweight plaster, 100 mm concrete block, 80 mm extruded polystyrene, 40 mm cavity and 100 mm brick outer leaf, meets current thermal insulation regulatory requirements. (See list of generic materials above for conductivity values. Use the following resistance constants: internal surface resistance 0·130 m²K/W, cavity resistance 0·180 m²K/W and external surface resistance 0·040 m²K/W.)
9. State whether a timber-frame cavity wall whose construction consists of 12·5 mm plasterboard, 50 × 140 mm timber studs at 400 mm centres, 140 mm sheep wool quilt insulation, 9 mm oriented strand board, 50 mm cavity and 100 mm brick. (See list of generic materials above for conductivity values. Use the following resistance constants: internal surface resistance 0·130 m²K/W, cavity resistance 0·180 m²K/W and external surface resistance 0·040 m²K/W.)
10. Ordinary Level, 2003, Question 6
    (a) Give two reasons why it is necessary to provide thermal insulation in a domestic dwelling.
    (b) Using notes and freehand sketches show a means of providing thermal insulation in each of the following locations:
        (i) Pitched roof,
        (ii) External wall,
        (iii) Concrete ground floor.
    (c) Suggest a suitable insulation material for each of the above locations.
11. Ordinary Level, 2005, Question 9
    Thermal insulation is widely used in the construction of domestic dwellings.

(a) Discuss in detail two advantages of using thermal insulation in the construction of a dwelling house.

(b) List two materials used to provide thermal insulation.

(c) Using notes and neat freehand sketches, describe a suitable location in a dwelling house for each insulation material listed at (b).

12. Higher Level, 2003, Question 5

A dwelling house built in the 1970s has external walls with uninsulated cavities. The owner has decided to insulate the walls of the house. The external walls have the following specification:

Outer leaf – 100 mm brick, Cavity – 75 mm (without insulation), Inner leaf – 100 mm concrete block with 16 mm plaster finish.

Thermal data of wall:

Conductivity of brickwork (k) 1·320 W/m °C

Conductivity of blockwork (k) 1·440 W/m °C

Conductivity of plaster (k) 0·430 W/m °C

Resistance of the external surface (R) 0·048 m² °C/W

Resistance of the internal surface (R) 0·122 m² °C/W

Resistance of the cavity (R) 0·170 m² °C/W.

(a) Calculate the U-value of the wall.

(b) The owner may choose either of the following methods to increase the insulation properties of the walls:

   (i) Filling the cavity with urea formaldehyde foam.
      OR

   (ii) Fixing insulated plasterboard sheeting to the inside wall surfaces. The insulated sheeting consists of 50 mm rigid urethane and 12·5 mm plasterboard.
      Calculate the U-value for each of the above options given the following thermal data:
      Conductivity of urea formaldehyde foam (k) 0·040 W/m °C
      Conductivity of rigid urethane (k) 0·023 W/m °C
      Conductivity of plasterboard (k) 0·160 W/m °C

(c) Evaluate both methods of insulation listed at (b) above, recommend a preferred method and give two reasons to support your recommendation.

13. Higher Level, 2005, Question 5

An extension to a dwelling house has a concrete flat roof with an asphalt finish. The total roof surface is 16 m² in area. The roof is constructed to the following specification:

(i) Concrete flat roof slab: Thickness 175 mm

(ii) Concrete screed: Thickness 60 mm

(iii) Layer of asphalt: Thickness 20 mm

(iv) Internal plaster to roof slab: Thickness 15 mm

Thermal data of roof:

Resistivity of asphalt 1·250 m °C/W

Resistivity of concrete screed 0·710 m °C/W

Resistivity of concrete roof slab 0·690 m °C/W

Resistivity of the plaster 2·170 m °C/W

Resistance of the internal surface (R) 0·104 m² °C/W

Resistance of the external surface (R) 0·413 m² °C/W

External temperature 11°C

Internal temperature 21°C.

(a) Calculate the U-value of the roof structure and the overall heat loss through the roof.

(b) Outline two design considerations that must be taken into account in the design of a roof for a domestic dwelling and describe, with the aid of notes and freehand sketches, the design detailing for each consideration outlined.

Sound is a form of energy transferred from a source to a person's ear. The transfer of energy can be through the air or through a solid object. The sound which travels through the air is called airborne sound; sound generated on an object is called impact sound. The sound energy causes vibrations of the air which the ear registers as a sound.

Sound energy may be transmitted directly or indirectly. Direct transmission of sound means the sound travels through a wall or floor from one of its sides to the other. Indirect sound transmission means that the sound travels around a wall through an adjacent wall or ceiling. Indirect sound transmission is also referred to as *flanking transmission*.

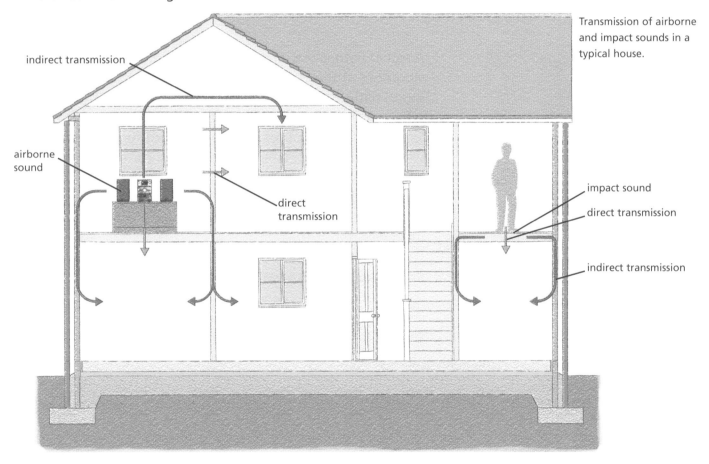

Transmission of airborne and impact sounds in a typical house.

Sound Waves

Sound energy travels in waves. The wave spreads out equally in all directions unless it is affected by another object in its path. Sound waves can travel through solids, liquids and gases but not through a vacuum. Sound travels faster in solids and liquids than it does in gases because of their higher densities. The particles of dense materials respond to vibrations more quickly and so convey the energy at a faster rate. Sound travels through air at 331 m/s, water at 1,498 m/s and granite at 6,000 m/s.

Room Acoustics

Room acoustics is about the study of sound within an enclosed space. Noise control from outside the room is an important element of acoustics and is examined later in this chapter. The general requirements for good acoustics include:

- adequate levels of sound,
- even distribution to all listeners,
- rate of decay (reverberation time) suitable for the type of room,
- background noise and external noise reduced to acceptable levels,
- absence of echoes.

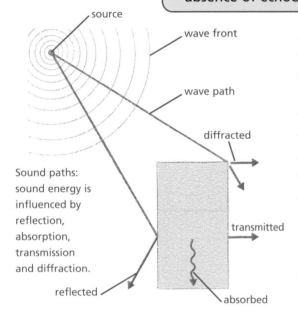

source
wave front
wave path
diffracted
transmitted
reflected
absorbed

Sound paths: sound energy is influenced by reflection, absorption, transmission and diffraction.

A sound path is the directional track made by the wave vibrations as they travel through the air. The behaviour of sound paths in an enclosed space are influenced by reflection, absorption, transmission and diffraction. Similarly to light, sound is diffracted at corners and this is why it is possible to hear sounds around corners. Sound absorption is a reduction in the amount of sound reflected from a surface. Absorption is very important to the quality of room acoustics – this is why the walls of cinemas are lined with fabric (curtains).

Reverberation is the term used to describe the continuation and enhancement of a sound caused by rapid multiple reflections between the surfaces of a room. Reverberation is not the same as echo because the

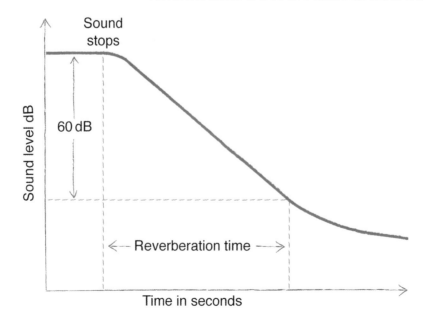

Reverberation time: the time taken for a sound to decay by 60 decibels from its original level.

reflections reach the listener too rapidly for them to be heard as a separate sound – instead they are heard as an extension of the original sound. The reverberation time of a sound is defined as the time taken for a sound to decay by 60 dB from its original level. This depends on the floor area of the room, the sound absorption of the surfaces of the walls and the frequency of the sound. The ideal reverberation time varies: speech requires short reverberation times for clarity (excessive reverberative sound will mask the next syllable) whereas longer reverberation times are thought to enhance the quality of music.

Another important concept in sound is the **Inverse Square Law**. This law states that the sound intensity from a point source decreases in inverse proportion to the square of the distance from the sound. In other words, if the distance to the sound source is doubled ($\times 2$) the sound intensity level is quartered $(1/2)^2 = 1/4$. This happens because the same energy is spread over a larger area as the sound spreads out.

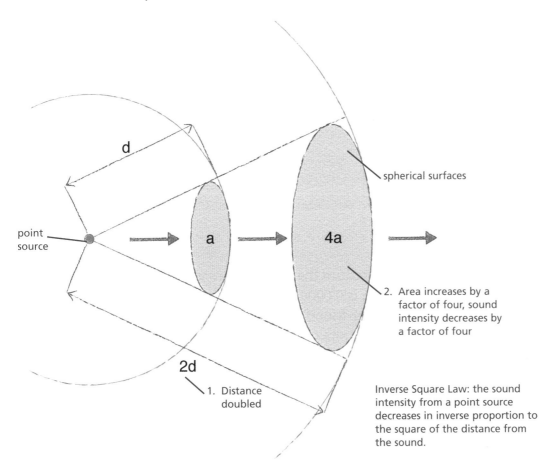

point source

d

a

4a

spherical surfaces

2d

1. Distance doubled

2. Area increases by a factor of four, sound intensity decreases by a factor of four

Inverse Square Law: the sound intensity from a point source decreases in inverse proportion to the square of the distance from the sound.

Sound Levels

Sound is measured in a number of ways including:

- Sound Power – The rate at which sound energy is produced at the source, measured in watts, W.
- Sound Intensity – The sound power distributed over unit area, measured in W/m^2.
- Sound Pressure – The average variation in atmospheric pressure caused by the sound.
- Sound energy levels are actually quite low – the effect relies on the sensitivity of human hearing. The threshold of hearing is the weakest sound that the average human ear can detect. The threshold of pain is the strongest sound that the average human ear can tolerate. While it is possible to specify the strength of a sound in terms of its intensity or pressure it is usually

Sound pressure in Pa	Sound level in dB	Typical Environment
200	140	Threshold of pain
	130	Aircraft take-off
20	120	Loud discotheque
	110	
2	100	Noisy factory
	90	Heavy lorry
0.2	80	High Street corner
	70	Vacuum cleaner
0.02	60	Normal conversation
	50	
0.002	40	Suburban living room
	30	
0.0002	20	Quiet countryside
	10	
0.00002	0	Threshold of hearing

Decibel scale.

inconvenient to do so because the numbers are very awkward. For example the threshold of hearing is 0·00002 Pascals (Pa) and the threshold of pain is 200 Pa. A much more user-friendly method is to describe the sound level as a ratio based on the threshold of hearing. Thus, we say that the threshold of hearing occurs at zero decibels (0 dB) and all other sounds are above this level. The decibel is defined as the logarithmic ratio of two quantities. This means that the sound level is calculated by getting the log (logarithm) to the base ten of the sound intensity level divided by the threshold of hearing intensity level. (Threshold of hearing: sound intensity level (I_0) = 1 $\times$ 10^{-12} W/m^2, sound intensity level = 10 log$_{10}$ (I/I_0).)

However, because this method is used, changes in sound level are not straightforward – an increase in sound level of 10 dB will sound twice as loud, whereas an increase in sound level of 20 dB will sound four times as loud. While a change of 1 dB may be detected by the human ear, a change of 3 dB is usually considered to be significant by the average person.

Principles of Acoustic Insulation

Acoustic insulation is the reduction of sound energy transmitted into an adjoining air space. The *sound reduction index* (SRI) is a measure of the level of insulation a material provides against the transmission of airborne sound. Sound reduction index values are established under controlled conditions in laboratories. These values can then be used by designers to calculate insulation requirements in buildings.

SRI examples:

- 100 mm concrete wall = 45 dB
- a closed window = 25 dB
- a slightly open window = 10 dB
- a door = 15 dB

Effective acoustic insulation of any structure depends on a number of factors, including:

- Heaviness – Heaviness is related to the mass law, which states that the acoustic insulation of a single leaf partition is proportional to its mass per unit area. This is because the greater the mass of the partition, the greater the sound energy required to set it in motion. In other words, a thick or dense wall requires more sound energy to cause it to vibrate – if it doesn't vibrate the sound energy cannot get through it. Acoustic insulation effectively increases by 5 dB when the mass is doubled.

215 mm block party wall plastered on both faces

Acoustic insulation: heaviness – the greater the mass of the partition, the greater the sound energy required to set it in motion.

- Completeness – This relates to the fact that airborne sound will travel through any gaps in the construction of a wall. The completeness of a structure depends on airtightness and uniformity. The sound reduction index of a 215 mm plastered brick wall will drop from 50 dB to 30 dB if a crack equivalent to only 0·1% of the area of the wall is present. Uniformity refers to the need to ensure that every element of a structure is properly sealed. For example, an unsealed door in a 100 mm blockwork wall will reduce the average SRI of the wall from 45 dB to 23 dB.

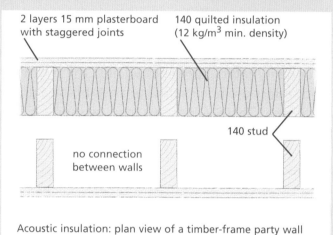

2 layers 15 mm plasterboard with staggered joints

140 quilted insulation (12 kg/m^3 min. density)

140 stud

no connection between walls

Acoustic insulation: plan view of a timber-frame party wall showing isolation of walls.

- Isolation – The idea behind isolation is that when sound energy has to transfer from one medium to another (e.g. blockwork to air) energy is lost. Therefore, discontinuous construction (e.g. cavity walls) is effective in reducing sound transmission.
- Flexibility – Flexibility is influenced by two factors: resonance and coincidence. Resonance occurs if the natural *frequency of the material* from which the partition is constructed is the same as the frequency of the sound. When this happens the partition will vibrate or resonate with the sound and its acoustic insulation performance will deteriorate. Resonance commonly occurs in thin rigid materials such as glass. For this reason, flexible materials are better for acoustic insulation. Unfortunately flexibility is not usually a desirable structural property in construction.

As noted above, the mass law suggests that the level of sound transmitted through a partition reduces with increasing partition thickness. However, at certain critical frequencies the performance of the partition drops and more sound is transmitted through it than we would expect. This drop in the performance of the partition is called the *coincidence dip*. Coincidence occurs when sound energy causes a partition to bend or flex inwards and outwards. When the *frequency at which the partition is bending* coincides with the frequency of the sound, the acoustic insulation of the partition deteriorates. For several octaves (multiples of the original frequency (e.g. 440 Hz, 880 Hz, 1,320 Hz) the acoustic insulation remains constant and less than that predicted by the mass law. Coincidence loss is greatest in double leaf constructions such as cavity walls.

The graph shows sound transmission loss for typical construction materials like plasterboard and plywood. These increase steadily, in good agreement with the mass law. (The higher the transmission loss, the less sound passes through the wall.) However, at higher frequencies, a clear dip occurs – the *coincidence dip*. This dip is centred at the coincidence frequency, which depends on the material's stiffness and its thickness.

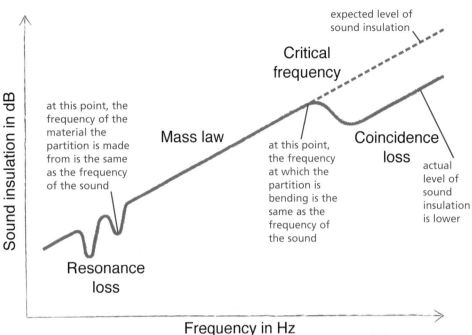

Acoustic insulation: resonance and coincidence loss.

This can cause problems in buildings because many everyday sounds occur at these frequencies which match the natural frequencies of construction materials (80–4,000 Hz). However, these problems can be overcome. For example, when two layers of material such as gypsum board are glued firmly together, they behave like a single thick layer, with an expected coincidence dip. However, if the layers are attached more loosely (e.g. with screws), the coincidence effect is reduced giving higher sound insulation near the coincidence frequency. The party wall of semi-detached timber-frame dwellings is lined with 2 layers of plasterboard (screwed or nailed together) to reduce sound transmission (and to prevent the spread of fire).

Acoustic Insulation within a House

Unwanted sound is referred to as *noise*. Noise in houses can be avoided by planning properly at the design stage. Strategies to limit noise in houses include:

- group noisy areas away from quiet ones (e.g. bathrooms should be kept away from bedrooms),
- have distance between different areas (i.e. distance from kitchen to living room),
- choose appropriate construction (heavy construction such as concrete absorbs sound energy).

The sound level in a room is influenced by a number of factors including:

- sound level outside building,
- transmission through building envelope (external),
- transmission within building (internal),
- noise generated within room,
- room acoustics.

For typical houses built in Ireland, the building regulations in relation to sound only apply to the party wall of semi-detached houses. In this case the building regulations require that the surface mass of the party wall of a semi-detached dwelling be a minimum of 415 kg/m². The surface mass is multiplied by the partition thickness to get its overall mass. For example, a 215 mm solid concrete block (density 1,900 kg/m³), 112·5 mm coursing, 12·5 mm lightweight plaster (density 600 kg/m³), on each face, would provide the required mass.

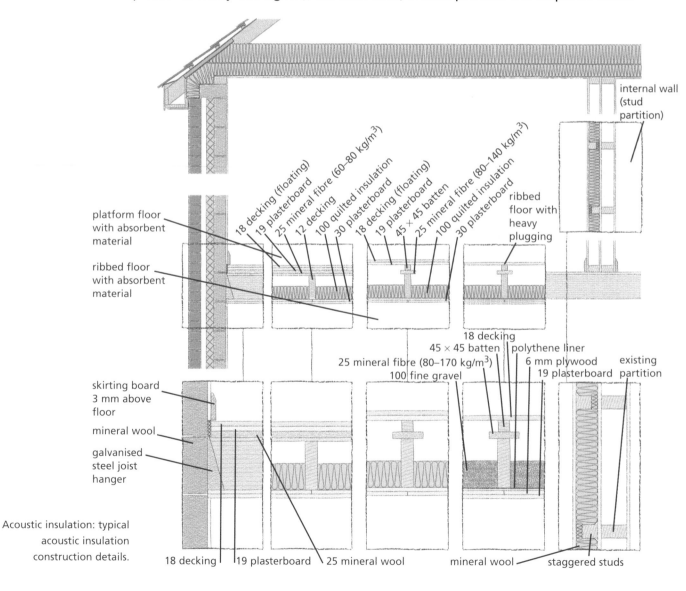

internal wall (stud partition)

18 decking (floating)
19 plasterboard
25 mineral fibre (60–80 kg/m³)
12 decking
100 quilted insulation
30 plasterboard
18 decking (floating)
19 plasterboard
45 × 45 batten
25 mineral fibre (80–140 kg/m³)
100 quilted insulation
30 plasterboard
ribbed floor with heavy plugging

platform floor with absorbent material

ribbed floor with absorbent material

18 decking
45 × 45 batten | polythene liner
25 mineral fibre (80–170 kg/m³) 6 mm plywood
100 fine gravel   19 plasterboard   existing partition

skirting board 3 mm above floor

mineral wool

galvanised steel joist hanger

Acoustic insulation: typical acoustic insulation construction details.

18 decking | 19 plasterboard | 25 mineral wool   mineral wool   staggered studs

We can check this, as follows:

Surface Mass
= (density of material) (thickness) (kg/m³)(m)
= (1,900)(0·215) + (600)(0·025)
= (408·5) + (15)
= 423·5 kg/m²

It is sometimes desirable to improve the acoustic insulation of a particular room in a house. This usually involves insulating the floor, internal walls and ceiling of the room. The usual principles of mass, isolation, completeness and flexibility are applied as shown above.

Revision Exercises

1. Outline and explain in your own words, the requirements for good acoustics.
2. Explain, using a neat annotated sketch, how sound is transmitted in a typical house.
3. Explain the factors influencing the sound level in a room.
4. Explain, using neat annotated sketches, the principles underpinning acoustic insulation of houses.
5. Discuss, in terms of direct and indirect transmission, the sound performance of your home. Use neat annotated sketches to illustrate your answer.
6. Describe, using neat annotated sketches, how one of the problems identified in your answer to question four could be solved.
7. Describe, using neat annotated sketches, how the acoustic insulation of your bedroom could be improved.
8. Higher Level, 2002, Question 6
    (a) Discuss sound insulation in buildings with reference to each of the following:
        (i)   Mass.
        (ii) Completeness.
        (iii) Isolation.
    (b) A living room is located on the first floor of a new house, directly above a bedroom. The floor consists of tongued and grooved flooring boards on wooden joists with a plasterboard ceiling beneath.
    Using notes and sketches, show two design details that will increase the sound insulation properties of the floor and minimise the transmission of noise to the bedroom beneath.

9. Higher Level, 2004, Question 9
    It is proposed to install a music system in the living room of a single-storey
    dwelling house. The house has a concrete floor and the living room is
    separated from an adjacent bedroom by a standard stud partition. The
    walls and ceilings have a smooth hardwall plaster finish. It is proposed to
    carry out renovations to improve sound insulation.
    (a) Using neat freehand sketches show two design details that would
        increase the sound insulation properties of the stud partition.
    (b) Explain in detail two sound insulation principles which would
        influence the design of the stud partition.
    (c) Using notes and sketches suggest two modifications which would
        improve the acoustic properties of the living room.

Electricity Basics

All material is made of atoms. An atom consists of three types of particles – protons, neutrons and electrons. These particles are held together by forces, one of which is an electric property called *charge*. There are two kinds of charge – positive and negative. Protons, which are found in the middle (nucleus) of an atom, have a positive charge. Electrons, which circulate around the outside of the nucleus (like planets orbiting the sun), have a negative charge. The negative electrons are held in orbit by their attraction to the positive neutrons.

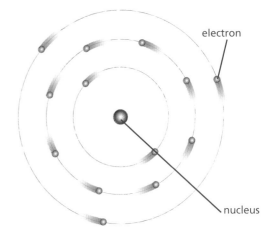

Electricity: simple model of an aluminium atom showing the electrons orbiting the nucleus.

The electrons in the outer orbits are held by relatively weak forces so outer orbits can sometimes lose or gain electrons. When electrons move from one place to another they transfer their electric charge. If electric charge transfers through a material then an electric current is said to flow. The amount of electric current is described by the quantity of charge which passes a fixed point in a given time. Thus, we can define electric current as, the rate of flow of charge in a material. Electric current is measured in units called amperes (symbol: A).

For a current to flow there must be a difference in charge between two points. This is similar to the pressure difference that must exist for water to flow

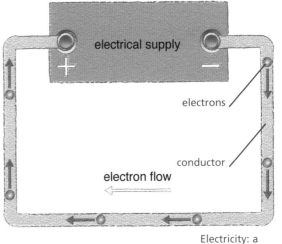

Electricity: a simple circuit.

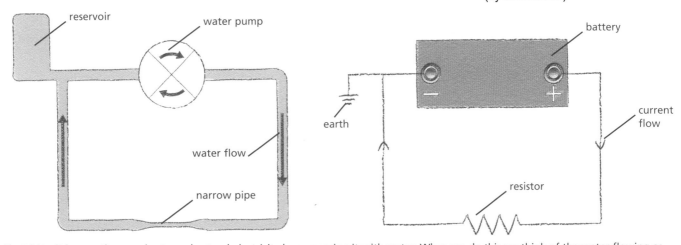

Electricity: it is sometimes easier to understand electricity by comparing it with water. When we do this we think of the water flowing as the electrical current flowing. The pressure of the water is the equivalent of voltage; the water pump is like a battery and the storage reservoir is like the earth. Where the water pipe narrows is the equivalent of a resistor.

in a pipe. This difference in charge is called potential difference and is measured in units called volts (symbol: V). Potential difference is sometimes referred to as voltage drop. In order to produce a potential difference there must be a source of electrical *pressure* acting on the charge. This source is called an electromotive force (EMF) and is also measured in volts. A battery is a typical source of EMF. Electromotive force is sometimes called voltage.

Materials that are good at transferring electricity are called *conductors*. Examples include metals such as copper and aluminium. Materials that are poor conductors are called *insulators*. Examples of insulators include synthetic rubber, plastics and wood. This is why electrical cables and cords are usually wrapped in PVC.

One of the most important effects of the flow of current is the heating effect. When current flows in a conductor heat is generated. This effect is used in many electrical appliances such as kettles, showers and immersion heaters.

Electrical energy and power have the same meaning as other forms of power and energy. For electrical energy, power is equal to current multiplied by potential difference.

Power = current × voltage

Electrical power is the measured in watts (symbol: W). The power rating of an electrical appliance is usually used to describe the amount of electrical energy needed to operate it. An typical electric kettle uses 2,500 W, while an electric shower uses around 8,000 W. Of course what really matters is not the rating but the duration of use. The energy consumed by an appliance depends on the power and the time. Energy consumption is measured in joules, where a joule is equal to one watt per second. For higher levels of energy consumption, like those associated with a typical home, energy consumption is measured in kilowatt hours (symbol: kWh), where one kilowatt hour is equal to one thousand watts per hour. Electricity suppliers like the ESB and Airtricity measure their customers' energy usage in kilowatt hours.

Power Supplies: Generation and Transmission

Electrical energy is used in homes all over the country. For this to be possible, large amounts of electricity must be generated, transmitted and distributed. Electricity is produced in a generator. A generator is a large motor consisting of coils and magnets. When the coil is rotated in a magnetic field (or a magnet is rotated in a coil) a current flows in the coil. This phenomenon is called *electromagnetic induction*. So, in order to generate electricity a rotational force is need to turn the coil or magnet. This force can be harnessed in a variety of ways, most of which involve the rotation of turbines. A turbine is like a large fan.

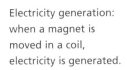

The generation of electricity in Ireland occurs in three types of stations:

Electricity generation: when a magnet is moved in a coil, electricity is generated.

- Thermal power stations – burn fossil fuels like coal, oil, gas and peat to boil water which produces steam to drive a turbine.
- Hydroelectric stations – use the energy of water flowing in rivers to drive a turbine.
- Wind power – uses wind energy to drive the turbine.

Wind powered and hydroelectric powered stations use renewable resources. Oil, coal, gas and peat are non-renewable resources – once they are burned they can no longer be used to generate power. The use of non–renewable resources has a negative effect on the environment because of the release of harmful greenhouse gases. When a typical household uses €100 worth (excluding VAT and charges) of coal-generated electricity this causes the release of 3,996 kg of CO_2, 46 kg of sulphur dioxide and 14 kg of nitrous oxide. These gases are by-products of coal burning and contribute to the greenhouse effect and global warming. Another way to think of it is that purchasing the same €100 of energy from a renewable energy source would be the equivalent of taking one car off the road for one year.

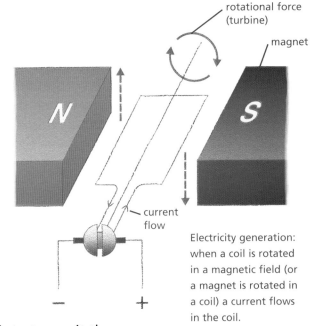

Electricity generation: when a coil is rotated in a magnetic field (or a magnet is rotated in a coil) a current flows in the coil.

Once the electricity has been generated, the next stage is to transmit the power around the country. This is done on a network of cables called the national grid. The benefits of a national grid include:

- Power doesn't have to be generated close to population centres.
- Hydroelectric stations are often in rural areas and need an efficient long-distance transmission system.
- Having all power stations on the one grid allows flexibility and efficiency.
  - Power can be redirected to where it is needed (e.g. when there are outages and/or cables down).
  - Power can be selected from cheapest stations (cost of fuels fluctuate).

Electricity is transmitted at very high voltages (400 kV, 220 kV, 110 kV). This is done to reduce the heating effect in the cables and thereby the amount of energy lost during transmission. The amount of heat energy produced in a cable is proportional to the square of the current. In other words, if

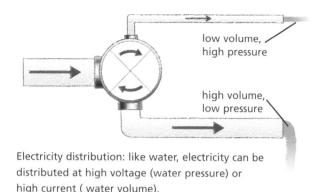

Electricity distribution: like water, electricity can be distributed at high voltage (water pressure) or high current (water volume).

the current is doubled the heat generated in the cable is quadrupled. In order to reduce the current the voltage is increased. The easiest way to understand this is to go back to the water analogy. It's like pumping a small volume of water at high pressure instead of a large volume at low pressure. So the current (volume) is decreased and the voltage (pressure) is increased.

When the electricity reaches the general area where it is needed the voltage is reduced at a transformer station. A transformer station is a station that reduces (transforms) the voltage of the electricity to lower levels. Typical transformer stations reduce the voltage to 38 kV, 20 kV and 10 kV. The electricity is then transmitted to local substations (or transformer drums on poles) where it is reduced to the 230 V level used in homes.

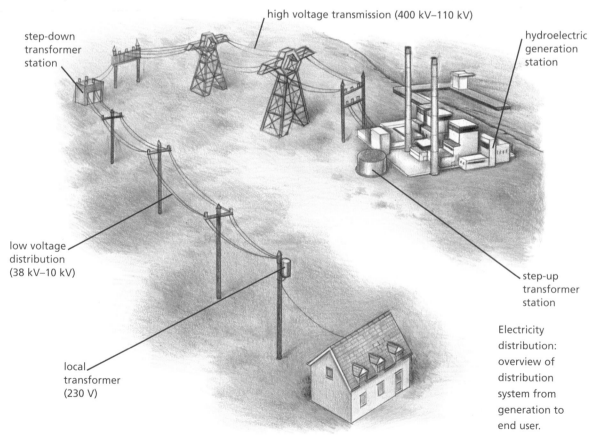

high voltage transmission (400 kV–110 kV)

step-down transformer station

hydroelectric generation station

low voltage distribution (38 kV–10 kV)

step-up transformer station

local transformer (230 V)

Electricity distribution: overview of distribution system from generation to end user.

Domestic Intake of Electricity

The intake of electricity into a house can occur above or below ground. In either case the supply is connected to a meter. The meter is usually housed externally in a purpose-made plastic meter box to facilitate reading by the supplier. In older houses the meter may be indoors.

The supply is then fed to an internal consumer unit or distribution board. This unit contains an isolating switch and connections to the various circuits in the home. Each circuit is connected to the distribution board via a circuit breaker.

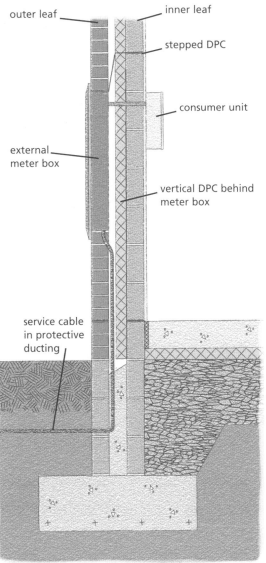

Domestic intake of electricity: typical intake detail.

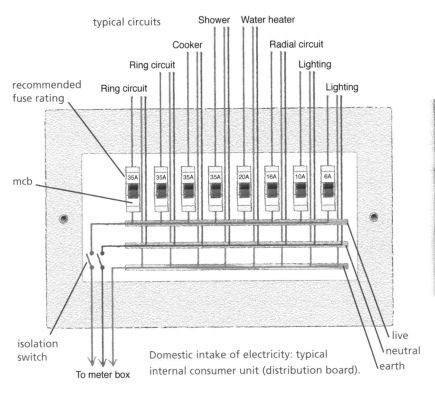

Domestic intake of electricity: typical internal consumer unit (distribution board).

Distribution of Electricity within the Home

Functions
The function of the electrical distribution system is to safely conduct electrical power to the various outlets and electrical appliances in a home.

Design Features
The system for distributing electricity must be designed to meet a number of performance criteria including:

- sufficient capacity for the needs of the home,
- minimum wastage of current in the cables,
- prevention of electrical injury,
- prevention of fire,
- means of isolation (cut-off supply),
- compliance with regulations and best practice.

A number of separate circuits are used to distribute electricity around a house, including:

- Radial Circuits
  - lighting circuits (usually two or more, depending on the floor area of the house),
  - cooker circuit (a separate circuit for an electric cooker),
  - shower circuits (a separate circuit for each electric shower),
  - immersion circuit (a separate circuit for the hot water immersion heater).
- Ring Main Circuits
  - socket circuits (usually two or more, depending on the floor area of the house).

Radial Circuits

A radial circuit is used to supply power in two ways:

- in a loop-type circuit to lighting,
- directly to a single item such as a cooker, shower or storage heater.

A radial lighting circuit is commonly referred to as a *loop-in* system because the live, neutral and earth supplies are looped from one ceiling rose (or other

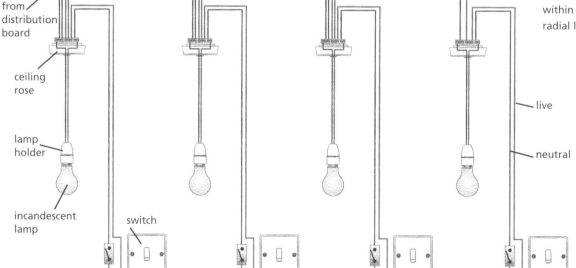

to distribution board

from distribution board

ceiling rose

lamp holder

incandescent lamp

switch

Distribution of electricity within the home: typical radial lighting circuit.

live

neutral

outlet) to the next. The ceiling rose contains separate brass terminals for the live, neutral and earth circuits, to which a number of connections can be made. The live supply is interrupted by a switch, so that only when the switch is thrown will the power be able to flow to the light fitting. The alternative to a loop-in system would be to run a separate circuit for each light fitting – however, this would require a lot of wiring and take much longer to install.

A two-way switching arrangement is usually used to control the light fittings in the hallway of a bungalow or the upstairs landing of a typical house. This allows the light fitting to be switched on or off from either end of the hallway (downstairs or upstairs). To achieve this, additional wiring is run between the switches to allow the current to flow when either is thrown.

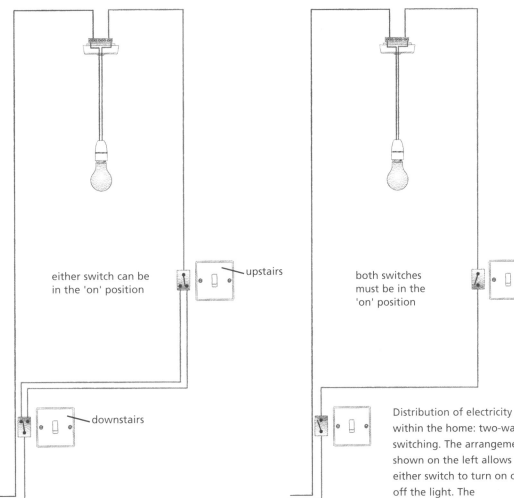

either switch can be in the 'on' position

upstairs

both switches must be in the 'on' position

downstairs

Distribution of electricity within the home: two-way switching. The arrangement shown on the left allows either switch to turn on or off the light. The arrangement on the right would be less effective because both switches would have to be in the same position for the light to illuminate.

A maximum of ten outlets per radial circuit (lighting) is recommended.

Ring Main Circuits

Commonly, a ring main circuit is used to supply power to the sockets in a house. The sockets can be installed at any point on the circuit which forms a *ring*, because it is connected to the distribution board at each end. The difference then, between a radial and a ring circuit, is that the last outlet of a ring circuit is connected back to the consumer unit. This means that current is supplied to the sockets from both ends of the ring main. This allows the wiring to be of lower current rating and smaller diameter than is required for a simple radial circuit connected at one end only. To prevent overloading, the following guidelines apply to the design of ring main circuits:

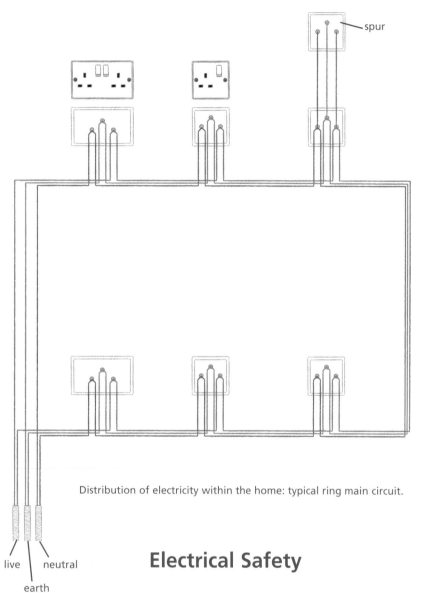

spur

- a separate ring main circuit must be provided for every 100 m² of floor area,
- a 2·5 mm² twin and earth double-insulated cable must be used,
- no limit to the number of sockets,
- radial spurs are allowed from 50% of sockets,
  – maximum of two sockets per spur.

Distribution of electricity within the home: typical ring main circuit.

live   neutral

earth

Electrical Safety

There are three design principles that ensure protection against electric shock in domestic installations:

- Isolation devices – fuses and circuit breakers that cut off the supply of electricity in the event of a fault.
- Earthing – appliances (e.g. kettles, cooker bodies) are connected to the earth to prevent injury if they become live.
- Isolation of live parts – e.g. every cable, cord, lighting and socket outlet is designed to protect the user against contact with live parts.

Isolation Devices

Mini Circuit Breakers (MCBs)

A mini circuit breaker is a safety device that isolates (cuts off the flow of electricity) a circuit if a fault occurs. There are two main types of faults:

- Overload – This occurs when an appliance is subjected to a greater load than that for which it was designed. Electrical overload can be caused by short circuit, incorrect installation or by misuse (e.g. running a high-powered appliance off a low-power extension cable).
- Short circuit – This occurs when a connection is accidentally made between the live and neutral conductors or between the live and earth. For example, if a nail or screw is driven into a wall and pierces the insulation on both live and neutral creating a *bridge* across them. A short circuit results in a very-high current flowing, which in turn trips the MCB.

Mini circuit breakers (MCBs) are designed to permit a specific amount of electrical current to flow. This is indicated on the MCB in amperes. If a fault develops in the circuit (or in an appliance connected to it) and the permitted flow rate is exceeded, the MCB will *trip*, isolating the circuit. Unlike a fuse which operates once and then has to be replaced, a circuit breaker can be reset manually once the fault has been repaired.

Residual Current Devices (RCDs)

A residual current device (RCD) provides protection against electrical currents flowing through a person's body or through wiring or electrical appliances which could cause a fire. A residual current device is used to provide backup protection in case of a fault. For example:

- a breakdown in insulation between a live part and metalwork,
- exposure to live parts when changing lights on a Christmas tree,
- cutting of a cable with an electric lawnmower.

When something like this happens an RCD can provide an additional level of protection against electric shock. The RCD monitors the current flowing in the live and neutral conductors. Normally the current flowing in both of these conductors will be the same. If a fault develops, the current balance will be upset and the RCD will isolate the supply. RCDs are available in a variety of operating levels for different applications. For shock protection in housing installations a 10 mA or 30 mA is used (depending on the specific installation). Higher rated RCDs (100–500 mA) are used in commercial installations for fire prevention.

Residual current device is a generic term used to describe a range of RCD products, including:

- Residual Current Circuit Breaker (RCCB) – A mechanical switch which can be manually opened and closed, and which is fitted with an RCD function to enable it to automatically disconnect power in the event of an earth fault. Available with trip current ratings of 10–500 mA.
- Residual Current Breaker with Overcurrent Protection (RCBO) – A two-in-one device combining the functions of a mini circuit breaker (MCB) and a residual current circuit breaker (RCCB). Available with trip current ratings of 10–500 mA.
- Socket Residual Circuit Device (SRCD) – A socket outlet which is fitted with an RCD function to enable it to automatically disconnect power in the event of an earth fault. Available with trip current ratings of 10 or 30 mA.
- Portable Residual Circuit Device (PRCD) – A portable plug or adaptor which is fitted with an RCD function to enable it to automatically disconnect power in the event of an earth fault. Available with trip current ratings of 10 or 30 mA. Used when working outdoors with power tools.

Earthing

Earthing is based on the principle that current always takes the path of least resistance. If a live wire touches the body of a metal appliance (e.g. kettle) and if a person touches the live body (of the kettle) the current will flow through them! To prevent this type of situation, an alternative route must be provided – this route is called the *earth*.

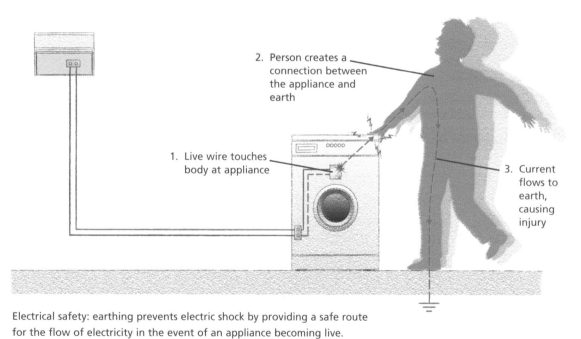

2. Person creates a connection between the appliance and earth

1. Live wire touches body at appliance

3. Current flows to earth, causing injury

The earth cable is connected from the distribution board to a steel bar that is in firm contact with the soil. This bar is copper or galvanised (2 m long, ⌀20 mm) buried in the ground

Electrical safety: earthing prevents electric shock by providing a safe route for the flow of electricity in the event of an appliance becoming live.

outside the home. A 16 mm² cable connects the rod to the distribution board. Copper piping, when used in central heating and water supply systems, is also connected to the earth system to prevent it becoming live in the event of a short circuit.

Isolation of Live Parts

Cables

Cables consist of conductors of low resistance (copper) covered by insulators of high resistance (PVC). Each conductor has a separate colour-coded PVC sheath. These three are protected by an outer PVC sheath. In this way the cable is said to be double-insulated. Cables are sized according to the cross-sectional area of the conductor (copper) and given a rating in amps. This is the maximum electrical load the cable can take without overheating. Different sizes are used for different applications (see ratings below). The conductor in a cable is a single wire strand. This makes cables quite stiff – although they are flexible enough for wiring through wall and floor cavities. Cables are usually used to conduct electricity from the consumer unit to sockets and lighting outlets. A typical cable has a grey outer sheath and colour-coded inner sheaths as follows:

- Live – brown.
- Neutral – blue.
- Earth – green and yellow.

Distribution of electricity within the home: typical cable – various sizes are used for wiring of radial and ring main circuits.

Cords

A cord is made by winding together a large number of very light copper strands. Cords are typically double-insulated as described above. However, in this case the conductor has a smaller diameter and the PVC sheathing is lighter. This gives the cord greater flexibility. A cord is sometimes referred to as a *flex*.

Distribution of electricity within the home: typical cord used to connect appliances to sockets.

Sockets and Switches

A socket or switch outlet consists of a galvanised steel box to which a front plate is screwed after the cables have been connected to the terminals. These galvanised steel boxes are made with circular knock-outs, which can removed, to allow cable access. The boxes are usually fixed within the wall, so that the front plate sits flush on the wall surface. Alternatively, for example when retro-fitting, a wall mounted plastic (polyester) box can be used. Modern socket outlets are fitted with shutters to prevent accidental contact (especially by children). These shutters are a single spring-loaded device that will only open when the earth pin on the plug pushes the shutter open – this is why the earth pin is longer than the others.

Plugs

Some appliances, like cookers, are permanently connected to the supply. Where this isn't the case, plugs are used to connect an appliance to the ring main circuit at sockets.

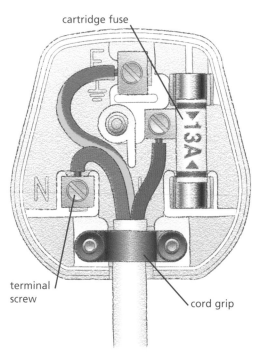

cartridge fuse

terminal screw

cord grip

Distribution of electricity within the home: typical plug showing correct wiring and a 13 Amp cartridge

A plug has a built-in safety device called a *cartridge fuse*. This fuse is rated to suit the power rating of the appliance to which it is connected. Cartridge fuses provide better protection for small appliances with thin cords. A variety of fuse ratings (3 A, 5 A, 13 A) are available and a suitable fuse should be fitted to allow the normal operating current while protecting the appliance and its cord as much as possible.

To calculate the fuse required for an appliance, the power is divided by the voltage to determine the actual current being drawn by the appliance, as shown in the following example.

Kettle with a power rating of 2,500 W:
- power   =   voltage × current
- 2,500   =   230 × current
- 2500/230 =   current
- 10·86 A   =   current flowing under normal use

Therefore, a fuse cartridge of 13 A is used (because 11 A and 12 A fuses are not made).

With some loads it is normal to use a slightly higher-rated fuse than the normal operating current. For example on 500 W halogen floodlights it is normal to use a 5 A fuse even though a 3 A fuse would carry the normal operating current. This is because halogen lights draw a significant surge of current when switched on because their cold resistance is far lower than their resistance at operating temperature.

Wiring a plug is an important everyday skill. A number of steps are followed:

1. Prepare the plug:
   (a) Put the plug on the table (don't wire it while holding it, in case the screwdriver slips).
   (b) Loosen the screws of the cord grip and remove the cover.
   (c) Loosen the cover screw and remove the cover.
   (d) Loosen the screws from the brass terminals.
2. Prepare the cord:
   (a) Measure the flex against the plug and strip the outer covering back as far as the cord grip, taking care not to cut through the coloured insulation on the wire strands.
   (b) If the coloured cables have to be trimmed, ensure that they are long enough to reach the terminals without straining – allow for a little slack on the green and yellow earth wire.
   (c) Strip about 10 mm of the coloured insulation from each core.
   (d) Twist the exposed wire strands of each core between the finger and thumb so that there are no loose strands, then double each twisted wire back on itself for about 5 mm.
3. Feed the cord under the loosened cord grip.
4. Fixing the flex to the plug:
   (a) Use this easy to remember guide for the colour codes:

   > bRown (live) to the **R**ight terminal
   > bLue (neutral) to the **L**eft terminal
   > **green** and yellow (earth) **in between** to the top terminal

   (b) Insert each core fully into the hole in the appropriate terminal.
   (c) Tighten the terminal screw firmly on the wire.
5. Tighten the cord grip screws, making sure that the cord grip is clamped on the full outer covering of the cable and not on the inner cores.
6. Calculate the appropriate cartridge fuse, fit into holder.
   (a) A blue 3 A or red 5 A fuse for lights and small appliances.
   (b) A brown 13 A fuse for larger appliances and heaters.
7. Securely replace the cover.

Note: If an appliance has a moulded plug (cannot be opened) that is damaged or faulty, it should be cut off and a new plug fitted.

Ratings

The following list is a guide to the conductor sizes and fuse ratings suitable for domestic installations:

- Lighting Circuits
  - size of cables: 1·5 mm²,
  - flexible cords between ceiling rose and lighting fittings: 0·5 mm² minimum,
  - MCB fuse rating: maximum 10 A (6 A recommended).
- Radial Circuits (other than lighting circuits)
  - size of cables: 2·5 mm²,
  - MCB fuse rating: maximum 20 A (16 A recommended).
- Ring Circuits
  - size of cables: 2·5 mm²,
  - MCB fuse rating: 35 A.
- Cooker Circuits 10 kW
  - size of cables: 6 mm² fixed wiring, 4 mm² flexible cable,
  - MCVB fuse: 35 A.
- Flexible Cords
  - size of cords for general use: 0·75 mm² minimum.

Revision Exercises

1. Explain, in your own words, how an electricity generator works.
2. Discuss briefly the benefits of renewable energy sources.
3. Explain why electrical energy is transmitted at high voltages.
4. Describe the performance criteria which a typical domestic electricity distribution system should meet.
5. Generate a neat sketch of a typical radial lighting circuit for the upper floor of a three-bedroom house.
6. Generate a neat sketch of a typical ring main socket circuit for the upper floor of a three-bedroom house.
7. Describe how a mini circuit breaker provides protection against electrical injury.
8. Describe how the various types of residual current device provide protection against electrical injury.
9. Explain, in your own words, why it is important to earth electrical appliances.
10. A plug is to be fitted to a 1,000 W hairdryer. Calculate the appropriate rating for the cartridge fuse.
11. Higher Level, 2001, Question 9
    (a) Using notes and sketches, show the correct wiring for two sockets in a ring main circuit of a domestic electrical installation.
    (b) Describe, using notes and sketches, the principles of earthing in a domestic electrical installation.
    (c) List and explain three safety procedures regarding the use of electrical power tools outdoors.

Rendering is the term used to describe the application of a surface finish to an external wall. Plastering is the term used to describe the application of a surface finish to an internal wall. Render and plaster have both an aesthetic and practical function. Aesthetically, the surface finish of a wall conveys a variety of messages, such as strength, stability, weather resistance, warmth and beauty. The surface finish can also say a lot about a building generally – e.g. a building might look sleek, welcoming and spacious like a glass-fronted clothing store or traditional, strong and secure like a bank.

Internal plastering is used to provide a smooth surface finish to the walls. This is important as it allows the home owner to decorate the walls to suit their taste. If the surface of the walls is uneven and rough then the only option would be to cover them with something like wood-chip wallpaper.

Render and plaster also serve practical functions. They improve the thermal and acoustic insulation of a house as well as improving the durability of the walls. For example, external concrete block walls must be rendered because the concrete is quite porous and without the protection of an external render, wind-driven rain might penetrate the fabric of the building causing dampness.

External Rendering of Blockwork

External rendering must meet the following performance criteria:

- Durability – The durability of a render depends on the degree of exposure of the building, the proportions of the mix used, the bond between the rendering and the wall and the quality of the workmanship.
- Resistance to Moisture – This depends mainly on the mix used. Surprisingly, stronger mixes (high cement content) are prone to cracking. This is because stronger mixes are less flexible and therefore cannot withstand the thermal and moisture movements associated with changes in the weather.
- Uniform Weathering – This depends upon the texture of the finished surface, the consistency of the mix used and structural details such as window cills. This is important because if the render does not weather uniformly it will not have a pleasing appearance and it may be susceptible to cracking.
- Aesthetics – Workmanship should be of a high standard and the finished render should be appropriate to the style of house and the locality.

Plain Rendering

Render is applied to a concrete blockwork wall in three coats:

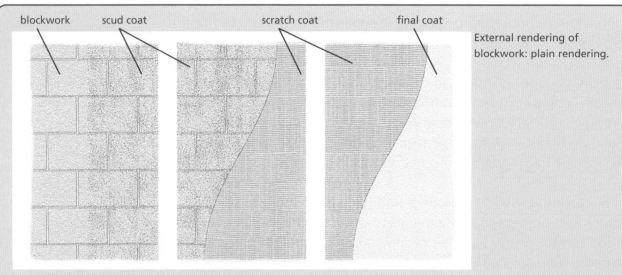

External rendering of blockwork: plain rendering.

blockwork   scud coat   scratch coat   final coat

- Scud Coat – A mixture of one part cement to 1½ or 2 parts sharp sand mixed with just enough water to give the consistency of a thick slurry. The scud coat is thrown onto the wall using a hand scoop to a thickness of 3 mm. The surface should be dampened periodically until hardened and then allowed to dry. The purpose of the scud coat is to provide a key or grip for the next coat of render.
- Scratch Coat – A mixture of one part cement to ½ part lime to 4 or 4½ sharp sand. The scratch coat is applied to a thickness of 8 to 10 mm and allowed to set firm. When the scratch coat is set firm (but before it hardens) it should be combed or scratched to provide a key for the following coat, but not so deep as to penetrate through to the scud coat.
- Final Coat – A mixture of one part cement to ½ part lime to 4 or 4½ sharp sand. Before the final coat is applied the scratch coat should be fully hardened and sufficiently dry to provide adequate suction (absorption). If the scratch coat is bone dry it should be dampened to prevent excessive suction. The final coat should be applied using a timber float because a steel float brings water and the finer particles of cement and lime to the surface, which on drying out, shrink and cause surface cracks. A timber float leaves the surface with a coarse texture which is less likely to crack. The final coat should be thinner than the scratch coat. Final coats are usually about 6 mm thick. The overall thickness of the combined coats should not exceed 20 mm.

It is important to note that the scratch coat and final coat are of similar strength and that the final coat is thinner than the scratch coat. This is to prevent cracking due to movement caused by changes in temperature and moisture. To prevent crazing, the final coat should not be allowed to dry out too quickly, especially in warm weather.

crazing

cracking

External rendering of blockwork: crazing and cracking of plain rendering.

Finishes

A finish is a decorative coat or texture applied to plain rendering. As well as the plain finish described previously, textured or thrown finishes may be applied. A textured finish is applied by scraping a pattern into the surface of the final coat before it has fully hardened. A thrown finish is applied by throwing a wet mixture onto the wall. Thrown finishes are generally more reliable under exposed conditions because they offer greater weather resistance, durability and resistance to crazing and cracking.

There are two main types of thrown finishes – wet and dry. Wet finishes are commonly referred to as roughcast or wetdash. Wetdash is a mixture of cement, lime, sand and coarse aggregate. The proportions used depend on the exposure. For areas of severe exposure (anywhere to the west of a line drawn from Cork city to Galway to Derry) a mix of $1:\frac{1}{2}:3:1\frac{1}{2}$ (cement:lime:sand:coarse aggregate). For moderately exposed areas (rest of the country) a 1:1:3:2 mix is recommended. A roughcast finish is achieved by throwing the final coat of rendering as a wet mix and leaving it un-trowelled. The roughness of the finish is determined by the shape and size of the coarse aggregate in the mix.

Wetdash.

A drydash finish is produced by throwing crushed chippings or pebbles on to a freshly applied mortar layer using a small shovel or scoop and leaving it exposed. The final coat should be laid on about 8 mm thick, depending upon aggregate size, and a rule or float should be passed lightly over to straighten it. The aggregate should be well washed and drained and thrown wet onto the final coat while it is still plastic (i.e. soft).

Drydash.

The whole surface is completed in one continuous effort, without interruptions, to ensure a uniform appearance. Again, the proportions used depend on the exposure. For areas of severe exposure a mix of $1:\frac{1}{2}:4$ (cement:lime:sand) is recommended. For moderately exposed areas a 1:1:5 mix is recommended.

Plastering

Plastering is a general term used to describe the process of applying a surface finish to the internal surfaces of walls. These might be the internal surfaces of external walls or the surfaces of internal walls such as stud partitions. There are two general approaches to this: the first involves applying a number of coats of plaster (creamy mixture of gypsum plaster and water) directly to the surface of the wall – the second involves fixing sheets of plasterboard to the wall surface and then applying a finish to the board surface.

Materials

The plasters used in internal plastering are gypsum-based products. Gypsum is a chalk-like mineral that is available as natural gypsum, which is mined all over the world, and as a synthetic by-product of industrial processes, such as fossil-fuelled power stations. Gypsum is a crystalline combination of calcium sulphate and water. During the manufacture of plaster products the gypsum is ground into powder form and heated, removing the water. A number of types of plaster are produced depending on: the temperature reached, the amount of water driven out of the raw material and the additives used. These include:

- Hemihydrate Gypsum Plaster – A quick drying plaster used for creating cornice moulds, coving and other decorative features. This product is traditionally known as *Plaster of Paris*.
- Retarded Hemihydrate Gypsum Plaster – This plaster has a retarding agent added during manufacture to slow the setting time to around two hours. This allows enough time for spreading and levelling. This plaster is commonly used as a finish coat to gypsum undercoats and plasterboards. Alternatively, it is mixed with a lightweight aggregate, such as expanded perlite or vermiculite, for use as an undercoat.
- Anhydrous Gypsum Plaster – This is commonly used as a thin finish coat over cement-based undercoats. A sulphate accelerator is added during manufacture to decrease its setting time, which otherwise would be so slow as to make it unsuitable for use as a finish coat plaster. A characteristic of this plaster is that it can be reworked by sprinkling the surface with water.

A variety of terms are used to describe these materials. These are often brand names of products which have become popular in the construction industry. For example, terms such as, *bonding plaster*, *finish plaster*, *skim coat* and *board finish* are all, in fact, brand names.

Plasterboard consists of a gypsum plaster corebonded between two sheets of heavy paper. The paper protects and reinforces the core, making it possible to cut neatly and install without it crumbling away. Most standard plasterboard has one ivory face and one grey. The paper on the ivory face is specially designed for finishing and should be visible after installation. Plasterboard is available in a range of thicknesses from 9·5 mm to 19 mm, although 9·5 mm and 12·5 mm are most commonly used in house construction. Board widths vary from 900 mm to 1,200 mm, while lengths range from 1,800 mm to 3,000 mm. A commonly used size is 1,200 mm × 2,400 mm. The edge of plasterboard sheets are manufactured with square, rounded and tapered edges, depending on the finish to be used.

When curved surfaces are required a glass-fibre reinforced gypsum board can be used. This board does not have a paper surface, instead it is made with layers of glass fibre immediately below each surface. This enables the boards to be easily bent for use on curved structures.

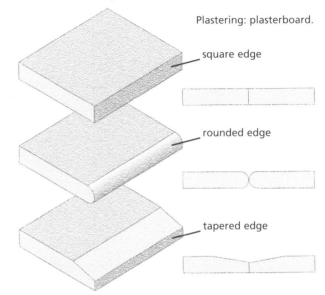

Plastering: plasterboard.

square edge

rounded edge

tapered edge

Storage
All plaster products should be stored in a dry, weatherproof, frost-free, enclosed shed or building with a dry floor (if the floor is concrete it should be stored on a timber platform). Any contact with moisture shortens the setting time and may reduce the strength of the set plaster.

Mixing
Plaster is mixed using clean water drawn directly from a tap or hydrant. The consistency is judged mainly by experience – a creamy plastic mix that spreads well is the norm. Cleanliness is essential and any mix left in containers from previous mixes should be washed out before use. Mixing old and new batches will shorten the setting time and may reduce the strength of the plaster when set.

Internal Finishing of Blockwork

The function of internal plastering is to provide a smooth, hard, level surface for decoration with emulsion paint or wallpaper. Internal plastering also provides a degree of thermal and acoustic insulation (see chapter 22, Sound and Acoustic Insulation, completeness) and fire resistance (gypsum is non-combustible). Where the walls are constructed from concrete blocks there are two methods commonly used to achieve a good quality finish – plastering and dry lining.

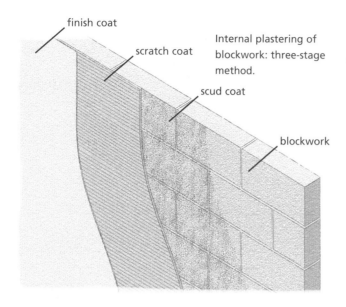

finish coat

scratch coat

Internal plastering of blockwork: three-stage method.

scud coat

blockwork

Internal Plastering of Blockwork: Three-stage Method

1. A scud coat 3–5 mm thick is applied, ruled to a reasonably flat finish and deeply scratched to provide a key (mix 1 sand : 2 cement), and allowed to harden.
2. A scratch coat 10–16 mm thick is applied and lightly scratched to provide a key (mix 1 cement : 1 lime : 6 sand).
3. When the scratch coat has hardened a finish coat of a gypsum plaster 2–3 mm thick is applied and finished smooth with a steel float. It is very important to allow drying shrinkage to occur in sand and cement base coats before the application of a gypsum finish coat. If the base coats have not fully dried before the gypsum top coat is applied, subsequent shrinkage of the base coat may cause cracking of the finish coat.
4. Skirting boards etc. fitted.
5. Paint and wallpaper applied.

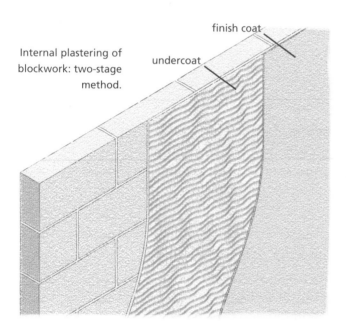

finish coat

Internal plastering of blockwork: two-stage method.

undercoat

Internal Plastering of Blockwork: Two-stage Method

1. A gypsum undercoat plaster 9 mm thick is applied and deeply scratched to provide a key.
2. When the undercoat has set, a finish coat of a gypsum plaster 2–3 mm thick is applied and finished smooth with a steel float.
3. Skirting boards etc. fitted.
4. Paint and wallpaper applied.

Dry Lining

Dry lining is a reliable approach to the internal finishing of masonry walls. It developed from the need to create an internal wall surface that was not affected by moisture or dampness seeping through the external walls. In older houses, where the external walls were of single leaf construction, there was always the risk that wind-driven rain would penetrate to the inner surface of the walls during a long spell of particularly bad weather. Dry lining was developed to overcome this problem by creating an internal wall surface that is not in contact with the external wall. Today, it is used for this reason in older houses, but also in newer houses to retain heat or to conserve heat energy. The procedure involves:

- fixing pressure-treated timber battens (or proprietary metal linings) to the wall,
- placing thermal insulation between the battens,
- fixing a vapour barrier across the insulation and battens – to inhibit the penetration of moisture generated inside the house (e.g. during cooking),
- screw fixing plasterboard sheets (explained below) to the battens.

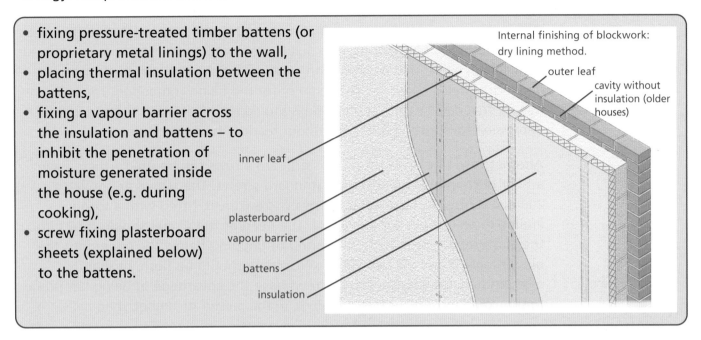

Internal finishing of blockwork: dry lining method.

outer leaf

cavity without insulation (older houses)

inner leaf

plasterboard

vapour barrier

battens

insulation

Alternatively, the plasterboard sheets may be secured using dabs of plasterboard adhesive. This approach is often used when the plasterboard sheets are supplied bonded to an insulant, such as expanded polystyrene or polyurethane. In this case, the procedure involves:

- applying dabs of plasterboard adhesive onto the wall to coincide with the edges and centre line of each sheet,
- applying a continuous seal of plasterboard adhesive to the perimeter of the wall,
- applying a continuous seal of plasterboard adhesive to the perimeter of any opes (e.g. door and/or window),
  - alternatively, the perimeter seal may be of pressure-treated timber battens,
- pressing the sheets firmly into place,
  - screw-fixing perimeter edges to timber battens (if used).

Once the adhesive dabs have set, two mechanical fixings (screws) are installed along the horizontal centre line of the boards approximately 15 mm in from the edge.

Installing Plasterboard to Internal Walls and Ceilings

As well as its use in dry lining, plasterboard is used in ceilings and stud partitions. Plasterboard is fixed to ceiling joists and studs using nails or screws. There is a relationship between the spacing of the ceiling joists (or studs) and the plasterboard dimensions. For example, 9·5 mm plasterboard: 450 mm joist and stud spacing, and 12·5 mm plasterboard: 600 mm joist and stud spacing. Thin plasterboards require greater support because they are not as strong and stiff as thicker boards. The spacing of joists and studs is also related to the board width: 600 mm spacing suits 1,200 mm board width. The boards are fixed with 40 mm galvanised nails with an annular ring shank (i.e. ridged surface) and a cross-hatch head, to provide a key for the filling compound. The nails are usually driven at 150 mm centres. Nails should be driven so as to leave a slight depression for spotting (filling), without tearing the paper surface.

Alternatively, zinc plated steel plasterboard screws can be used – these have a finer pitch (more threads) than standard screws. The screws are driven at 200 mm centres using a suitable power screwdriver, fitted with a torque limiter, or adjustable depth control. The head of the screws should recess into the plasterboard face, without tearing or popping below the paper into the plasterboard gypsum core. Once fixed, screws should be flush, or slightly recessed to allow easy spotting. Screws have the advantage of being less prone to shrinkage movement (i.e. popping), in the plaster or timber, than nails.

It is essential that plasterboards are supported along their edges. For this reason, nogging pieces are installed to support board edges where necessary. The board ends should be staggered to avoid a continuous joint. This is done by starting every second row with half of a board.

Once the boards are securely fixed, the joints are filled. This is necessary to prevent unsightly cracks, caused by shrinkage or structural movement, developing over time. The use of tapered edge plasterboards makes this possible. The tapered edges create a shallow trough that makes it possible to fill the joints without creating a *hump*. The procedure typically involves the following steps:

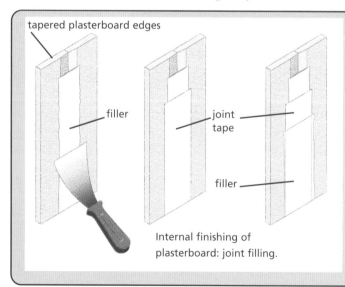

tapered plasterboard edges

filler

joint tape

filler

Internal finishing of plasterboard: joint filling.

- Filler is applied to the full length of the joint.
- A continuous band of joint tape is pressed firmly into the filler (ensuring no air bubbles are trapped).
- the filler is allowed to set (approx. one hour).
- A second layer of filler is applied – over the tape – and smoothed flush with the surface of the board.
- A damp sponge is then used to remove excess filler from the outer edges, taking care not to disturb the main joint filling.

Joint paper has chamfered edges and a centre crease which makes it easy to fold into internal angles (corners). External angles are usually protected by beading to prevent damage during day-to-day use. A variety of external angle beads are available, including:

- galvanised steel mesh or perforated steel,
- paper tape laminated to thin galvanised steel strip,
- extruded uPVC profile – allows for curving around arches and reveals down to 250 mm radius.

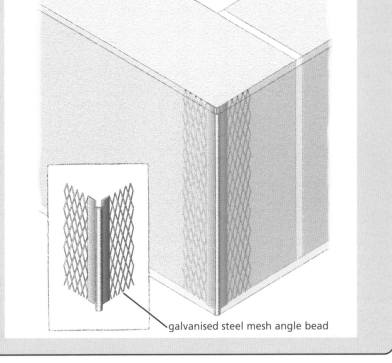

Internal finishing of plasterboard: external angle bead used to protect corners.

galvanised steel mesh angle bead

Nail and screw heads are spotted (spot filled) flush with the board surface using the same filler.

Finishing of Plasterboard

There are two systems which can be used to apply a finish to plasterboard – these can be referred to as *wet* or *dry* systems. The wet system, which is the traditional approach, is a popular method because it can achieve a high-quality surface finish. However, the mixing and application of the plaster is labour intensive. The speed and convenience of the dry system, coupled with improvements in plasterboard products (e.g. ivory facings) has seen an increase in the use of the dry system.

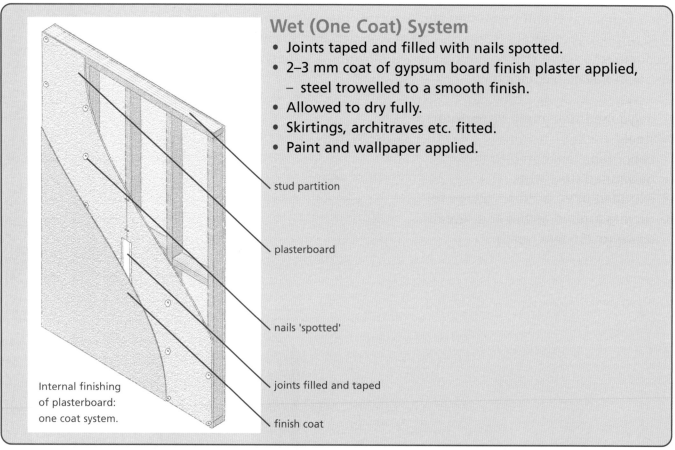

Wet (One Coat) System

- Joints taped and filled with nails spotted.
- 2–3 mm coat of gypsum board finish plaster applied,
  - steel trowelled to a smooth finish.
- Allowed to dry fully.
- Skirtings, architraves etc. fitted.
- Paint and wallpaper applied.

stud partition

plasterboard

nails 'spotted'

joints filled and taped

finish coat

Internal finishing
of plasterboard:
one coat system.

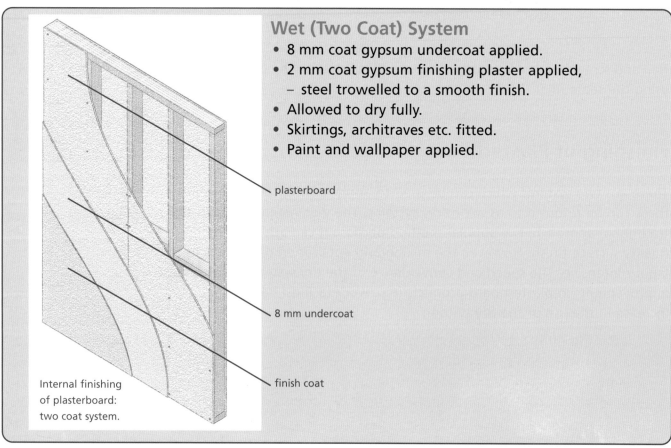

Wet (Two Coat) System

- 8 mm coat gypsum undercoat applied.
- 2 mm coat gypsum finishing plaster applied,
  - steel trowelled to a smooth finish.
- Allowed to dry fully.
- Skirtings, architraves etc. fitted.
- Paint and wallpaper applied.

plasterboard

8 mm undercoat

finish coat

Internal finishing
of plasterboard:
two coat system.

Dry System

- Joints taped and filled, nail holes filled.
- Joint finish applied.
  - This is a filler designed for joint filling and nail spotting when no further plaster is to be used – it is applied in a 250 mm-wide band to create a smooth finish over the entire joint area – the edges are feathered out with a slightly damp sponge.
- Sealing agent applied to plasterboard surface to allow a good bond with paint and wallpaper.
- Skirtings, architraves etc. fitted.
- Paint and wallpaper applied directly to ivory face of plasterboard.

2. Tape

1. Filler

3. Joint finish applied

Revision Exercises

1. Outline and explain the performance requirements of external rendering.
2. A small group of holiday homes are being built in an exposed area on the west coast of Ireland. The houses have a concrete block outer leaf. Suggest, giving reasons, a suitable render for the houses and describe the procedure for applying the render.
3. A single-storey house situated in the countryside is being built in the vicinity of a number of traditional thatched cottages. The house has a concrete block outer leaf. Suggest a suitable render for the house and describe the procedure for applying the render.
4. Explain, using neat annotated sketches, how plasterboard is installed on an internal stud partition.
5. Describe the wet and dry systems of finishing plasterboard. Discuss the advantages and disadvantages of each approach.
6. Ordinary Level, 2004, Question 6
   Plasterboard and insulation are to be fixed, as a dry lining, to the inner surface of an external block wall.
   (a) Using notes and neat freehand sketches, show one method of fixing the dry lining to the wall.
   (b) Give two reasons why it may be necessary to fix dry lining to the external walls of an old house.
   (c) Describe the preparation of the surface of the dry lined wall prior to the application of a paint finish.
7. Ordinary Level, 2005, Question 2
   A non-load-bearing timber stud partition with plasterboard finish

separates two bedrooms on the first floor of a house.

(a) Using notes and neat freehand sketches, describe the construction of the stud partition. Indicate on the sketches the names and sizes of all parts.

(b) Describe two methods of providing a surface finish to the plasterboard prior to painting.

(c) Recommend one of the finishes described at (b) and give two reasons for recommending this finish.

8. Higher Level, 2000, Question 6
    Using notes and sketches, describe the application of:
    (a) A smooth plaster (plain render) finish to an external block wall.
    (b) A gypsum-based finish to an internal block wall.
    (c) A gypsum-based finish to an internal stud partition wall.
    Give details of materials, thickness and sequence of undercoats for each finish.

9. Higher Level, 2004, Question 3
    (a) Using notes and freehand sketches, describe the application of an external render to the walls of a new house of concrete block construction. Give details of materials, mix proportions and sequence of coats required.

    (b) The original external render of an old house is to be removed to reveal solid stone walls of random rubble construction, as shown on the sketch below. The owner has the option of either leaving the external stonework exposed or of re-plastering the walls. Outline two reasons in favour of each option listed above. Recommend a preferred option and give two reasons to support your recommendation.

    (c) If the house is to be re-plastered, a 1 lime : 3 sand mix is recommended for the external render. Give two reasons why such a mix is recommended for this house.

Paint was first used for decorative purposes in the caves of France and Spain as early as 15,000 BC. Many centuries later, Asian cultures developed hard, clear varnishes to add lustre to their craft work. Since then, the need for paint has broadened to include functional as well as aesthetic requirements. Well-finished surfaces (e.g. walls, furniture) are attractive to look at, easy to clean and hygienic.

Functions of Finishing

Finishes perform a number of functions including:

- Prevent Decay – Timber and metals will deteriorate if not protected from exposure to the elements. Consider, for example, the wide variation in temperatures and moisture that a typical hardwood front door is exposed to in a single year. If the wood is not properly protected it will not last very long.
- Maintain Hygiene – Porous surfaces such as raw timber or concrete blockwork may absorb moisture and provide an opportunity for bacteria to develop especially in warm, moist environments such as kitchens and bathrooms.
- Improve Appearance – A painted finish is an excellent way of improving the surface of unattractive materials, whereas a varnish finish will allow the natural beauty of wood to be appreciated for a long time

Types of Finishes

There is a wide range of finishes available today and the classification of them has become less clear cut. Traditionally, a finish could be identified by the application. For example, paints and varnishes that were used for finishing wood could be described as *oil-based* – these were tough, durable finishes that usually gave a shiny or gloss finish. These paints were easily identified by the fact that white spirit or turpentine would have to be used for thinning (diluting) or brush cleaning. Paints used for metals were similar. Most other paints, such as those used for painting walls, were *water-based*. Today, however, paint manufacturers have developed a much broader range of finishes (the industry term is *coatings*). In particular, the growth in the do-it-yourself market has led to the development of quick-drying, water-based varnishes. In fact, more and more finishes are being manufactured in *water-based* format. This means that it is no longer possible to classify a finish by application. The only way to be certain about the constituents of any product is to read the information on the label.

Paints

Paints have three major components – pigments, binders, and solvents.

The main function of a pigment is to add colour to the paint. Organic and inorganic pigments are available in a wide array of colours. Organic pigments, though often more expensive than inorganic colours, offer a wider range of shades. Inorganic pigments are derived from various metallic ores. The most commonly used of these pigments is white titanium dioxide. Other pigments include carbon black, red lead, chrome yellow, molybdate orange, zinc yellow, and iron oxides.

Pigments are dispersed in binders (also called resins) which provide the protective and mechanical properties of the paint film. As it dries, the binder (resin) forms a film that allows the paint to adhere to the surface. Traditionally, binders were made from natural materials such as linseed oil. While these are still used in oil-based paints, synthetic resins now make up more than ninety per cent of binders. Synthetic binders include alkyd resins made from acids, anhydrides and alcohols as well as acrylic binders. Vinyl, epoxy, urethane and other specialty resins are used when chemical resistance or extra durability and adherence are needed. Acrylic resins are a common component in glossy latex paints.

The pigment and binder are mixed and dissolved in a solvent, which controls the consistency of the paint and evaporates after the paint is applied. Without a solvent, the pigment-binder mixture would be too thick to spread easily and uniformly over surfaces. For non-water-based paints, oily solvents such as toluene and xylene are used. Some solvents are derived from alcohols, while others, including mineral spirits and naphthas, are distilled from petroleum. Pungent-smelling esters are found in clear lacquers, and ketone solvents are used to make and remove paint. Until synthetic resins were introduced, tree-derived turpentines were the standard paint solvents.

Besides pigment, binder and solvent, paints also contain other additives. They may have thickening extenders such as calcium carbonate and talc, dispersing and drying agents, fungicides and anti-mildew agents. Specialty additives include anti-settling agent, anti-coagulants, anti-skinning agents, flattening agents, deodorants and flame retardants.

Varnishes

Varnish is a general term used to describe a clear or coloured finish applied to wood, usually by brush. Varnishes are generally more flexible than paints and are capable of withstanding movement in timber which might cause paints to crack or peel. The type of varnish used depends on a number of factors, including:

- material – raw timber or previously varnished,
- interior or exterior use,
- clear or coloured finish,
- time available – quick drying.

There are three main types of varnish:

- Oil Borne – Oil varnishes are made by combining drying oils such as linseed oil with resins and then thinning the mixture with a solvent. Drying agents are included to speed up the drying process.
- Spirit (or *Solvent*) Borne – Spirit varnishes consist of resins dissolved in solvents such as methylated spirit. Drying occurs through evaporation rather than by means of drying agents.
- Water Borne – These relatively new products are usually sold as *quick drying* varnishes for interior and exterior applications. This is because they are usually touch dry in approximately one hour. This can be helpful in a number of situations (e.g. for floors and exterior work).

Polyurethane varnishes are commonly used for timber finishing. These consist of a polyurethane resin dispersed in a solvent or in water.

Surface Preparation

Surface preparation is necessary to ensure the correct adhesion between the surface and the applied finish. If a surface is not properly prepared the finish will likely peel, crack or flake off within a few weeks of application. Surface preparation is a potentially hazardous activity. The generation of dust and the use of chemicals are common. For this reason appropriate personal protective equipment should be worn at all times. This includes:

- eye protection (e.g. safety goggles),
- respiratory protection (e.g. face mask),
- skin protection (e.g. gloves),
- clothing protection (e.g. overalls).

It is also essential to ensure that adequate ventilation (i.e. fresh air) is available. Solvent-based finishes should not be applied in

Personal protective equipment, including protection for the eyes, ears, lungs and skin should be worn when working with paint products and paintbrush-cleaning chemicals.

confined spaces. Generally, the use of water-borne finishes is less hazardous than solvent-based finishes. Always follow the safety precautions indicated on the product packaging.

Timber

All timber surfaces to be finished must be clean and dry, with dust, dirt, wax and grease removed. The timber should be allowed to acclimatise to its end-use environment. The moisture content should not exceed fourteen per cent prior to coating. Traditionally a knotting agent is used to seal knots so that resin does not leak out through the finished surface. However, many manufacturers do not recommend the use of knotting agents as they are not always fully effective in sealing the resin. In addition, the presence of unsightly knots is often highlighted and adhesion of coatings can be impaired. Any indentations in the surface of the timber may be filled with a filler specifically designed for use with timber. General or all-purpose fillers are not suitable, particularly on external areas, as they cannot cope with timber movement.

New Timber

Any nails or pins should be punched below the surface of the timber. The holes (and any indentations) are then filled. Some timber species contain high levels of natural oils and some softwoods can be highly resinous. Resinous deposits should be removed with a scraper. Any remaining residues should be removed using a lint-free cloth dampened with methylated spirits. The surface is then sanded smooth using a medium-grit paper working down to a fine grade. The timber surface should then be degreased by wiping with a cloth dampened with methylated spirits. The methylated spirits should be allowed to evaporate fully before the finish is applied.

Previously Finished Timber

When working with previously finished timber any loose, flaky or bubbling coating should be removed by use of a scraper and abrasive paper. In cases where the existing finish has deteriorated significantly, a heat gun or chemical stripping agent may be used. These are both potentially hazardous operations and must be carried out safely. A heat gun causes the finish to soften sufficiently to make it possible to scrape it away. A chemical stripping agent will have a similar effect. It is essential that the manufacturer's instructions are followed when working with these products.

In damp areas (e.g. kitchens and bathrooms), superficial mould or mildew may be present. This must be removed using a suitable fungicide. Surfaces should be washed with water and a mild detergent to achieve a clean surface. The surface should then be rinsed thoroughly and allowed to dry completely. Timber with more substantial decay will have to be removed and replaced and a suitable preservative applied. Clear preservatives, that do not affect the colour to the wood, may be used. Any bare patches should be primed with one coat of finish. Finishes in a poor condition should be removed completely and the surface thoroughly sanded and then treated as new timber.

Walls

External masonry surfaces require similar attention, including:

- removal of all dirt and loose paint with a stiff brush,
- mould and algal growth to be dealt with using a fungicidal wash,
- cracks and holes should be filled with a suitable flexible filler,
- any surfaces that remain powdery or chalky should be sealed with a suitable primer.

New masonry will need to be sealed with a bonding or sealing agent. If this is not done the masonry surface will be too absorbent and require excessive use of paint. Also, using a sealing agent will improve the adhesion of the paint to the masonry surface.

Internal walls which have been freshly plastered need to be checked to ensure they are smooth and free from plaster splashes or other damage. Any defects should be filled and sanded as required to achieve a smooth finish.

Application of Finishes

There are three principal methods of applying a finish:

- brush,
- roller,
- spray.

The quality of the equipment and tools used will have a significant impact on the quality of finish achieved. A good quality brush is always worth the extra expense. Cheap brushes tend to lose bristles and generate a poor surface finish. It is

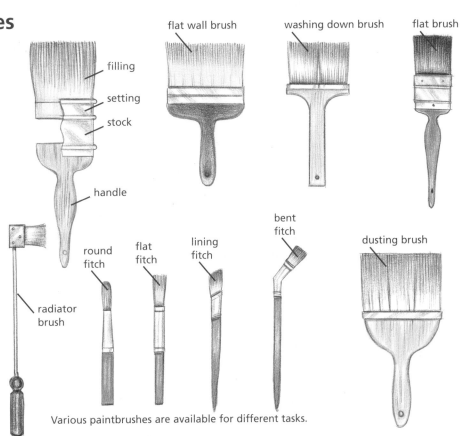

Various paintbrushes are available for different tasks.

also essential that brushes are kept clean during use. A dropped brush will quite likely pick up grit and other dirt and should be thoroughly cleaned before resuming work. Similarly, all equipment should be thoroughly cleaned after use. The cleaning procedure for every product is described on the container. All waste materials should be disposed of in an environmentally sound manner. There are many *bring centres* around the country to enable proper disposal of paint and varnish cans.

The sequence of application and the number of coats required will depend on the substrate (material to be finished), the environmental conditions and the product used.

Substrate	Environment	Finish Coat	Coats Required	Application Method
Wood	Interior	Solvent-based paint (e.g. gloss paint)	• Primer • Undercoat • Finish coat * sand lightly between coats	Brush
Wood	Interior	Oil (e.g. linseed oil)	• 3 coats of oil * sand lightly between coats	Brush or lint-free cloth
Wood	Interior	Wax (e.g. beeswax)	• 3 coats of wax * buff vigorously between coats	Lint-free cloth
Wood	Interior	Varnish (e.g. polyurethane)	• 3 coats of varnish * sand lightly between coats	Brush
Wood	Interior	Cellulose lacquer	• Sealer • Undercoat • Finish coat * sand lightly between coats	Spray gun

Substrate	Environment	Finish Coat	Coats Required	Application Method
Wood	Exterior	Solvent-based paint (e.g. gloss paint)	• Primer • Undercoat • Finish coat *a preservative may be required in exposed areas (e.g. fascias) *sand lightly between coats	Brush
Metal	Exterior	Cellulose-based paint (e.g. Hammerite®)	• Primer * primer is used on galvanised steel, aluminium or very rusty metals • 1 coat	Brush and also available as aerosol spray can
Metal	Exterior	Solvent-based paint (e.g. gloss paint)	• Primer • Finish coat	Brush
External render	Exterior	External masonry paint (e.g. Dulux Weathershield®)	• 2 to 3 coats	Brush or spray
Internal plaster	Interior	Emulsion paint (e.g. matt, eggshell, soft sheen and other emulsion paints)	• 2 to 3 coats	Brush, roller or spray
Plasterboard	Interior	Emulsion paint	• Sealer • 2 to 3 coats of emulsion	Brush, roller or spray

Revision Exercises

1. Explain, in your own words, why it is necessary to apply a finish to many materials.
2. Outline the functions of the various components of paint.
3. Describe the main types of varnish for the finishing of timber around the home.

4. Describe how a timber casement window, which was last maintained three years ago, would be prepared for finishing.

5. Describe how plasterboard, which has been prepared using the *dry* system would be finished with an emulsion paint.

6. Describe how external walls, which have a cement-based plain render are prepared for painting.

7. Ordinary Level, 2000, Question 4

   Explain in detail the sequence of operations required when preparing and painting the following surface (assume that the surfaces have not been previously painted):

   (a) External wrought iron gate.

   (b) External smooth plastered wall.

8. Ordinary Level, 2002, Question 8, (a) and (b)

   (a) Explain, in detail, the sequence of operations required for preparing and painting new wood for an external use such as the fascia board shown in the diagram below.

   (b) List and explain two safety precautions that should be observed when applying preservatives to wood.